WITHDRAWN

Ecology
of

Ecology
of
Natural Resources

François Ramade

Professor of Ecology and Zoology at the Université de Paris-Sud (Orsay)
Member of the IUCN Commission on Ecology (Gland, Switzerland)

Translated from the French by W. J. Duffin, University of Hull

JOHN WILEY & SONS
Chichester . New York . Brisbane . Toronto . Singapore

First published under the title Ecologie des Ressources Naturelles by François Ramade.

© Masson, Editeur, Paris, 1981

Copyright © 1984 by John Wiley & Sons Ltd

Library of Congress Cataloging in Publication Data:

Ramade, François.
    Ecology of natural resources.

    Translation of: Écologie des ressources naturelles.
    Bibliography: p.
    Includes indexes.
    1. Ecology.   2. Natural resources.   I. Title.
QH541.R2513 1985   333.7′2   84–3678

ISBN 0 471 90104 0 (cloth)
ISBN 0 471 90625 5 (paper)

British Library Cataloguing in Publication Data:

Ramade, François
    Ecology of natural resources.
    1. Natural resources   2. Ecology
    I. Title   II. Ecologie des ressources
    naturelles. *English*
    333.7   HC55

ISBN 0 471 90104 0 (cloth)
ISBN 0 471 90625 5 (paper)

Typeset by Pintail Studios, Ringwood, Hampshire
and printed by The Pitman Press, Bath, Avon

Contents

viii CONTENTS

Foreword

Reader, do you remember how, barely 10 years ago, futurologists and economic prophets were announcing to us all in a blaze of publicity that the major problem of tomorrow's citizens would be the occupation of their leisure time? And how the few pioneers in ecology who put forward quantitative evidence opposing this anaesthetizing myth were met with nothing but a disdainful silence from traditionalists on all sides?

Now the wind has changed. Shifts in the balance of world power and the resultant redrawing of the geopolitical map, the energy 'crisis' and upheavals in the Third World have put an end to the complacence of many of those holding the above opinion. We have passed from a smug optimism to a chronic pessimism, trying hard to ignore the future for the better enjoyment of the present. Après nous la fin du monde!

Such an attitude is just as irresponsible as the previous one, and this new book of François Ramade clearly demonstrates that, while the situation is serious, it is certainly not desperate—as long as we are willing to look beyond the end of our noses and draw the necessary conclusions in time. There *are* solutions to our problems but they must be willingly adopted or they will be violently imposed from outside. If the necessary measures are not taken in time, 'the Third World will only be able to advance by shattering the world's social order in one way or another' (Fernand Braudel, *Civilisation matérielle, économie et capitalisme*, Vol. 3, 1979, p. 469). At best, this would bring about a progressive collapse and decline of the civilization of which we are so proud; at worst, it would provoke a great march northward of the starving, and the dawn of the new Middle Ages.

What are these measures that seem forced on us by the inescapable laws of physics and biology?

First of all, it is more than ever necessary to put an end to the world's population explosion. Particularly urgent is the need to think at once about reducing the pressure in already overpopulated regions by freely accepted birth control: a measure that will undoubtedly require a colossal effort of education.

In parallel with that, the enormous wastage of renewable resources indulged in so freely by the West's consumer society should be reduced as far as is humanly possible and, at the same time, everything that can be recycled should be recycled.

The third, equally urgent, measure is the conservation of the potential biological productivity of land, sea and inland waters, and its rehabilitation in those environments where fertility has been thoughtlessly reduced through activities working against nature. This action must be accompanied by the preservation of the world's genetic stock of plants and animals, the fruit of hundreds of millions of years of evolution.

The final objective—and by no means the easiest to attain—is the need to learn how to share the available resources so as to reduce the most glaring inequalities between the 'haves' and the 'have nots'. Alas, no political movement of any importance dares for one moment to face this problem, because its solution necessarily implies some reduction in the standard of living of the present most wealthy countries.

It is greatly to be feared, therefore, that the inevitable unpopularity of many of the measures needed for an immediate response to the four requirements I have just outlined means that the necessary decisions will be taken too late and under the pressure of a 'crisis' that will have become endemic.

Moreover, things are not made any easier by the obsession of too many Third World leaders with a blind copying of models of development which were valid in other places and at other times, but which are ill-adapted to their current situation. Nor does the moral abdication on the part of too many 'elites' in wealthy countries help in any way. At a time when egocentricity, hedonism and laxity seem to have become the three Western pillars of wisdom, how can we hope to arouse in ourselves the great effort of imagination and research, the altruistic spirit and the sheer crusading attitude needed to take rapid decisions?

I hope that the pages which follow will nevertheless cause many future leaders of our society to reflect on such matters, and help to give their lives a new sense of direction. For all those who are not satisfied with the dullness of daily existence, there are as many exhilarating tasks to be undertaken today as there were yesterday. Knowing how to respond quickly and efficiently to the appeals of thousands of millions of the underprivileged presents a challenge that we must have the courage to accept as soon as possible.

François Bourlière,
Honorary President of the International Union for the Conservation of Nature; President of the Société Nouvelle de Protection de la Nature and of the International Association for Ecology (INTECOL)

Acknowledgements

It gives me great pleasure to thank Professor J.-M. Pérès, Membre de l'Institut de France, who kindly offered to read the chapter on marine resources and suggested various modifications; Professor F. Bourlière, President of the International Association of Ecology (INTECOL), who was kind enough to contribute the foreword; and Professor M. Lamotte, Director of the Zoology Laboratory at the ENS, who allowed me to have access to a large volume of documentation on research into the ecology of the savannas carried out at Lamto.

I should also like to express my sincere thanks to my friends and colleagues, Y. Gillon, J. C. Moreteau and F. Terrasson, who have contributed to the illustration of this book by allowing me to use certain photographs from their personal collections.

Introduction

Like all other living beings, humans require matter and energy. It is not being too sceptical about possible advances in science and technology to say that our species could obviously not set itself free from such requirements, whatever future progress might be envisaged.

Human beings are animals and are therefore heterotrophic organisms.[1] As such, their metabolic requirements are met by the air they breathe and by the water and organically derived food they ingest. Like other living species, they depend on the cosmic system from which they have descended: essentially the sun together with the ecosphere, the superficial part of our planet where the environmental conditions exist that make life possible.

However, continual advances in technology have caused other needs to appear in addition to those resulting from natural biological processes. The development of an ever more complex industrialized society in which the production of manufactured objects is incessantly growing entails a continually increasing use of primary energy and of organic and inorganic raw materials.

Just as metabolic processes involve the discharge of mineral and organic excreta, so the activities of a technological civilization release waste products into the environment. In both cases, the discharged material does not simply disappear from the environment that receives it: instead, it circulates in biological systems which can cease to function properly if their homeostatic mechanisms are overstretched.

Another feature of the contemporary world that arises from human activity and has a considerable impact on the ecosphere is the explosive growth in population. This, along with the unending increase in *per capita* consumption of manufactured goods, puts great pressure on nature and natural resources.

This takes us a long way from one of the fundamental concepts of industrial civilization, that of *Homo ecônomicus*, which regards humanity as the owner of all mineral and biological resources and thinks of these resources as inexhaustible and to be disposed of as we please. According to this idea, humanity is extraneous to the ecosphere and thus independent of any ecological changes in it that might be produced by human activity.

The very concept of a natural resource is worth some reflection. A resource can be defined simply as any form of energy or matter necessary to satisfy the physiological needs of humanity or to sustain all the various activities leading to production. The flow patterns of such resources through human civilization are very complex and so can be studied from several different angles.

Between the stage at which the resource is extracted and that of its use by a consumer, it undergoes many transformations, and these often have an impact on the overall functioning of the ecosystems in which the processes occur. A classic distinction is frequently made between non-renewable and renewable resources. Potential sources of energy such as hydrocarbons and fissile materials clearly come into the first category, but for other types of resource the distinction is often difficult to make. Even minerals could be allocated to the second category since they can theoretically be recycled from both domestic and industrial waste and this would circumvent the problem of their exhaustion.

Water and all resources of a biological origin are usually classified as renewable. Even when polluted, water is not chemically modified in any way by being used and so can be recycled after purification. Plant and animal resources, on the other hand, although potentially renewable, are very often so overexploited that the possibility of regeneration in many parts of the world has been greatly reduced and sometimes completely compromised by the destruction of the ecosystems on which they depend.

In reality, the natural resources of the ecosphere are being wastefully consumed at an increasing rate under the combined effect of population pressure and the dramatic increase in industrial production. The current rate of use takes absolutely no account of the real size of available reserves of minerals or fossil fuels, nor does it concern itself with the rate of renewal of plant or animal resources. The needs of future generations are similarly ignored. In addition to that, malnutrition is spreading in the Third World and, in a future that is closer than some people think, the industrialized and overpopulated countries of Europe and other continents will no longer be protected from shortages of animal protein.

---

1. See glossary, p. 222.

Another form of damage to the ecosphere which has very worrying ecological consequences is the over-exploitation of the Earth's plant cover. The degradation of forests, quite apart from the decrease in timber production it brings, causes irreversible changes in climate and soil. The extension of deserts into grassland or open forests; the erosion of soils in mountainous terrain or in areas where fragile land has been irresponsibly cultivated—these all demonstrate the extent of the upheaval stemming from the overexploitation of natural resources indulged in by humanity.

Animal resources are even more threatened. During the last two centuries, more than 600 species of birds and mammals have become extinct because of human activity. As for marine ecosystems, whose biomass was long considered inexhaustible, these are now also showing disquieting signs of overfishing through excessive landings of the main species of economically valuable fish.

In what follows, therefore, I intend to analyse the ways in which the main categories of natural resource are exploited and to examine the major principles of methods for the rational management and conservation of such resources. Now, more than ever before, is the time for implementing a world strategy for the protection of nature in the cause of a lasting and stable world. The aim of this book is to explain the scientific foundations and methodological approach on which such a strategy must be based.

Chapter 1

Ecological Concepts Related to Nature and Natural Resources

1.1 Ecosphere and Biosphere

The word *nature* is used in a good many senses and it includes ideas that vary widely according to the educational background of those using it and the amount of scientific training they have had, especially in biology. The concept of nature generally held by the non-specialist, the politician or the technocrat is undoubtedly only distantly related to that held by the ecologist. For that reason, I prefer in the following pages to use instead the terms *biosphere* and *ecosphere*, since these indicate the two regions of our planet which contain everything ordinarily understood as 'nature'.

The *biosphere* can be simply defined as that part of the Earth in which life is permanently possible and which contains all living organisms. It consists of the terrestrial oceans and the surfaces of the continents, together with the adjacent atmosphere (that is, the troposphere), with the exception of the polar ice caps and the higher mountain slopes above the snow line. These latter regions, described as parabiospheric, are included along with the biosphere itself in a larger system, the *ecosphere*, which also embraces the upper layers of the lithosphere and the whole of the atmosphere above the troposphere.

The aim of *ecology* is to investigate the relation of living organisms to each other and to their surroundings and it thus provides, more than anything else, an essential basis for any rational approach to the study of the biosphere. Since the ecosphere is the origin of all natural resources except solar energy, it is easy to appreciate the importance of ecological science to the understanding of problems caused by the consumption of such resources in our present technological civilization.

1.2 Fundamental Ecological Variables

The living world can be studied in ecological terms at various levels of organization having greater and greater complexity. The effects of the main factors typical of any given environment can be considered (a) on an individual taken in isolation (autecology or eco-physiology), (b) on the population of a given species (population ecology or demo-ecology), or (c) on the entire community of living organisms subject to the same environmental conditions (ecology of the eco-system or synecology).

The concept of an ecological factor

Whatever the level of organization being studied, we are always led to consider the effects of a certain number of fundamental parameters called *ecological factors*: these are the typical physico-chemical or biological characteristics of each environment that are capable of acting *directly* on the living organisms. Thus, in a continental environment, the temperature and rainfall are ecological factors whereas the altitude is not: it can be divided up into a series of parameters (such as temperature, amount of daylight, atmospheric pressure, etc.) through which it exerts its influence and produces its effects.

Ecological factors can be classified in several different ways. First of all, we can distinguish between *abiotic factors* (climate, physico-chemical composition of the environment, for example) and *biotic factors* (such as parasitism, predation, food supply, etc.).

Then, some ecological factors can be said to be *independent of the population density* because they exert the same influences on the individual whatever the density of the population to which it belongs (temperature is an example). Other factors, such as the quantity of food available or the effect of predators, may be *dependent on the population density*.

The last method of classifying ecological factors, that of Mondchasky, takes into account the effect of cosmic phenomena on them. It is quite obvious that annual, seasonal and daily fluctuations connected with the Earth's movement in its orbit could have an influence on a whole series of ecological factors which would then vary cyclically, sometimes in a pronounced manner, sometimes less so.

1

Table 1.1   Classification of ecological factors (adapted from Dajoz, 1971)

Abiotic factors	**Climatic factors** temperature amount of daylight humidity rainfall others (wind, etc.) **Non-climatic physico-chemical factors** *Aquatic environment* pressure concentration of mineral salts concentration of dissolved oxygen *Edaphic environment* particle size chemical composition	Factors independent of population density	Primary periodic
			Secondary periodic
			Secondary periodic or aperiodic
	Trophic factors concentration of inorganic nutrients available food supply	Factors dependent on population density	Secondary periodic
Biotic factors	**Biotic factors** *Intra-specific interactions* *Inter-specific interactions* competition predation parasitism etc.		Secondary periodic or aperiodic

Mondchasky distinguishes *primary periodic factors* (such as temperature and amount of daylight, for example), where the connection with the Earth's movement is direct, from *secondary periodic factors*, where the cyclic variations are derived from those of the primary factors (like atmospheric humidity, plant food supply, etc.). In addition, there are *aperiodic factors*, some of which exhibit fluctuations that are both violent and unpredictable (an exceptional drought or a volcanic eruption are good illustrations). Also classified as aperiodic are those factors having such a slow variation that they can be considered as virtually constant (the concentration of nutrient minerals in soils is an example).

Table 1.1 shows the relation between these various methods currently used for the classification of ecological factors.

The concept of a fundamental ecological variable

According to Watt (1973), the action of the many ecological factors can be reduced to a limited number of *fundamental ecological variables*, five in all: matter, energy, space, time and diversity. Any ecological phenomenon, no matter how complex, can be explained by the interplay of these five variables and, conversely, they must all be involved in the interpretation of such a phenomenon. I consider each of them in turn.

1.2.1   Matter

All living organisms consist of a certain number of chemical elements essential for the building of their biological molecules. In decreasing order of the amounts found in the human body, these are: oxygen, carbon, hydrogen, nitrogen, phosphorus, sulphur, etc., as indicated in Table 1.2. The table also shows that this order is very different from that of the relative abundances in the lithosphere and other large regions of the ecosphere. In all, about 40 chemical elements are found in living organisms, most of them being truly biogenic: in other words, they play a significant and precise role in the structure and/or the functioning of the organisms.

There are several fundamental principles governing the way matter is used by living organisms and the way that it circulates in ecological systems:

The law of tolerance

This law states that for each element there exists a range of concentrations, called the *interval of tolerance*, in which all physiological processes involving that element can take place normally. Consequently, it is only within this interval that life is possible for any particular plant or animal species. There is an optimum concentration in the interval, called the *preferendum*, for which metabolic processes occur at

Table 1.2 Comparative proportions of principal chemical elements present in living matter and in the ecosphere generally

Element	Atomic number	Lithosphere, atmosphere, hydrosphere (%)	Human body (%)
Hydrogen	1	0.95	9.31
Carbon	6	0.18	19.37
Nitrogen	7	0.03	5.14
Oxygen	8	50.02	62.81
Fluorine	9	0.10	0.009
Sodium	11	2.36	0.26
Magnesium	12	2.08	0.04
Aluminium	13	7.30	0.001
Silicon	14	25.80	negligible
Phosphorus	15	0.11	0.64
Sulphur	16	0.11	0.63
Chlorine	17	0.20	0.18
Potassium	19	2.28	0.22
Calcium	20	3.22	1.38
Manganese	25	0.08	0.0001
Iron	26	4.18	0.005

maximum speed. At concentrations less than the lower limit of the tolerance interval, death of the organism occurs through deficiency of the element concerned, while beyond the upper limit death occurs through excess (Figure 1.1).

Take as an example the case of nitrates in the soil. They are essential for the majority of plants, providing the source of inorganic nitrogen. The optimum growth of a plant like maize will occur for some definite concentration of these salts in the soil, other things being equal. Below a certain concentration, nitrogen deficiency will prevent the development of the plant, while at the other extreme an excess of nitrogen in the form of nitrates will cause inhibition of growth and even death through phytotoxicity.

The law of the minimum

This law, discovered by Liebig in the last century, states that the growth of a plant is only possible as long as all essential elements are present in sufficient quantities; and that speed of growth is controlled by the essential element that is present in the lowest concentration: growth ceases if only one essential element is present in the environment in insufficient amounts.

This law can be generalized to cover all living organisms and the whole extent of the interval of tolerance. Clearly, at the other extreme, an element present in excessive amounts will be enough to prevent development because the threshold of toxicity has been reached. Similarly, there will be an optimum value for the concentration of *each* essential element at which the speed of growth will be a maximum.

If, now, *all* the elements necessary for the development of a given species are taken into account, it is possible to define what is called the *ecological niche* in matter space. If the interval of tolerance for each essential element is represented along a different axis as in Figure 1.2, the effect of all the various concentrations is combined in a multidimensional volume, or hypervolume, within which the organism will grow. The total extent of this ecological niche will in fact be represented by a hypervolume of $n + 1$ dimensions as shown in Figure 1.3, where n is the number of physiologically essential elements for the species being studied. In Figure 1.3, $n = 2$.

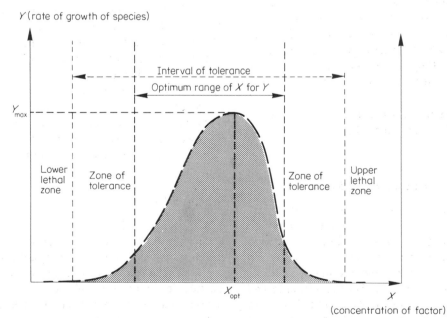

Figure 1.1 Diagrammatic illustration of the law of tolerance. X_{opt} is the preferendum, the concentration producing the maximum growth rate

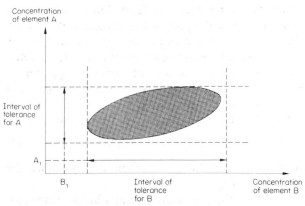

Figure 1.2   Application of the law of the minimum in the case of two elements considered simultaneously. The organism in question cannot grow or even survive with the concentrations denoted by A_1 or B_1, even if the other element has a sufficient concentration

The law of the conservation of matter

In ecological systems, matter never becomes a 'waste product' accumulating indefinitely in the surroundings. There is in fact an almost perfect and permanent recycling of matter, alternating between organic and inorganic forms, which is brought about by the three categories into which living organisms can be assigned: the primary photosynthetic producers (which are autotrophic), the animal consumers and the animal decomposers (both of which are heterotrophic). The cycle is illustrated in Figure 1.4.

1.2.2   Energy

All living systems, from the cell to the most complex ecological community, are energy converters more than anything else. At all levels of biological organization, processes exist which channel energy into the various activities of the living system in such a way that the flow of the energy is adapted and controlled. Not only that, but the utilization and the conversion of energy by biological systems are both quite remarkable in that they occur with the most extraordinary efficiency, greatly exceeding that of the most perfect machines of human design.

Energy is essential for all the vital processes at every level, from that of the most elementary cellular mechanisms to that of the entire biosphere. The body temperature of a mammal, the total number of organisms populating a biocoenosis, their speed of development, their rate of reproduction: all of these

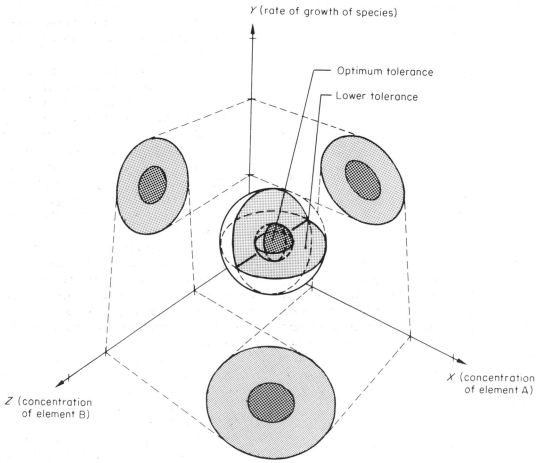

Figure 1.3   Representation of an ecological niche as a hypervolume in a space of three dimensions and its projections on the three planes which they define. (From Blondel and Bourlière, 1979, p. 350)

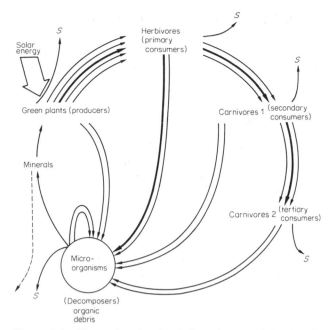

Figure 1.4   Illustrating the circulation of matter (thick lines) and the flow of energy (thin lines) in an ecosystem. S = entropy. (Adapted from Lemée, 1967)

depend on the amount of energy available. It is important to realize, however, that the flow of energy in ecological systems is subject to the laws of thermodynamics as stated by Carnot and others, and we take a brief look at these before proceeding.

The first law: the principle of the conservation of energy

Energy can be neither created nor destroyed, but only changed from one form to another. This principle can be expressed by the equation:

$$\Delta E = \Delta H - \Delta w \qquad (1)$$

where ΔE is the increase in the energy of the isolated system, ΔH the increase in heat (or enthalpy) of the system, and Δw the amount of work done by the system on its surroundings.

Since the law is universally true, it certainly applies to all biological reactions, including those involving photosynthesis (the transformation of light energy to biochemical energy), muscular work (the transformation of biochemical energy to mechanical energy), neural conduction (the transformation of chemical energy to electrical energy) and so on. It also applies to processes involving energy flow that are appropriate to whole communities of living organisms, such as primary production and secondary production.

The second law: the principal of the degradation of energy

The second law of thermodynamics states that no

processes involving a transformation of energy can take place without a partial degradation of the energy, which passes from a concentrated, ordered form to one that is dilute and unusable (in other words, to heat at a low temperature). The loss of useful energy from a system during such processes is proportional to a quantity called the entropy, S.

The principle can be expressed by the equation

$$\Delta G = \Delta H - T\,\Delta S \qquad (2)$$

where ΔG is the change in usable energy (called the free energy)

ΔH is the exchange of heat with the surroundings

ΔS is the change in entropy

and T is the absolute temperature at which the process occurs.

A corollary to this principle is that no biological process can take place with 100 per cent efficiency.

The unicyclic flow of energy

Energy is quite different from matter in that it can only pass once through any given trophic level of the food chain. During such a single cycle of energy flow in an ecosystem, the second law shows that it is degraded as it progresses, so that it is gradually dispersed and lost to the surroundings in a non-usable form (entropy), as is illustrated in Figure 1.4.

The energy upon which all ecological systems depend has only one external source: the input from the sun. As a result, the structure and functioning of communities of living organisms inhabiting a particular region on the Earth's surface will be determined by the absolute values of the flux of solar radiation at each point, together with the size of their annual fluctuations. The distribution of large ecological units (macroecosystems or biomes) over the biosphere thus depends essentially on the characteristics of the solar radiation at different latitudes.

A *law of optimalization in the use of energy* exists both at the level of species and at that of communities. All species that occupy a particular ecological niche use the available energy more efficiently than other species having similar requirements but less well adapted to the environmental conditions appropriate to that niche.

In the same way, whole ecosystems tend to evolve naturally towards a structure consisting of communities that use the available energy most efficiently. This is expressed in terms of the ratio of the biomass B of the community to the amount of energy $\mathscr{E}$ entering the biotope per unit surface area per unit time. Thus the biomass supported per unit of energy flow, $B/\mathscr{E}$, will increase as the ecosystem develops towards maturity. $B/\mathscr{E}$ ratios are low for cultivated land and natural grassland but become greater for shrublands and reach a maximum for ancient forests and, most clearly, for primeval forests and coral reefs.

Lindemann's law

An important consequence of the second law of thermodynamics concerns the transfer of energy in ecosystems: only a fraction of the energy reaching a given trophic level in a community is transmitted to a higher trophic level. As an example, if we consider the food chain:

plant → herbivore → carnivore 1 → carnivore 2

then as a general rule no more than 10 per cent of the energy corresponding to the herbivore level passes to that of carnivore 1, and so on. . . .

In reality there are wide variations in the efficiency with which animals use energy. Their rate of consumption of energy per unit mass is generally higher, other things being equal, if they are smaller. This is because the loss of energy by radiation is directly related to the surface area S of the body and this becomes greater in relation to the mass M as the linear size l decreases.

To show this more clearly, we have that

$$S = kl^2 \quad \text{and} \quad M = Kl^3$$

where k and K are constants. Thus

$$\frac{S}{M} = \frac{k/K}{l}$$

so that the ratio (surface area)/(body mass) increases as l becomes smaller, thus leading to the higher energy loss.

It can be shown experimentally that the mass M of an animal and its metabolic rate of energy consumption E are related by

$$M = AE^{3/2}$$

where A is a constant for the species under consideration.

1.2.3   Space

This is a fundamental ecological variable because the total mass of living organisms which can populate an ecosystem depends directly on the space available. In the first place, the available area determines the intensity of the competition within and between species, something that is particularly obvious in the case of cultivated plants. Here, the relation between the dry weight P of a plant and the density d of sowing (or the space available to each individual S, where $d = 1/S$) is

$$\log P = a - b \log(1/S) = a - b \log d$$

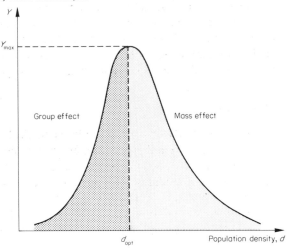

Figure 1.5   Illustration of Allee's principle. The respective domains of the group effect and the mass effect are shown on either side of the optimum

where a and b are constants. From field measurements that effectively yield values for a and b, we obtain the general relations:

$$P = k'd^{-3/2} = k'S^{3/2}$$

where k' is another constant.

In practice, the speed of growth, the individual weight, the fertility rate and other physiological factors do not vary monotonically as a function of population density either in plants or animals. There is a certain value of the density for which these factors are a maximum, and on both sides of which they consequently show a decrease. This condition is embodied in *Allee's principle*, according to which the density of population can be a *limiting factor* by being either too high or too low (Figure 1.5). On the other hand, in some cases the available space per individual can produce favourable effects if it is greater than or equal to the optimum (the effect of the group), and unfavourable effects when too low (the effect of the mass).

In the same way as with the overall density of population, the *distribution of the individuals* in each species over the available space also plays a fundamental ecological role. Classically, a distinction is made between distributions that are *random, clumped* or *uniform*. The degree of non-uniformity in a habitat is one of the chief factors governing the extent of its occupation by each living species. It has been noted, for example, that plant-eating forest insects proliferate much more easily in a homogeneous distribution of their host species than in discontinuous afforestation. Forests that are planted uniformly and monospecifically offer the most favourable conditions for population explosions among phyllophagous and xylophagous species.

The edge effect provides another example of the influence of spatial structure on populations, and an uneven topography, too, generally induces variations in the occupation of space by plants: this multiplies habitats favourable to those animal species which can then more readily find protection against predators.

1.2.4 Time

This is quite clearly another essential ecological variable. It is concerned in the evolution of every ecological system towards a state of maturity (characterized by the optimum accumulation of biomass—and therefore of energy—per unit surface area). Moreover, the normal functioning of any ecosystem implies a tendency to harness the maximum amount of energy per unit time.

During an annual cycle, the length of time during which ecological factors have tolerable values determines the types of community capable of occupying any given environment. The presence of any particular species or any particular community in a given biotope depends above all on the length of the favourable period during which sufficient biomass ('reserves') can be accumulated for living through the unfavourable season.

Again, a limiting factor of prime importance is the time available for the accomplishment of some essential biological process, such as the germination of a seed, the capture of prey by a carnivore, or the encounter with a mate for reproduction. Thus, if the density of prey diminishes, the increase in time necessary for their capture inevitably causes the expenditure of extra energy which can be fatal for the predator. In arid regions, similarly, an unusual shortening of the period during which the soil contains enough water can prevent the germination of seeds of annual plants or stop their growth.

1.2.5 Diversity

This is the last of the fundamental ecological variables and it is one that, more than any of the others, characterizes a whole community of living organisms. The concept of diversity is related to what naturalists call, rather subjectively, the *richness* of a population.[1]

The *species diversity d* can be expressed mathematically most simply in terms of two variables: the number N of individuals present in a given biotope and the total number s of species to which they belong. Then

$$d = \frac{s}{\sqrt{N}} \quad \text{(Menhinick, 1964)} \quad (1)$$

However, a deeper analysis of the structure of populations leads us to take into account the number of individuals in each species. For example, it is easy to see that if 5 per cent, say, of the species contained 99 per cent of the total population of a community, its species diversity would be much lower than if all the species were present with equal numbers of individuals.

The species diversity, then, apparently involves the number of ways that individuals can be combined to form groups (the species) within each of which the individuals are essentially similar. In addition, the greater the number of species in a given community, the greater is the degree of saturation of potential niches and thus the greater is the structural complexity of the food web. Finally, the amount of information contained in an ecological system becomes larger as its species diversity increases.

These various considerations have led certain ecologists to propose more elaborate expressions than (1) as an indication of diversity, expressions which involve the application of information theory. Several formulae exist for *general information diversity D*, such as that proposed by Margaleff in 1956:

$$D = \frac{1}{N} \left(\log_2 N! - \sum_{i=1}^{s} \log_2 n_i! \right) \quad (2)$$

where N is the total number of individuals in the community, s is the total number of species present and n_i is the number of individuals in the ith species.

Shannon's *index of general diversity* also uses logarithms to the base 2. It corresponds to the total number of questions with yes/no answers that are needed to identify an object drawn at random from a finite collection of different objects. The unit of information is the *bit* (short for *binary digit*), which corresponds to the case of two types of object present in equal proportions (only one alternative). The Shannon–Weaver index $\bar{H}$ is given by

$$\bar{H} = -\sum_{i=1}^{s} \left(\frac{n_i}{N} \right) \log_2 \left(\frac{n_i}{N} \right) \quad (3)$$

where the ratio $n_i/N = p_i$ expresses the probability of encountering the ith species in the total population N.

Although these indices are calculated on samples, the index $\bar{H}$ is in practice independent of sample size because it is mainly determined by the dominant species—in all cases the most abundant—and the error introduced by the omission of rare species is negligible.

The expression for the structural diversity of a population also depends on the level of the study being carried out. In applying the Shannon index, three types of diversity can be distinguished:

— type α, $\bar{H}_\alpha$, called the *intrabiotopic* diversity or the *microcosmic* diversity (i.e. within one community)

---

1. The species richness of a population corresponds to the total number of species present in it.

— type γ, $\bar{H}_\gamma$, called the *sectorial* or *macrocosmic* diversity: this is calculated by taking into account the whole collection of mixed biotopes contained in a given geographical region

— type β, $\bar{H}_\beta$, which compares the differences of population between two adjacent biotopes 1 and 2, and which is given by

$$\bar{H}_\beta = \bar{H}_{a_{12}} - 0.5(\bar{H}_{a_1} + \bar{H}_{a_2}) \qquad (4)$$

Using logarithms to the base 2 once more, $\bar{H}_{a_1}$ is the Shannon index for population 1, $\bar{H}_{a_2}$ for population 2, and $\bar{H}_{a_{12}}$ for both together. Hence $\bar{H}_\beta$ represents an *index of similarity* between biotopes. Blondel (1979) gives some excellent examples of such indices (see Figure 1.6).

Another important concept in analysing the structure of communities is that of *equitability* or *evenness*. This is expressed as the ratio between the observed diversity $\bar{H}$ and the maximum theoretical diversity $\bar{H}_{max}$ (calculated by assuming that the species present in a given sample are all represented by equal numbers). The equitability E is then given by the simple relation

$$E = \bar{H}/\bar{H}_{max} \qquad (5)$$

There is theoretically a minimum diversity $\bar{H}_{min}$ in which all the species but one would be represented in a sample by a single individual and the one species by all the rest of the individuals. A *coefficient of redundancy* R can then be calculated using the relation

$$R = \frac{\bar{H}_{max} - \bar{H}}{\bar{H}_{max} - \bar{H}_{min}} \qquad (6)$$

so that $R = 0$ for $\bar{H} = \bar{H}_{max}$ and $R = 1$ for $\bar{H} = \bar{H}_{min}$. Just as with equitability, a monitoring of changes in

this coefficient allows us to tell whether some of the species in a population are proliferating more than others.

The structure of a community or of a particular group can also be represented by the relative frequencies of the various classes chosen according to the taxonomy of the population. Curves showing how the relative abundances of species are distributed, or the mathematical formulae corresponding to the curves, are both quantitative ways of expressing the species diversity (Figure 1.7).

The species diversities of ecosystems depend not

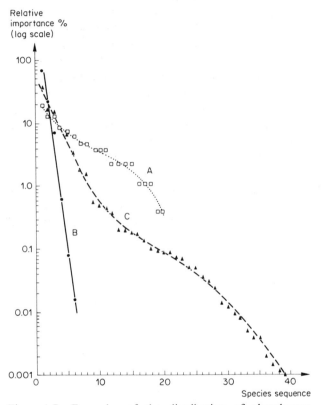

Figure 1.7 Examples of the distribution of abundances within a population. In the data shown for three natural communities, the relative importance of each species is plotted against its position in the ordered sequence from highest to lowest importance values (importance is measured by quantities such as *density* for animal species or *surface covered by foliage* for plant species). The curves A, B and C are those resulting from three different hypotheses as to the way space is occupied by various species in a community, thus demonstrating that different hypotheses are applicable to different populations:

A The MacArthur distribution (from the random niche-boundary hypothesis). Data for nesting bird pairs in a deciduous forest.

B The geometrical series distribution (from the niche pre-emption hypothesis). Data for vascular plant species in a sub-alpine fir forest.

C The log-normal distribution (occupied niche spaces determined by large numbers of independent variables). Data for vascular plant species in a deciduous cove forest.

(From Whittaker, 1975)

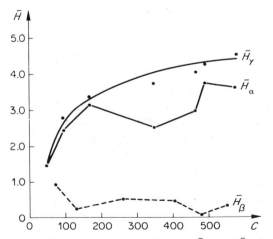

Figure 1.6 Shannon indices of diversity $\bar{H}_\alpha$ and $\bar{H}_\gamma$ for an assortment of biotopes in Provence plotted against an index of complexity C for the vegetation in the regions studied. Also plotted is the index of similarity $\bar{H}_\beta$ for pairs of adjacent biotopes. (From Blondel, 1979)

only on their area, but on the stability of the habitats they contain and on their degree of geographical isolation.

If we look first at stability, it is noticeable that in general the diversity decreases when passing from an environment with stable conditions regarding climate, soil, etc., to one where conditions undergo fluctuations that are irregular or totally aperiodic. Thus if a course is traced along a transect of increasing height (Figure 1.8) or decreasing rainfall, the diversity is observed to decrease: during the annual cycle, mountainous or arid environments are subject to irregular fluctuations of large amplitude in their conditions.

A high species diversity is only one consequence of stability in climatic factors over a long period. In fact, such stability produces a high value of a much wider 'ecological' diversity in the spatial distribution of habitats and species (*pattern diversity*, Pielou, 1966).

There is also a strong correlation between the age of an ecosystem and the magnitude of its species diversity: a field of cereals or a plantation of young spruces show much lower values than those of natural grassland or a climax community of mature spruces. In the same way, coral reefs among oceanic biomes and tropical rain forests among terrestrial ones have the largest species diversities of all ecosystems. Their great age—millions of years or even tens of millions—demonstrates the great stability of the environments in which they have developed and is the fundamental reason for their extraordinary diversity.

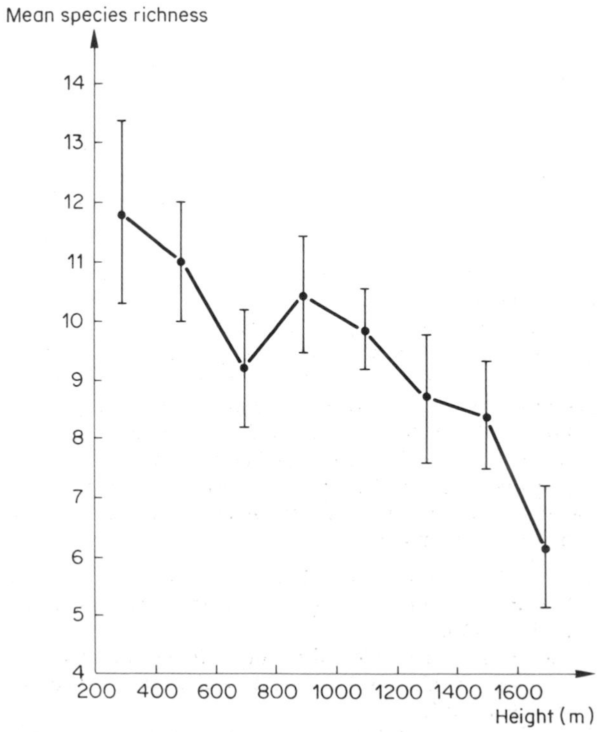

Finally, geographical isolation lowers the species diversity: other things being equal, an island will harbour fewer plant or animal species than a continental region of the same area.

1.3 Ecological Principles Governing the Use of Natural Resources

Certain ecological principles must be respected if the resources of the biosphere are to be protected and used rationally, principles which involve phenomena occurring at all the various levels of organization: population, community, macro-ecosystem and finally the entire biosphere (in increasing order of complexity).

Definition of a resource

A resource can be most simply defined as a form of energy and/or matter which is essential for the functioning of organisms, populations and ecosystems. In the particular case of humans, a resource is any form of energy or matter essential for the fulfilment of physiological, socio-economic and cultural needs, both at the individual level and that of the community.

When the functioning of natural ecosystems or of technological civilizations is analysed, it is found that their usage of natural resources involves a permanent transformation of matter (in living organisms, through their metabolic activity; in human societies, through industrial processes). This transformation is the result of a continuous flow and consumption of energy (originating from the sun as regards the biosphere or from fossil fuels as regards technological civilization).

Such considerations lead us to another definition of the term 'resource'. As already mentioned, a resource may consist of one of the various forms of primary energy present in the ecosphere. In addition, however, it may be defined as anything needed by a living organism such that an increase in its availability leads to an increase in energy flow through the organism, and thus a greater rate of energy conversion. In this definition, the term 'living organism' may be interpreted at any level of complexity: from the individual, through a given population, to a complete ecosystem.

The five fundamental ecological variables already discussed are thus included among the natural resources, since it is quite easy to see that energy, matter, space, time and diversity all fit the definition just given. As a result of this, all theories concerning the use of natural resources must depend on the laws which govern variations in these five quantities.

Figure 1.8 Variation of the species richness with height: showing the decrease in the mean number of species of nesting birds as height increases on the slopes of Mont Ventoux, Provence. (From Blondel, 1979)

Renewable and non-renewable resources

This is a classic distinction in which all resources are subdivided into two separate groups. However, it is not always easy to see how to make such a distinction, since it will depend on the time-scale adopted and on the physical extent of the resource under consideration.

Non-renewable resources are those dependent on a finite stock and not reproducible—like fossil fuels, or raw materials whose usage involves dispersal (e.g. phosphate fertilizers). Renewable resources are those that are reproducible: in other words, obtained from the biomass of living organisms.

Solar energy is a special case, and although it has a fixed rate of flow (the intensity of solar radiation) it can be classified roughly as a renewable resource inasmuch as solar 'reserves' are inexhaustible on a human scale.

1.3.1   Principles concerning energy

(a) The production of biomass in a given ecosystem, and thus the availability of plant and animal resources, is governed in the first place by *solar radiation*. Indeed, the potential biological productivity in every ecosystem is determined by the mean intensity of this radiation more than by anything else. In the final analysis, the whole of the energy flow in a given community depends on the amount of solar energy captured per unit area per unit time by the leaf systems and stems of green plants.

(b) Energy is transformed in *food webs* from one form to another but is neither created nor destroyed, in accordance with the first law of thermodynamics.

Both in primary producers and in animal consumers (= secondary producers), part of the energy available is stored in the biomass, with the remainder being used up in metabolic processes. Thus in autotrophic plants, only a fraction of the incident solar radiation is converted into biochemical energy through photosynthesis. After that, immediate photorespiration and other forms of respiration will reutilize a large proportion of the solar radiation already converted into biochemical substances. Ultimately, the energy P_N stored in the biomass through the processes of growth and reproduction is related to the gross intake of energy P_B by the expression

$$P_N = P_B - R_a$$

where R_a is the total respiratory consumption by the plants necessary for the maintenance of growth and reproduction.

In animals, the energy flow undergoes a greater number of 'partitionings' than it does in plants. Here, in addition to the energy needed for basal metabolism and for growth and reproduction, there is a certain amount required for locomotion and, in homoiotherms, the consumption needed to maintain a constant internal temperature. Figure 1.9 illustrates schematically the energy flow in an arbitrary animal species. Some losses occur when food is ingested since it is never completely assimilated (faeces). After that, there is further consumption of energy, in the shape of other excreta and of radiated heat, associated with metabolic processes occurring at all levels of organic activity.

The efficiency of energy conversion in animal biomass (secondary productivity) is really quite low and the quantity stored at the end of a metabolic chain by means of growth and reproduction only represents a small fraction of the amount taken in by the organism through feeding.

Because energy is a basic resource, all living species—including humans—make optimum use of it by adopting an *energy strategy* which becomes evident when the partitioning is examined at the level of the whole population of the species concerned.

One category of plant and animal species (annual grasses, most insects, rodents) devotes a high proportion of its energy intake to reproduction, the rest being used for the growth of the individuals. Thus, with the field mouse *Microtus arvalis*, females devote around half of the energy to their growth and the rest to gestation. Here the energy strategy favours reproduction because 30–53 per cent of the energy converted into biomass is incorporated in the offspring and the remainder (70–47 per cent) is stored in the tissues of the adult female.

At the other extreme, a second category consisting of large, long-lived organisms (humans, elephants, whales) stores the majority of the available energy in the growth and reproduction of the adult biomass, only a small fraction going to the offspring. Thus in the blue whale *Balaenoptera musculus* 97 per cent of the energy ends up in the biomass of the adult female and only 3 per cent in the young.

The first category is said to adopt a demographic strategy of type r, which is characterized by a very high intrinsic rate of natural growth. The second category consists of species having a low fecundity, a high capacity for surviving unfavourable ecological periods and a tendency to display a maximum biomass depending on the capacity of their environment to support it. This is called a strategy of type K.

Another aspect of the energy strategy adopted by animal species concerns their mobility in the environment. As a general rule, r-type strategies are observed to devote a smaller proportion of their available energy to locomotion than do K-type strategies. In this way, r-strategies leave a greater amount of energy for reproduction, but this is offset by their low mobility which prevents them from escaping out of an environment when conditions have become unfavourable.

The efficiency of energy conversion in ecosystems is always low and it can under no circumstances

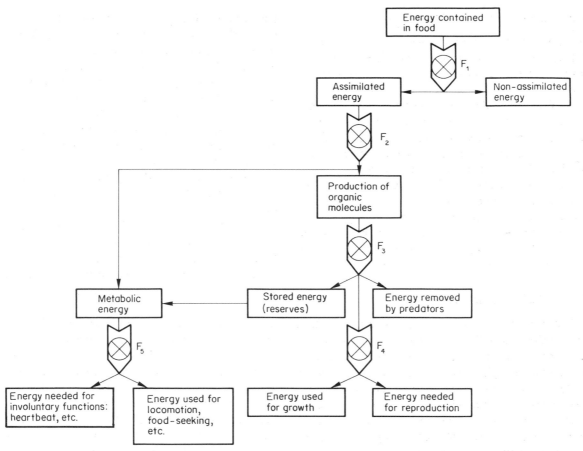

Figure 1.9   Flow-chart for the available energy in an arbitrary animal species. F_1, F_2, ... indicate the various levels at which the energy flow is partitioned, while the rectangular boxes show the physiological processes through which the energy passes

approach 100 per cent. This is a consequence of the second law of thermodynamics which, as we have already seen, shows that all transformation of energy is accompanied by a loss of quality in some of the initial amount present. Thus, at all stages in food webs or food chains, there will be a loss of useful energy, reflecting what is observed during typical metabolic processes in each individual (see above).

A classic example of the very low efficiency of ecosystems and of the low yield from energy conversion in food chains has been given by Odum (1959). He calculated the various energy transfers that occurred in the theoretical case of a child fed strictly for a year on meat from a calf which itself ate only lucerne. The food chain is

solar radiation → lucerne → calf → child

Odum calculated that only 0.24 per cent of the incident solar radiation is used for growth by the cultivated plant, only 8 per cent of that growth is used by the calf, and barely 0.7 per cent of the energy associated with the biomass of the calf contributes to the growth of the child. Overall, little more than a millionth of the initial solar radiation is converted into carnivoral biomass, in this case the biomass of the child.

Out of the total solar radiation incident on the earth of about 10^{18} kcal per day, some 10^6 kcal per day is taken up by producers, around 10^4 kcal per day passes to herbivores and only 10^3 kcal per day goes to carnivores of the first order. That is an excellent example of Lindemann's law and it illustrates why food chains cannot be of unlimited length.

In terrestrial environments, the longest chains are generally of the type:

trophic level:

I   II   III   IV

plant → herbivore → carnivore 1 → carnivore 2

As an example, we have in a continental environment:

grass → field mouse → weasel → eagle owl

and most common of all:

grass → cattle → humans

In an aquatic environment, we can have:

phytoplankton → zooplankton → microphagous fish → predator fish → superpredator fish

and one of the longest food chains in an oceanic

environment but also involving humans is that of the tunafish:

trophic level:

I   II   III

phytoplankton → zooplankton → anchovy →

IV   V   VI

mackerel → tuna → humans

This last example provides another illustration of Lindemann's law since the amount of energy stored in the biomass of the tunafish represents at best only 10^{-5} of that contained in the biomass of the phytoplankton used to produce it! It is clear that humans use oceanic resources with a derisory energetic efficiency: because the fish that are harvested commercially are mostly to be found in trophic levels IV and V, this would be equivalent on land to our being fed only on lion or vulture meat, as has been forcefully pointed out by Watt.

A corollary of this principle shows that the overall energy efficiency of a food chain is the greater the shorter it is. As a result, it would be in the interests of

the nutritional needs of the human race to be placed at the level of a herbivore rather than at that of a carnivore, since that would greatly increase the amount of energy obtainable from plants. However, increases in crop production do not follow the same pattern as those of human populations. . . .

1.3.2 Principles concerning matter

(a) In the light of human needs, no natural resource can be said to exist in unlimited quantities, whether it be mineral or biological. The availability of a resource that seems on the face of it to be present in the ecosphere in enormous amounts is in practice restricted by various abiotic and/or biotic factors. Such factors as the spatial distribution of a mineral element or biogenic molecule (the case of water), the difficulties of transference from the lithosphere (the case of phosphorus) or from the atmosphere (the case of nitrogen), and many others, may combine in such a way that many chemical elements and mineral

Table 1.3 Primary productivity, net production and plant biomass of large biomes (expressed in tonnes of dry organic matter) (from Whittaker and Likens, in Lieth and Whittaker, 1975)

Ecosystem type	Area (10^6 km²)	Net primary productivity (g m^{-2} yr^{-1}) Normal range	Mean	Net production worldwide (10^9 t yr^{-1})	Biomass per unit area (t ha^{-1}) Normal range	Mean	Total biomass worldwide (10^9 t)
Tropical rain forest	17.0	1000–3500	2200	37.4	60–800	450	765
Tropical seasonal forest	7.5	1000–2500	1600	12.0	60–600	350	260
Temperate evergreen forest	5.0	600–2500	1300	6.5	60–2000	350	175
Temperate deciduous forest	7.0	600–2500	1200	8.4	60–600	300	210
Boreal forest (taiga)	12.0	400–2000	800	9.6	60–400	200	240
Woodland and shrubland	8.5	250–1200	700	6.0	20–200	60	50
Savanna	15.0	200–2000	900	13.5	2–150	40	60
Temperate grassland	9.0	200–1500	600	5.4	2–50	16	14
Tundra	8.0	10–400	140	1.1	1–30	6	5
Desert and semi-desert scrub	18.0	10–250	90	1.6	1–40	7	13
Extreme desert (sand), polar regions	24.0	0–10	3	0.07	0–2	0.2	0.5
Cultivated land	14.0	100–3500	650	9.1	4–120	10	14
Swamp and marsh	2.0	800–3500	2000	4.1	30–500	150	30
Lake and stream	2.0	100–1500	250	0.5	0–1	0.2	0.05
Total continental	149.0		773	115.0		123	1837
Open ocean	332.0	2–400	125	41.5	0.01–0.05	0.03	1.0
Upwelling zones	0.4	400–1000	500	0.2	0.5–10	2.0	0.008
Continental shelf	26.6	200–600	360	9.6	0.1–4.0	1.0	0.27
Algal bed and coral reef	0.6	500–4000	2500	1.6	0.4–40	20	1.2
Estuaries	1.4	200–3500	1500	2.1	0.1–60	10	1.4
Total marine	361.0		152	55.0		0.1	3.9
Full total	510.0		333	170.0		36	1841

Plate I   Limiting factors

1   The sun provides the only external source of energy for the biosphere. Thermo-nuclear reactions occurring in the solar interior (essentially a 'burning-up' of hydrogen) deliver a flux of energy to the ecosphere that has been remarkably constant for close on 5000 million years. Astrophysicists estimate that the sun will maintain its present level of activity for another 5000 million years before becoming a red giant.

2   A tropical rain forest on the slopes of Mount Kenya. In general, the availability of water is a prime limiting factor in continental ecosystems. In tropical rain forests, the high biological productivity and exceptional species diversity are explained by the combination of an abundant rainfall and a stable and favourable temperature.

3   Death Valley in the north of the Mojave Desert in California. Here, the lack of rain produces a low species diversity and a negligible biological productivity
(Photographs F. Ramade)

Plate II   Consumers and food chains

1   Herbivores form by far the largest proportion of the biomass of consumers. The photograph shows a herd of wildebeest (*Connochaetes taurinus*), a species of antelope dominant in the savannas of East Africa. This is a region in which food chains of the type grasses → ungulates → carnivores still exist undisturbed by human activity, at least in the nature reserves.

2   In the Masai Mara Reserve (Kenya), a lioness prepares to devour a wildebeest she has just killed, while vultures gather, awaiting their chance.

3   A colony of water birds on the banks of Lake Nakuru in Kenya. Cormorants, pelicans and other fish-eating birds are situated at the peak of complex food webs consisting of five or six superposed levels. This feature makes them particularly vulnerable to contamination of the environment by persistent pollutants.
(Photographs F. Ramade)

molecules which are abundant on the Earth's surface become limiting factors for primary and secondary production because of their unavailability. In the case of biological resources, even those whose biomass in a given ecosystem appears considerable, we must not lose sight of the fact that their true availability depends on the energy flow in that ecosystem: in other words, on the productivity.

Tables 1.3 and 1.4 give figures for the primary and secondary productivities of the large continental and oceanic ecosystems. It is noticeable that no proportional relation exists between total biomass and productivity. In particular, environments like tropical rain forests or animal species like whales and other cetacea, all with considerable biomass, do not exhibit a distinctly greater primary or secondary productivity than more modest ecosystems or species. Ignorance of that fact has led to some well-known miscalculations and sometimes to disasters: laterization of tropical forest soils, and the quasi-extinction of certain species of whale through overfishing are examples of this.

(b) A second principle that concerns matter is a generalization of Liebig's law and it takes the following

form: both the scarcity (or even just a relative lack of availability) of a mineral element and its excess through natural or human agency (pollution) can act as factors limiting the take-up of other natural resources.

The principle is well illustrated by the case of water in semi-arid climates and by that of phosphorus in land-based or aquatic environments. In the latter example, for instance, the concentration of phosphates in soils or water is a basic factor that determines the productivity of continental or oceanic ecosystems. Fresh-water environments illustrate this: a relative deficiency of phosphorus is responsible for the low productivity of oligotrophic lakes, while at the other extreme an excess of phosphorus stimulates so much primary productivity that it is the main cause of dystrophic lakes—in which, for example, salmon and trout breeding can be ruined because of the lowered secondary productivity.

(c) A third principle concerning matter is what we shall describe as the *law of the optimum*. This states that an increase in the availability of a nutrient, whether a mineral element or an organic molecule, does not produce an unlimited increase in productivity:

Table 1.4   Secondary production and productivity in the biosphere (from Whittaker and Likens, in Lieth and Whittaker, 1975)

Ecosystem type	Area (10⁶ km²)	Leaf-surface area (10⁶ km²)	Biomass of litter (10⁹ t)	Animal consumption (10⁶ t yr⁻¹)	Secondary production (10⁶ t yr⁻¹)	Secondary productivity of animal matter (kg ha⁻¹ yr⁻¹)	Animal biomass (10⁶ t)
Tropical rain forest	17.0	136	3.4	2600	260	152.9	330
Tropical seasonal forest	7.5	38	3.8	720	72	96.0	90
Temperate evergreen forest	5.0	60	15.0	260	26	52	50
Temperate deciduous forest	7.0	35	14.0	420	42	60	110
Boreal forest (taiga)	12.0	144	48.0	380	38	31.7	57
Woodland and shrubland	8.5	34	5.1	300	30	35.3	40
Savanna	15.0	60	3.0	2000	300	200	220
Temperate grassland	9.0	32	3.6	540	80	88.9	60
Tundra	8.0	16	8.0	33	3	3.8	3.5
Desert and semi-desert scrub	18.0	18	0.36	48	7	3.9	8
Extreme desert (sand), polar regions	24.0	1.2	0.03	0.2	0.02	0.008	0.02
Cultivated land	14.0	56	1.4	90	9	6.4	6
Swamp and marsh	2.0	14	5.0	320	32	160	20
Lake and stream	2.0			100	10	50	10
Total continental	149.0	644	111	7811	909	61	1005
Open ocean	332.0			16600	2500	75.3	800
Upwelling zones	0.4			70	11	275.0	4
Continental shelf	26.6			3000	430	161.7	160
Algal bed and coral reef	0.6			240	36	600	12
Estuaries	1.4			320	48	342.9	21
Total marine	361.0			20230	3025	83.8	997
Full total	510.0			28041	3934		2002

there is always either an asymptote or a maximum beyond which the productivity decreases.

A good illustration of this principle is provided by the *law of diminishing yield*, which is well known to agronomists. According to this, the yield from cultivated land never increases in proportion to the supply of fertilizer. Instead, when the latter grows in a geometric progression, the yield increases roughly in arithmetic progression—although this cannot be exact because as the fertilizer supply tends to infinity, the yield tends to an asymptotic limit.

In cases where the law of diminishing yield applies, the increase in availability of resources in the ecosystem is a *non-stimulatory effect*. At the other extreme, resources may induce *stimulatory effects*. Here, in an initial phase, an increase in the availability of a resource produces a growth in productivity (or other biological processes) that is more than merely proportional. Gradually, however, further increases in availability will produce limiting factors that slow down the growth, as in Figure 1.10.

There are two different ways in which the availability of a resource can affect ecological processes like productivity. In the first category are resources which produce a stimulatory effect, where the relationship is expressed by logistic or sigmoid curves as in Figure 1.10. After the initial rise, the effects eventually tend to an asymptotic limit as the availability tends to infinity. This type of response is observed not only for resources under the heading of 'matter' but also for energy.

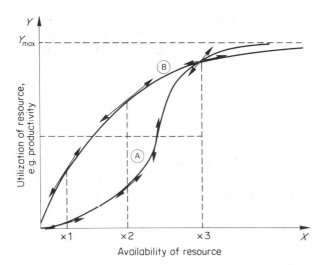

Figure 1.10 Difference between resource utilization of the stimulatory type (A) and non-stimulatory type (B). In the first case there is an intermediate region where the increase in availability of the resource stimulates further utilization. In the second case, which corresponds to the law of diminishing yield, the growth in availability of the resource is not accompanied by a proportional increase in its utilization. As the resource increases, the corresponding slope of the tangent to the curve decreases

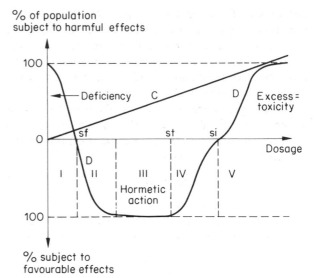

Figure 1.11 The curve D illustrates the dose–response relation for an ecological factor of the hormetic type. Here the availability of the substance in excess (toxicity) acts as a limiting factor just as much as deficiency does. The line C is the dose–response relation for the case of a toxic agent which does not have a threshold of toxicity. (From Ramade, 1979a)

In the second category are resources for which an increase in availability produces a growth in the effects up to a maximum, beyond which they begin to decrease. Here, excessive concentration of an element will have harmful effects and will become toxic or pollutant. This type of relationship between effect and concentration is observed with a large number of mineral and organic substances which are said to have *hormetic* properties. They are materials, for instance, which are indispensable to living organisms in small doses but which become toxic beyond certain concentrations: fluorine, cobalt and some vitamins are well-known examples (Figure 1.11).

1.3.3 Principles concerning space

(a) The space available to each biological entity, whether an individual, a species or a community, is always restricted because the area it occupies is smaller than that covered by the complete ecosystem to which it belongs. This proposition is just as true for humans as for other species: the amount of land capable of being cultivated in any country is always smaller than the total area because of the relief or the poor structure of some of the soil, or for bioclimatic reasons.

(b) The area available for ensuring that each individual is fed decreases more rapidly than the population density increases. In no case is the area *per capita* available for food production, s_a, related to the total available area S and the total population N by simple proportionality: $s_a = S/N$ does not hold.

For each animal species, a far from negligible

proportion of the territory is assigned to uses other than food supply, particularly to movement and to the nesting or living quarters. This is especially true for humans, where non-agricultural uses of the soil (roads, dwellings, industrial zones) consume a significant fraction of the available land.

It often goes unrecognized that the area *per capita* assigned to uses other than agriculture, s_n, is not constant but varies as a growing function of the population size according to the relation

$$s_n = kSN^a \qquad (1)$$

where k and a are constants, with $a > 1$ and k dependent on the units of area chosen. The area *per capita* available for food production is now given by

$$s_a = \frac{S - s_n}{N} \qquad (2)$$

so that (1) and (2) together give

$$s_a = \frac{S}{N} - kSN^{a-1} \qquad (3)$$

for the variation of s_a with the number of inhabitants N.

1.3.4 Principles concerning diversity

Of all the fundamental ecological variables, diversity is the one that best characterizes the biotic component of ecosystems. It embraces the whole collection of properties peculiar to renewable biological resources and it can, conversely, itself be regarded as a resource.

(a) The first principle concerns the relationship between diversity and the stability of ecological factors. The greater this stability, the greater the diversity of the ecosystem: biocoenoses which have evolved in stable environments or which have suffered only regular and foreseeable fluctuations show the maximum species diversities. I have already quoted coral reefs and tropical rain forests as examples of this.

In contrast to that, all human activity in a natural ecosystem, because it is unforeseeable and almost always acyclic, inevitably has the effect of decreasing the species diversity. Not only that, but the principle is applicable in many cases where adaptation by an unwanted species is prevented by deliberately unseasonal human intervention: the case of treatment by pesticides is an example.

(b) A second principle relates to the effects of another resource, time, on the species diversity, effects which are manifested as a fundamental ecological phenomenon called *succession*. The diversity of an ecosystem increases as a function of time from an initial state where it has a minimum to one where it has a maximum value, when the ecosystem is called a *climax community*.

One application of this principle is observed when humans cease to intervene in a given ecosystem. Examples are rangelands which have been overgrazed as a result of excessive numbers of domestic animals, and moorland or garrigue that has been produced by excessive clearing of primitive forest cover and/or by fire. In these cases it is noticeable that the ecosystem evolves continuously towards its earlier state characterized by a maximum species diversity.

(c) The biomass/productivity ratio of an ecosystem is proportional to its diversity.

The diversity is a measure of the degree of organizational complexity in ecosystems: in other words, their degree of negentropy. It is thus clearly linked to the amount of energy (B) stored in the biomass and to the flow of energy per unit time (P) which traverses the ecosystem (productivity). The mean length of time during which a given quantity of living matter (and therefore of energy) remains in the system is given by the expression

$$t = k\frac{B}{P} \qquad (4)$$

where k is a constant. The ratio B/P has the dimension of time because, for instance, B can be expressed in kcal per hectare and P in kcal per hectare per year.

However, it is also quite clear that the time taken for a quantity of energy to traverse an ecosystem will be the greater with greater diversity. Diversity is, after all, a measure of the complexity of food webs. The path of the energy from the producers through to the super-carnivores and the detritivores will quite obviously take longer in a food web comprising a large number of steps than in a short linear chain of the type producer → herbivore. Consequently, we can link the diversity $\bar{H}$ to the ratio B/P by the expression

$$\bar{H} = K\frac{B}{P} \qquad (5)$$

where K is a constant.

(d) The ratio B/P grows as a linear function of diversity and/or time up to an asymptotic limiting value. A rearrangement of (5) shows that

$$\frac{B}{P} = \bar{H}/K$$

so that B/P and $\bar{H}$ are proportional.

Experience shows that this principle is certainly valid. The B/P ratio is low for a field of cereal or artificial grassland, and is a maximum in primitive tropical forests. Moreover, during an ecological succession, the increase in diversity from the initial stage to the climax stage is accompanied by a growth in the B/P ratio as a function of time.

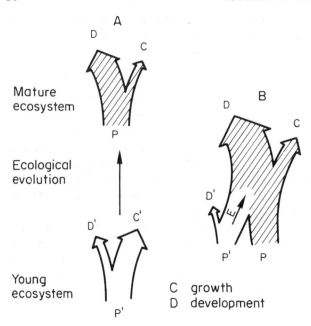

Figure 1.12 The distribution of energy absorbed by an ecosystem depends on its diversity level and on its degree of exploitation.
    Diagram A: In an immature ecosystem, most of the energy goes into the increase in biomass and only a small porportion into the development and maintenance of the various structures. The ratio is gradually reversed during ecological succession, a greater and greater proportion being devoted to development.
    Diagram B: The exploitation of an immature ecosystem by a mature one is shown by the partial or total injection of the increase in biomass of the first into the second, which then develops and grows faster than if it were limited to its own resources. The excess energy which was allowing the immature ecosystem to develop is consumed as it becomes available by the mature ecosystem. The exploited system thus has its development retarded, inhibited or even reversed

    (e) The most diversified ecosystems are the most stable. The capacity of a community for homeostasis will grow with its diversity: there will be a larger number of links in the food web and more organisms ·with redundant functions.
    By contrast, young ecosystems of low diversity will possess communities with few species, using r-type strategies, which will thus be subject to large fluctuations of population when environmental conditions change. In that case, the absence or rarity of species with equivalent functions will make a homeostatic regulation of the community uncertain or impossible.
    (f) Diversified ecosystems exploit less diversified ecosystems. In general, mature systems (climax communities) are observed to remove energy, matter and diversity from immature or less diversified systems that surround them.
    A classic example is provided by forest–savanna contact in tropical regions. Forest animals mostly seek their food in the neighbouring open ground where grass is the predominant ground cover. By doing that, they

prevent the savanna from evolving to a more diversified stage: the removal of seeds by birds, and the crushing of young trees, prevent the more advanced stages of succession from developing. The transition to tree-covered land is thus blocked by the exploitation and the savanna remains in a state of immaturity.
    A similar example is observed in a marine environment at the meeting zone between blue tropical waters and the immature ecosystem of the colder waters. Predators in the tropical zone seek their food in the cold waters of high productivity and maintain the latter ecosystem in a less advanced successional stage (Figure 1.12).

1.3.5 Principles concerning populations

The growth and regulation of natural populations are governed by a number of laws which will now be developed. Knowledge of these is essential for the rational exploitation of the resources in the biosphere.

The law of population growth in the absence of limiting factors

When an environment temporarily provides a superabundance of natural resources, the populations that live in it can grow without the restraint of limiting factors. Yeasts or insects, for example, can be placed under such conditions in the laboratory; and we ought not to forget that the human species is experiencing similar favourable conditions because of medical advances, agricultural production and so on, which have produced the population explosion of our times.
    When limiting factors are absent, the rate of growth of a population with time is observed to remain constant. Suppose N_0 is the population at time t_0, N that at time t, when this rate of growth *per capita* R will be given by the expression

$$R = \frac{(N - N_0)}{N(t - t_0)} = \frac{\Delta N}{N\,\Delta t} \qquad (1)$$

When $\Delta t \to 0$, let $R \to r$, so that (1) becomes

$$r = \frac{1}{N}\frac{dN}{dt} \qquad (2)$$

r is called the intrinsic rate of natural growth. From this

$$dN = rN\,dt \qquad (3)$$

and integration of this equation leads to the solution

$$N = N_0\,e^{r(t - t_0)} \qquad (4)$$

Taking t_0 as the origin of time means that $t_0 = 0$, so that

$$N = N_0\,e^{rt} \quad \text{(Figure 1.13, curve A)} \qquad (5)$$

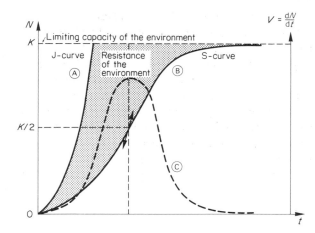

Figure 1.13 Curves giving the theoretical population in the absence (A) and the presence (B) of limiting factors. The curve C represents the gradient of B and thus shows the variation in the speed of growth

The speed of growth of the population V can be deduced from (3) and is

$$V = \mathrm{d}N/\mathrm{d}t = rN \qquad (6)$$

This shows that when a population is not subject to any limiting factors, its speed of growth increases exponentially with time!

A number of important parameters can be obtained from these expressions:

(a) *Calculation of r* In general, the rate of growth R (per year or per generation) is known. Thus for the human species at the present time, $R = 2$ per cent per year. This means that after 1 year, $N = N_0(1 + R)$ and we have from (5) since $t = 1$:

$$1 + R = e^r \qquad (5')$$

giving

$$r = \log_e(1 + R) \qquad (7)$$

For small values of R, r will be approximately equal to R, but this can no longer be assumed for high rates of growth. For example, in the case of a species that doubles its population each year, $R = 100$ per cent, whereas $r = \log_e 2 = 0.693$.

(b) *Doubling time* This is the time taken for the size of a population to double. Call this time t_2, at which $N_t = 2N_0$. From (5) we then have $2 = e^{rt_2}$ so that

$$t_2 = \frac{\log_e 2}{r} = \frac{0.693}{r} \qquad (8)$$

Take the case of a country like Mexico where the annual rate of growth is 3 per cent. We can put $r \sim R = 0.03$. In this case, using this value of r in (8), the doubling time will be of the order of 23 years. Since Mexico had 67 million inhabitants in 1979, this means that its population will approach 135 million by the beginning of the next century, unless there is soon a

complete reversal of its demographic policy . . . which would appear difficult, to say the least.

The law of population growth with limiting factors

In the great majority of cases, all living species will very quickly find themselves confronted by the *resistance of the environment*, which opposes population growth more strongly as the size of the population itself increases. Limiting factors, both extrinsic and intrinsic, will only allow the number of individuals to reach a ceiling called the *limiting capacity* of the environment, K in Figure 1.13.

The mathematician Verhuist, who in 1838 was the first to propose a theory of the growth of populations in a restrictive environment, considered that the rate of growth R is a linearly decreasing function of the population N:

$$R = r\left(1 - \frac{N}{K}\right) \qquad (9)$$

where r is the intrinsic rate of natural growth in the absence of limiting factors and K is the limiting capacity, or the maximum value that N can attain. It can be seen from (9) that at the beginning of the growth of a population the instantaneous rate of growth R is equal to r. Then, as $N \to K$, $R \to 0$.

Since in general R is equal to $(1/N)\,\mathrm{d}N/\mathrm{d}t$, the speed of growth is given by

$$\frac{\mathrm{d}N}{\mathrm{d}t} = RN = r\left(1 - \frac{N}{K}\right)N \qquad (10)$$

This can also be written in the forms:

$$\mathrm{d}N = N\left(r - \frac{rN}{K}\right)\mathrm{d}t \qquad (11)$$

and

$$\mathrm{d}N = N(a - bN)\,\mathrm{d}t \qquad (11')$$

where a and b are constants. The solution of this differential equation is of the form

$$N = \frac{K}{1 + c\,e^{-rt}} \qquad (12)$$

where

$$c = \frac{K - N_0}{N_0}, \qquad (13)$$

N_0 being the initial population at time $t = 0$. Equation (12) is represented graphically by a sigmoid or S-curve as shown in Figure 1.13.

The speed of growth of the population is given by the slope of the S-curve and is also shown in Figure 1.13. It varies with time in such a way that it increases to a maximum and then decreases. The point at which the maximum is attained can be calculated from (10), from which we obtain by differentiation

$$\frac{\mathrm{d}^2 N}{\mathrm{d}t^2} = r\left(1 - \frac{N}{K}\right) - \frac{rN}{K} = r\left(1 - 2\frac{N}{K}\right) \quad (14)$$

The rate $\mathrm{d}N/\mathrm{d}t$ is a maximum for $\mathrm{d}^2 N/\mathrm{d}t^2 = 0$, and from (14) we see that this condition is realized when $N = K/2$. We can thus see that in such a population the maximum speed of growth is attained when the size of the population is half the limiting capacity of the environment.

It should be noted that in Figure 1.13, two growth curves are plotted for the same initial population N_0 and the same intrinsic rate of natural growth r: one is an exponential curve, called the curve of *biotic potential*, expressing the growth in the absence of limiting factors, while the other is the sigmoid curve we have just calculated. For times near the origin, the two curves are practically coincident, but as time increases a greater and greater gap appears between them which corresponds to the resistance of the environment, $1 - N/K$ or $(K - N)/K$.

The tangents to the S-curve of Figure 1.13 give a graphical indication of the variations in the speed of growth as a function of time. The point of inflexion in the S-curve, at $N = K/2$, is at the point of maximum slope of these tangents. The curve C in the figure, giving the variation of the speed of growth, recalls the shape of the Gaussian curve.

In practice, it is observed in real natural populations approaching the limiting value K that the value of R does not fall according to $r(1 - N/K)$ given by (9). This is because the population can find enough resources in the environment to live on and to exceed the value K for a time. Ultimately, the resistance of the environment comes into play and causes a temporary collapse in numbers. There will then be oscillations about K having more or less regular periods (Figure 1.14). Various mathematical expressions have been proposed to account for such fluctuations in natural populations as their numbers approach the limiting capacity.

Stability of populations

In a stable environment subject only to small fluctuations in its ecological factors the populations themselves are stable.

We saw in section 1.3.4 that mature ecosystems exhibit a maximum diversity. Now the greater the complexity of a biocoenosis the greater will be its degree of homeostasis. High diversity carries with it very low levels of wastage (both of matter and energy) because of the great variety of niches occupied by the species forming the community.

Such high efficiency in the utilization of energy in mature and highly diversified ecosystems leaves no

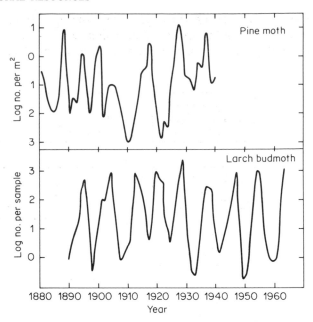

Figure 1.14   Examples of oscillations in a natural population about the limiting capacity. The data shown are for two species of moth whose larvae feed on the foliage of European conifers. Top: pine moth *Bupalus*. Bottom: larch budmoth *Zeiraphera diniana* in the Engadin. (From E. P. Odum, 1971)

room for wastage of resources and there is therefore no large surplus which would permit great fluctuations in species populations.

The relationship between stability and complexity can be illustrated very simply by thinking in terms of food webs. Suppose that a certain species of herbivore has its population controlled by six species of predator in one ecosystem, while in another it only serves as prey to a single carnivore. In the first case, any decrease in the efficiency of one predator will be largely compensated for by the five others, whereas a similar decrease in the second case means that the population of the herbivore will grow exponentially until other limiting factors come into play and cause it to collapse. The more complex food web in the first case clearly makes for greater stability.

This is by no means a special case: it is only in species belonging to ecosystems with very simple food chains (like lemmings in the tundra or xylophagous insects in monospecific forest plantations) that sudden explosions of population are experienced. They are unknown in highly diversified tropical ecosystems.

1.3.6   Interaction between principles

The various principles that have been put forward and discussed in this section show a strong degree of interdependence. For example, the utilization of energy, both at the level of populations and ecosystems, conditions the differentiation of ecological niches, the diversity, the productivity, the stability of populations, their spatial distribution and so on. Similarly, the principles concerning diversity interfere with those related to other resources, etc. The functional relationships existing between the various principles are illustrated in Figure 1.15.

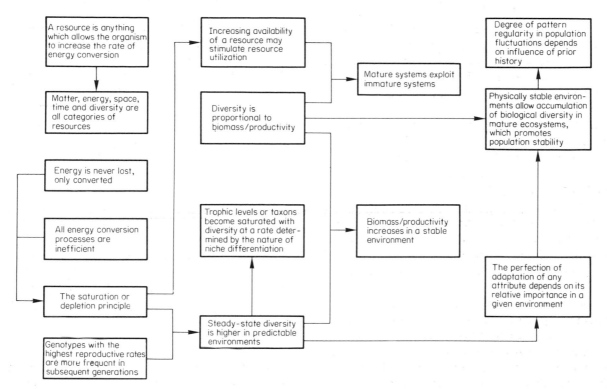

Figure 1.15   Chart showing the main interrelationships between principles governing the rational use of natural resources. (Reproduced with permission from Watt, *Principles of Environmental Science*, McGraw-Hill, 1973, p. 45)

Chapter 2

Energy and Mineral Resources

Sources of energy and supplies of minerals occupy a special place among the various natural resources exploited by humans. They are both essentially non-renewable[1] even where they have a biological origin, as in the case of fossil fuels or of certain metal-bearing deposits. They thus form a stock which is undergoing more or less rapid exhaustion according to the size of the various geological deposits and the speed with which contemporary technological civilization is exploiting them.

However, there is a fundamental difference between the sources of energy that are potentially contained in fossil or nuclear fuels and the supply of mineral elements. The energy contained in oil and coal or in uranium-235 is unavoidably lost in any useful form after combustion or fission, in accordance with the laws of thermodynamics. Mineral elements, on the other hand, do not disappear after use, whether they are metals or any other inorganic substance: iron remains iron (even when oxidized it only changes its chemical state), glass remains glass, and so on. We have even invented synthetic organic materials like plastics that resist almost any form of biogeochemical degradation!

Unlike raw materials from mineral sources, therefore, energy resources are essentially a limiting factor in our industrial civilization. The energy can be neither recovered nor recycled after use. In contrast to that, the exhaustion of high-grade metal-bearing deposits would not be particularly worrying to us if there were an inexhaustible and virtually cost-free source of energy available. With such a source it would then be feasible to extract important metals from rocks in parts of the lithosphere where they are widespread but in very low concentrations. In practice, the consumption of energy involved in isolating the copper in granites, for example, is so great that it precludes any such undertaking in the climate of the current energy crisis.

---

1. This preliminary discussion excludes the renewable energy resources mentioned later in the chapter since they form as yet so small a part of the present world utilization.

2.1 The Flow of Energy and the Cycle of Matter in the Human Ecosystem

An analysis of the way ecosystems function soon shows that the flow of energy and the cycle of matter cannot be dissociated from each other. Before going very far into the study of resources, therefore, ecological models will be used to look in more detail at the manner in which energy and matter are transported through anthropoecosystems (a term covering the various human societies with ever-increasing cultural levels that have succeeded each other through the ages).

The first examples of this type of analysis applied to the study of a particular human society or activity date back to the beginning of the last decade. However, the credit for having systematized the method belongs to H. T. Odum (1971) who had begun to use his basic flow-charts to describe the functioning of a natural ecosystem, the well-known Silver Springs in Florida, as long ago as 1957.

The whole history of the relationship between humanity and the biosphere is marked by two quite fundamental changes: the appearance of agriculture during Neolithic times and the advent of modern industrial civilization in the nineteenth century. It is important, therefore, to analyse the flow of energy and the cycle of matter in two types of human ecosystem: one dominated by agriculture and the other by industrial processes. The ecological models to be used for those analyses will have as their basis the diagram shown in Figure 2.1, which illustrates the functioning of a natural ecosystem not exploited in any way by humans.

In a primitive rural civilization of the Neolithic type, the only energy source available to humankind was solar radiation (Figure 2.2). The energy intake *per capita* in such societies depended entirely on photosynthetic conversion and was therefore determined by the amount of land under cultivation or used for rearing animals. The only fuel used for heating, cooking of food and primitive metallurgy was wood (using about 4000 kcal of energy per day per

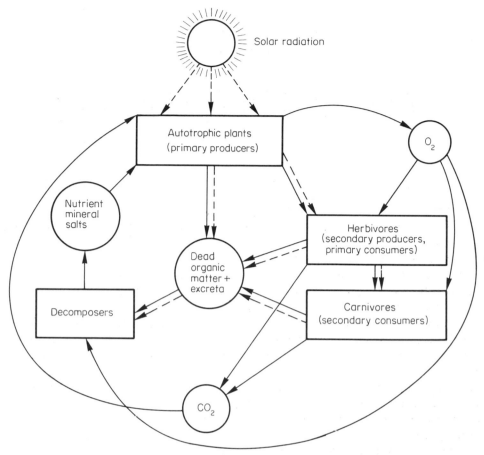

Figure 2.1   Diagram showing the flow of energy (broken lines) and the cycle of matter (continuous lines) in an ecosystem not exploited by humans

individual) and the only available source of mechanical work depended on animal traction (using approximately the same amount of energy). The amount of work performed by humans themselves would correspond to energy supplied by the individual's food intake (3000–4000 kcal per day per individual). The total of these three quantities shows that each individual in a Neolithic society would dispose of around 10 000–12 000 kcal of energy per day. The proportion of solar energy used by humans could not be increased beyond a certain limit fixed by the maximum area of land *per capita* that could be exploited with the modest techniques available. We can conclude that the overall energy equilibrium in the ecosphere was not upset by Neolithic civilization: some 4000 kcal per m^2 per day of solar energy reached the Earth's surface and a similar amount was radiated into space in the form of heat. With the low population densities of those times, only a minute proportion of that energy was under the control of human society.

Neither was the cycle of matter affected by such a primitive civilization. The only waste produced by human activity was fermentable organic matter and the various materials used (that is, wood, plant fibres, even certain metals like iron) were subject to more or less

short-term degradation by micro-organisms and other biogeochemical agents. As a result, there was no accumulation of waste products because all the matter used by humans was mineralized and dispersed in the environment in such a way as to ensure the recycling of elements, especially in cultivated soils. The Neolithic human ecosystem thus functioned in a way very similar to that of a natural ecosystem, in relation both to the flow of energy and to the cycle of matter.

Our present industrial civilization offers a contrast: the relationship between humankind and the ecosphere as regards energy flow has been profoundly changed both in degree and kind by our massive recourse to fossil fuels (Figure 2.3). Current world consumption is more than 10^{10} t.c.e. (tonnes of coal equivalent) per year, mostly in the form of the various organic fossils (coal, lignite, oil and natural gas).

Table 2.1 compares the energy flow in various ecological and technological systems. Examination of the figures shows that the present consumption of fossil fuels corresponds to an energy flow of 0.44 kcal per m^2 per day (assuming the energy to be uniformly distributed over the whole surface of the ecosphere) whereas the amount of solar radiation taken up by agricultural production is only 0.3 kcal per

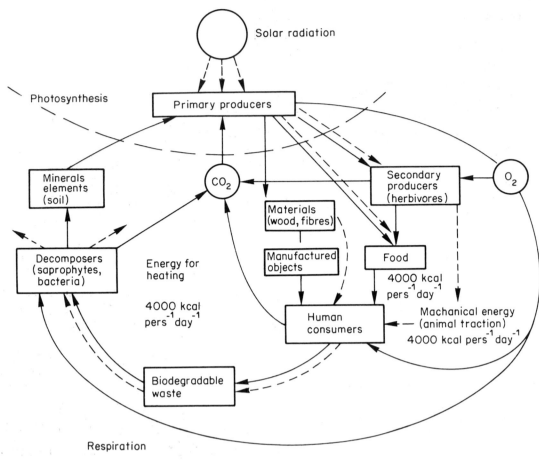

Figure 2.2   Diagram showing the flow of energy (broken lines) and cycle of matter (continuous lines) in a Neolithic civilization. (From Ramade, 1978a, p. 116)

Table 2.1   *Comparison of energy flows in various ecological and technological systems*
(From H. T. Odum, 1971)

System	Energy flow (kcal m^{-2} day^{-1})
Incident solar radiation	
Solar energy flux at the surface of the ecosphere	5110
Solar energy flux available to green plants	3400
Maximum converted by photosynthesis	170
Primary production	
In a tropical rain forest	131
Gross primary production of the biosphere	6
Primary production of world agriculture (related to the total cultivated area: 8.9% of the continental surfaces)	0.3*
Systems of consumers	
Respiration in a tropical rain forest	131
Neolithic village (100 m^2 per inhabitant)	30
World consumption of fossil fuels (related to the surface area of the ecosphere)	0.44*
Consumption of fossil fuels in the USA (related to the total surface area without Alaska)	7.95*
Animal community (oyster colony)	57
Consumption of fossil fuels in a large modern city	4000

* Figures updated to 1979.

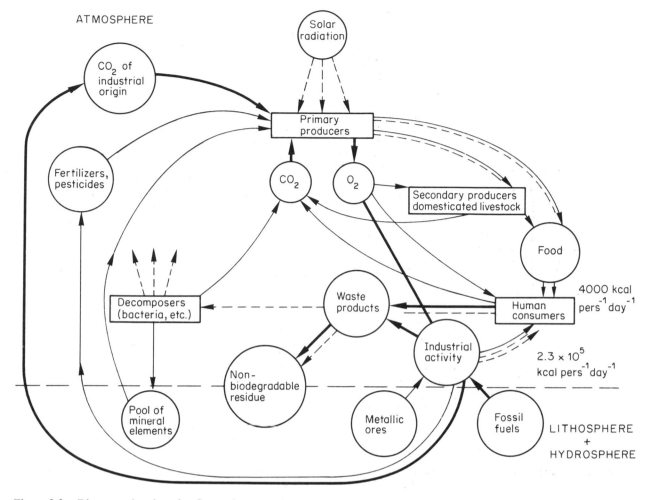

Figure 2.3   Diagram showing the flow of energy (broken lines) and the cycle of matter (continuous lines) in modern industrial civilization. (From Ramade, 1978a)

Table 2.2   *Growth of daily energy consumption* per capita *throughout human history* (from E. Cook, 1975)

Period and type of civilization	Energy consumption (kcal per s^{-1} day^{-1})				
	Food*	Domestic consumption	Industry and agriculture	Transport	Total
10^6 years ago— early Palaeolithic	2 000				2 000
10^5 years ago— middle Palaeolithic	3 000	2 000			5 000
10^4 years ago— early Neolithic	4 000	4 000	4 000		12 000
600 years ago— end of Middle Ages (N.W. Europe)	6 000	12 000	7 000	1 000	26 000
100 years ago— early industrial society	7 000	32 000	24 000	14 000	77 000
AD 1970— technological society	10 000	66 000	91 000	63 000	230 000

* These quantities are the energy content of the plant food needed to produce the meat and dairy products consumed by humans.

m^2 per day (also assuming uniform distribution). Another point to notice is that the consumption of fossil fuels in a large modern city is equivalent to an energy flow of as much as 4000 kcal per m^2 per day: a value greater than the average incoming flux of solar energy (3400 kcal per m^2 per day)!

Table 2.2 shows how consumption has increased through the ages. Today, the energy consumption *per capita* has reached very high levels and can exceed 300 000 kcal per day in certain regions of Western Europe and North America (the value of 230 000 kcal per person per day was an estimate of the mean consumption in the USA in 1970).

In primitive rural civilizations, humans could be said to have lived on the 'interest' from their energy stock produced by cultivated plants and forests. This 'interest', 'trapped' by photosynthesis and consumed by humans through primary plant production, represented only a small fraction of the total incident solar radiation. In complete contrast, today's technological societies are living on their energy 'capital'. The table shows that the proportion of individual energy consumption arising directly from solar radiation (i.e. the food needed by humans plus animal traction) has continually fallen since Neolithic times and, from the end of the eighteenth century, has formed only a minor part of the overall consumption: fossil fuels have become the preponderant source of energy.

As a result, present technological society is slowly but surely exhausting our energy capital existing in the form of fossil fuels derived from the carbon content of the lithosphere. The stock of such fuels represents, in the last analysis, an accumulation of solar energy throughout geological periods after transformation to biochemical energy by fossilized plants. Our present massive dependence on oil, if prolonged for several decades, will lead to the exhaustion in less than a century of a quantity of fossil hydrocarbons that has taken more than 100 million years to form!

The development of nuclear energy also represents consumption from a stock and not from a flow. It is based on the utilization of 'capital' in the form of fissile material which collected when the planetary system condensed some 4700 million years ago. The use of fast breeder reactors instead of conventional ones will merely delay the eventual exhaustion of fissile and fertile materials in the Earth's crust.

The massive use of fossil fuels is causing a considerable amount of ecological disturbance. The ejection of tens of thousands of millions of tonnes of CO_2 gas per year into the atmosphere could modify terrestrial climates by increasing the greenhouse effect. More generally, the cycle of matter is being completely disrupted, as has been shown by a great deal of research into cases like the biogeochemical carbon cycle just referred to. It has been established, for instance, that the atmospheric concentration of CO_2, a perfectly stable gas over millions of years, has increased from 280 p.p.m.[1] at the beginning of the industrial era to 335 p.p.m. today, with a current annual increase of 1 p.p.m.

In a similar way, there is increasing disruption of the nitrogen and phosphorus cycles produced by the massive use of chemical fertilizers. In fact, it is no exaggeration to say that modern industrial society has considerably upset the whole cycle of matter in the biosphere. On the one hand, there is the discharge of various toxic substances into the environment, particularly into water, which decreases or even destroys its self-purifying capabilities. At the same time, there is also an increasing quantity of persistent and even indestructible material being dispersed throughout the world: non-corrosible metals, plastics, chlorinated hydrocarbon pesticides, long-lived radioactive elements and so on.

As a result of all this, we are taking part in a linear process with, at one extremity, resources of raw materials in shorter and shorter supply and, at the other, an ever-growing mass of non-biodegradable waste products. The formidable problems of pollution that we face are nevertheless the very consequences of the way modern technological society functions in that no account is taken of ecological models. Pollution itself, which is due to a disruption of the cycle of matter, leads to a very considerable wastage of natural resources, not only because those responsible for industrial activity generally reject all forms of recycling but also because the pollutants contaminating soils and water destroy biological resources even before they can be utilized.

In fact, the apparent prosperity of industrial countries is in a large measure artificial if the plundering of scarce and non-renewable natural resources on which it is based is taken into account.

Yet the inefficient use of energy and the wastage of raw materials is completely in line with even the most trivial activities in what is called the 'consumer society': a society that persuades people to discard many everyday objects (made, for instance, from plastics, i.e. from oil) not only when they are just a little worn but even when they are slightly outdated. Such waste has been implicitly established and almost institutionalized in Western countries since the beginning of industrial production and it has brought a 'throw-away' state of mind to our civilization that is well summed up in the phrase 'no deposit, no return'. Worse still, limited resistance to wear is in many cases deliberately incorporated into the manufacturing process, whether it is of a small tool or of a motor car.

---

1. p.p.m. = parts per million.

As Labeyrie (1971) has written so forcefully:

> Even those that are resistant to the blandishments of the advertisements in the mass media have no chance of maintaining equipment in good condition if it is unrepairable and designed for rapid renewal. It is explained to us that such a policy increases the Gross National Product, but ecologists consider it to be a scandalous squandering of the wealth inherent in human work, in matter and in energy.

All this will bring us more quickly to the day when technological civilization will be plunged into an insurmountable crisis over resources because of the exponential nature of growth. In the end, it is by no means improbable that shortages which become firmly established could lead inevitably to political tensions and thus to a third world conflict. . . .

Thus, at a time when the availability of energy and raw materials is a burning issue, it is certainly not a useless exercise to consider both the actual quantities of each of these resources that exist and the nature of the various solutions for their replacement that can be foreseen in the medium and long term.

2.2 Energy Resources

In this section two main categories of energy resource are distinguished. The first includes those that occur in the form of a non-renewable, and therefore exhaustible, stock: fossil fuels, fissile nuclear fuels and, to some extent, geothermal energy. These might be called sources of 'potential' energy. The second category consists of 'free' sources or natural flows of energy that are essentially inexhaustible: solar radiation and gravitational energy (in the form of tides).

2.2.1 Potential sources of energy, 1: fossil fuels

It was not until 17 October 1973 that the public at large—and the politicians of Western Europe—began to be aware of an essential factor in today's environmental crisis: the relative scarcity of certain fossil fuels and their wasteful utilization. That this should have been so seems somewhat paradoxical when we think of the formidable growth in the consumption of energy, mostly in the form of hydrocarbons, that everybody in industrialized countries must have witnessed during the last few decades.

The growth in energy consumption: its speed and nature

It is obvious that indefinite quantitative growth is an absurd idea if a little thought is devoted even to the general idea of an exponential increase. Any activity

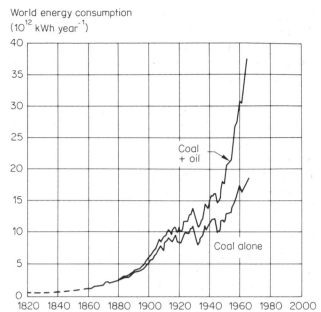

Figure 2.4   Growth in world consumption of energy from the middle of the nineteenth century to the present. (From Hubbert, 1971)

whose size is regularly doubled after a definite period of time, say every 10 years, achieves a magnitude after n such periods that is greater than the sum of those at the ends of *all* previous periods (just as the last term in the series 2^0, 2^1, 2^2, . . . , 2^{n-1}, 2^n is greater than the sum $1 + 2 + 4 + . . . + 2^{n-1}$).

To give a more concrete example of the same point: doubling the energy consumption per year every 10 years means that during such a period it is necessary to bring into service at least as many new oil wells, to construct at least as many new power stations, etc.,[1] as have been brought into operation since hydrocarbons and electricity began to be used. An example in quite another area (cf. chapter 5, p. 134): the doubling of the human population in the next 35 years means that in that time world food production must grow by an amount equal to the total development made during the course of the 10 000 years that separate us from the early Neolithic period!

Nevertheless, as Figure 2.4 shows, there *has* been an enormous growth in the world consumption of energy, mainly in the form of fossil fuels, since the dawn of the technological era (which we might fix with good reason in 1859, the date of the sinking of the first oil well by Colonel Drake at Titusville in Pennyslvania). Even though the world population has quadrupled since that date, the growth in consumption is principally the result of technological progress: it is the *per capita* increase that has been the essential cause of the un-

---

1. In fact, more than this are needed since the older, outdated installations and oil wells that have become exhausted must also be replaced.

Table 2.3   Growth in world energy consumption over the last 60 years

	1925	1938	1950	1960	1968	1975
World consumption per year (in 10^6 t.c.e.)	1484	1790	2610	4196	6306	8490
World population (in 10^9)	1.89	2.16	2.50	3.00	3.48	3.97
Consumption *per capita* (in t.c.e. per year)	0.785	0.826	1.042	1.403	1.810	2.140

ending growth in primary energy needs, a growth that has been particularly spectacular in the last few decades as is shown in Table 2.3.

The total quantity of coal consumed from the time of its first use in the seventeenth century until 1860 is estimated at 7000 million tonnes. Between 1860 and 1970, some 133 000 million tonnes have been consumed, so that in 110 years 19 times more coal has been used than in the preceding seven centuries!

Another example that might provide a clearer appreciation of the fantastic growth in energy consumption: it is estimated that between its origins some 1 million years ago and the year 1860, humanity consumed some 35 000 million tonnes of coal equivalent, an amount that is approximately what is called 1 Q.[1] Between 1860 and 1970, about 10 Q were used, while in the single year of 1980, world consumption rose by 0.28 Q. It is estimated (with a margin of error of ±25 per cent) that our technologicol society will need 20 Q by the end of the century.

In practice, world consumption increases more quickly than the simple exponential law would predict. If P is the amount of energy consumed up to a time t, and P_0 is the initial energy at $t = 0$, then exponential growth would mean that

$$P = P_0 e^{rt} \qquad (1)$$

where

$$r = \log_e(1 + R) \qquad (2)$$

R being the annual rate of increase and both r and R being constant. However, studies of the world increase in energy consumption as a function of time, as in Figure 2.4, show that the two parameters r and R themselves vary as an increasing function of time, so that energy needs eventually increase at a rate that is greater than that of an exponential function. Thus, the annual rate of increase was about 2 per cent per year in the middle of the nineteenth century, 3.5 per cent at the beginning of the twentieth century, rose to 5.6 per cent in the 1960s and was even higher in 1973. . . . Remembering how explosive an increase is entailed by the normal exponential growth, these figures largely explain, and justify, the use of the phrase 'energy crisis'.

If present trends are extrapolated into the future, they show that the world would need between 200 and 400 Q of energy between now and AD 2050 (Felden, 1976). As we shall see later, the total amount available from fossil fuels contained in the lithosphere is estimated to be 250 Q! It is thus quite obvious that the current rate of consumption can only continue without major problems for a few more decades. . . .

Turning to the *nature* of world energy consumption, we see that the relative proportions in which the various sources have been used have also changed profoundly over the last hundred years, especially since the Second World War. At the beginning of the century when 'coal was king' the part played by hydrocarbons like oil and natural gas was very small, but they have dominated the scene since 1950, particularly in the Western world.

During the hundreds of thousands of years between the discovery of fire and the dawn of the modern era, the only fuel in use was wood, and even in 1850 it still constituted some 90 per cent of the energy supply in the USA and continental Europe. However, by around 1900 its contribution had fallen to 20 per cent in North America, where more than 70 per cent of the total requirements were being met by coal: worldwide this new fuel was meeting some 90 per cent of total needs at the same date.

The decline of coal began during the 1920s. By 1950, oil and natural gas were providing 60 per cent of US energy needs and by 1975 the same fuels constituted 67 per cent of world consumption—in spite of the 'energy crisis' created by the Arab–Israeli war of 1973. Between 1929 and 1971, world production of coal only grew by 70 per cent, whereas that of oil increased by 1000 per cent.

Table 2.4 shows how world energy consumption has grown in the last quarter of a century, and a notable feature is that the proportion provided through primary electricity production (hydroelectric and nuclear) is still under 10 per cent overall. This is partly because of the standstill in the development of hydroelectric power in industrialized countries where most of the favourable sites have already been exploited. In addition, although nuclear power emerged as a significant source in the middle of the 1960s, its proportion of the total remains at less than 5 per cent in spite of the size of nuclear programmes in some countries.

1. 1 Q 36.62 × 10^9 t.c.e. = 10^{18} British thermal units = 2.52 × 10^{17} kcal = about 10^{21} J or 3 × 10^{14} kWh.

Table 2.4   Growth of world energy consumption for various primary sources (from Felden, 1976)

| Region | Year | Proportion of total consumption (%) | | | | Total |
		Coal and lignite	Oil	Natural gas	Primary electricity (hydroelec. + nuclear)	(10^9 t.c.e.)
World		60	26	10	4	2.7
USA		40	34	22	4	1.2
USSR	1950	76	20	3	1	0.4
EEC (9)†		70	23	1	6	0.55
France		67	24	6.5	3.5	0.09
World		50	30	14	6	4.5
USA		24	39	32	5	1.7
USSR	1960	62	25	10	3	0.66
EEC (9)†		64	28	2	6	0.78
France		55	31	4	10	0.09
World		29	43	22	6	8
USA		19	43	33	5	2.7
USSR	1973	39	32	25	4	1.3
EEC (9)†		23	57	15	5	1.4
France		17	66	9	8	0.26
World		20	45	25	10	14
USA		20	40	27	13	4.1
USSR	1985*	27	38	27	8	2.5
EEC (9)†		17	49–51	18–23	16–19	2.2
France		9–12	40–46	13–16	31	0.36–0.39

* Predictions made from figures for growth to 1973.
† Data for the nine countries which were member states of the EEC between 1973 and 1981.

Table 2.4 also reveals the dominant part played by hydrocarbons in supplying the energy requirements of Europe as a whole and particularly of France, where they represent 75 per cent of the total. More generally, it is clear that the various derivatives of fossilized carbon provide 95 per cent of world energy needs, a proportion that showed no sign of decreasing in the 1970s and that shows every sign of being maintained into the future if the overall growth continues.

Figure 2.5 shows the structure of primary energy supplies in the major industrialized regions during 1975. In the Eastern bloc, coal is still predominant, providing more than half the total. In contrast, it is the hydrocarbons that play the major role in the West, a feature that can be easily explained by the influence exerted on national energy policies by multinational oil companies. The choice of an oil-based industrial economy has placed not only France but Europe as a whole and even the USA in a state of almost absolute dependence on the OPEC countries. The nature of the problem is emphasized by Table 2.5, which shows the proportion of national energy needs provided by internal sources and indicates the difficulties which will be faced in the future. Only the USSR is self-sufficient in energy, while Western countries, particularly Japan and European nations, have watched their levels of self-sufficiency fall continuously over the last 30 years.

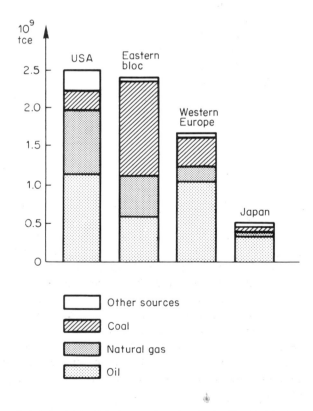

Figure 2.5   The structure of energy supplies in industrialized countries in 1975

Table 2.5   Proportion of energy needs provided by internal production in industrialized countries

	1960(%)	1973(%)	1976(%)
USSR	100	100	100
USA	95	89	77
Japan	65	14	—
EEC (9)*	69	38	—
France	61	24	23

* Data for the nine countries which were member states of the EEC between 1973 and 1981.

The geopolitical problems that arise from this situation, coupled with the sheer size of current fuel consumption, leads us to consider the full extent of the energy crisis. One of the first questions usually asked concerns the real magnitude of the total fossil fuel resources contained in the lithosphere. This is because, given the inertia inherent in all technological, and hence economic, development, such resources will continue to provide the basic energy requirements of industrial countries until the end of the century, even if there is some increase in the proportion provided by nuclear reactors and by natural sources such as solar radiation.

Theory of the production cycle of a non-renewable resource

At the end of the 1960s, the insatiable appetite of so-called developed countries for energy and raw materials was leading some authorities in the field to consider questions about the total reserves of the various natural resources of strategic importance: reserves, that is, which were available and which could be technically exploited. The answers to such questions can be arrived at by combining the use of several mathematical, statistical and geological methods of investigation, a complete description of which lies far outside the scope of this account. In what follows, therefore, I shall attempt to give only a general outline of the principles involved.

A fossil fuel or any other material in the Earth's crust occurs as a collection of discontinuous deposits, whose exploitation as time goes on can be illustrated diagrammatically in various ways. Let us suppose that the total amount of material produced from the beginning of exploitation until a certain time t is Q_p. Clearly Q_p increases as time goes on and will tend towards a maximum value Q_∞, say, as the deposit is gradually exhausted. The *rate* of production

$$P = \frac{\Delta Q_p}{\Delta t} \tag{1}$$

represents the quantity of fuel or mineral ΔQ_p produced in a given interval of time Δt (1 year, for example). If the rate of production is plotted graphically against time, from the beginning of extraction

until the complete exhaustion of deposits, as a smooth curve, the result represents the variation of

$$P = \frac{dQ_p}{dt} \tag{2}$$

Such a curve shows an initial growth of P that is quasi-exponential (Figure 2.6) until its rate of increase

$$V_p = \frac{dP}{dt} = \frac{d^2 Q_p}{dt^2} \tag{3}$$

reaches a maximum at the point of inflection. V_p then decreases and becomes zero at the time when P itself is a maximum. After that, P decreases in a manner symmetrically related to the ascending part of the curve, while V_p becomes negative, at first greater in magnitude and then falling away to zero (Figure 2.6). As $t \to \infty$, the total amount extracted or produced tends to Q_∞, the total mass of the resource in the deposit that is ultimately recoverable, so that

$$Q_\infty = \int_0^\infty P \, dt \tag{4}$$

which is the total area between the P curve and the time axis.

The *useful period* of a single deposit—or of the total resources of a material if the argument is extended to a world scale—is defined as the duration of exploitation needed to produce 80 per cent of the total material. This period is considered to begin after the production of 10 per cent of the total and to end after that of 90 per cent: in other words, the useful period is the time taken for the extraction of that 80 per cent of the total lying between $0.1\, Q_\infty$ and $0.9\, Q_\infty$.

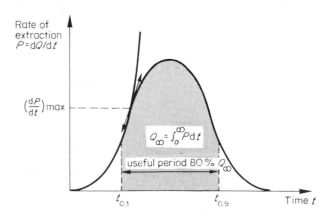

Figure 2.6   Production cycle of a fossil fuel or mineral resource illustrated by a curve of the rate of extraction against time. The quantity Q_p of the material extracted up to any time t is the area under the curve from 0 to t. The total mass of fuel or mineral contained in a deposit, Q_∞, is the total area between the curve and the time axis. The times $t_{0.1}$ and $t_{0.9}$ are those after which 10 and 90 per cent respectively of Q_∞ have been extracted

Equally important are (a) the curve giving the variation in time of the rate of *discovered* resources

$$D = \frac{dQ_d}{dt} \qquad (5)$$

where Q_d = the total quantity of the resource discovered up to time t, and (b) the curve showing the variation in the quantity of proved reserves (that is, resources that are known but not yet exploited)

$$Q_r = f(t) \qquad (6)$$

Mathematical and statistical considerations show that the *cumulative* masses of both discovered resources Q_d and extracted resources Q_p exhibit time variations represented as in Figure 2.7 by a logistic or sigmoid curve. At any time t

$$Q_d = \int_0^t D \, dt \qquad (7)$$

$$Q_p = \int_0^t P \, dt \qquad (7')$$

while Q_d, Q_p and Q_r are related by

$$Q_d = Q_p + Q_r \qquad (8)$$

It is obvious that when $t \to \infty$, $Q_p \to Q_d$ while $Q_r \to 0$. Moreover, the Q_d and Q_p curves are separated by an interval of time Δt, which is a corollary of (8) since, to put it more concretely, exploration precedes exploitation.

A noteworthy feature of the D and P curves shown in Figure 2.8 is their intersection situated at the midpoint between the maximum rate of discovery and the maximum rate of exploitation. After this point in time, the rate of extraction continues to grow while the rate of discovery is already declining. The intersection

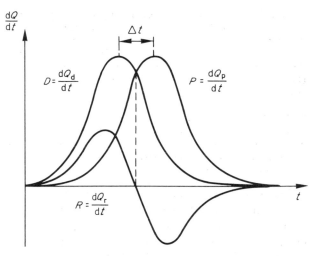

Figure 2.8 Variation with time of the rate of discovery $D = dQ_p/dt$ and the rate of extraction $P = dQ_p/dt$. Also shown is the variation of $R = D - P$, the rate of change of unexploited reserves. (From Hubbert, 1969, p. 168)

also marks the time at which the maximum quantity of proved reserves Q_r occurs.

Another feature of the $P = dQ_p/dt$ curve in Figure 2.6 that is worth noting is the point at which the production rate attains its maximum value (i.e. where $d^2P/dt^2 = 0$ at the point of inflection). From this time onwards, there is an increasing difference between the demand of the market (the exponential curve) and the actual rate of production. If maintenance of exponential growth is considered desirable, then this time marks the beginning of the search for means of compensating for the deficit, either by recourse to a different source of energy or to a substitute for a raw material. On the other hand, it may be thought desirable to make the consumption level off at a point corresponding to the maximum rate of production in order, for instance, to curtail energy growth. In that case, the changeover to other sources should be started well before the maximum of the P curve is attained because of the large time lags inherent in developing new systems for energy production in an industrial civilization like ours.

Data for the initial parts of the Q_d, Q_p and Q_r curves (and hence for those of their derivatives) are provided by the statistics for prospecting, exploration and exploitation of any given resource. The same statistical data also allow Q_∞ to be evaluated for each type of fuel or raw material provided information is available up to the maximum of the dQ_d/dt curve: failing that, geological information about the extent of the terrain containing potentially productive deposits is used. The value for Q_∞ thus obtained is an estimate of the total available resources contained in the lithosphere. The duration of the useful period for the various resources and the time at which production will reach its maximum can also be obtained with reasonable accuracy.

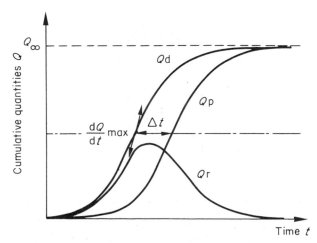

Figure 2.7 The variation with time of (a) the cumulative total of energy produced or resource extracted, Q_p; (b) the cumulative total of resources already discovered, Q_d; (c) Q_r, the unexploited reserves. (From Hubbert, 1969, p. 172)

Application of the theoretical treatment just outlined has enabled specialists in the field to estimate the total recoverable reserves and the useful periods for the main non-renewable energy sources. Tables 2.6 and 2.7 summarize the data for the estimated Q_∞ values in the case of the various fossil fuels, the most reliable calculations in this area being those of Hubbert and of Ryman for oil and natural gas and those of Averitt for coal.

The tables demonstrate several important facts. In the first place, Table 2.6 shows that coal is by far the largest in quantity of all the fossil fuels with 85 per cent of all the available reserves. The crushing preponderance of coal is emphasized by the fact that the majority of its Q_∞ corresponds to proved reserves whereas with oil and natural gas most of the reserves are only guesses in so far as there is still no idea of their exact location. In the case of oil, for example, out of a Q_∞ of 320×10^9 tonnes only 90×10^9 tonnes are proved reserves. It can well be understood that, unlike coal, there is great uncertainty about the total cost of extracting all the estimated oil reserves, even if we confine ourselves to those that are defined as 'conventional' (see note to Table 2.6).

Oil reserves

In 1973, oil provided some 43 per cent of the world energy supply. It is thus the largest of all energy sources at the present time, a situation made all the more paradoxical by the fact that current proved reserves appear to be particularly low: their estimated total of about 100×10^9 t represents 35 years of world consumption at the 1973 level (2.85×10^9 t per year). In fact, even that calculation is faulty since it does not take into account the growth in demand which stood at a rate of 12 per cent per year at the beginning of the 1970s. Over the whole period 1890–1970 the mean rate of world oil production rose by 6.94 per cent per year, but even at this more modest growth rate—corresponding fairly exactly to a doubling of production every decade—the proved world oil reserves would be exhausted at the latest by 1999.

Fortunately, estimates of the total ultimate resources

Table 2.6   Estimates of ultimately recoverable resources (Q_∞) for the principal types of fossil fuel in the Earth's crust (adapted from Hubbert, 1969)

Type of fuel	Q_∞ (tonnes)	Q_∞ (10^{15} kWh thermal)	% of total in Earth's crust
Coal and lignite	7.6×10^{12}	55.9	85.1
Natural gas*	645×10^9	5.1	7.8
Oil†	320×10^9	3.85	5.9
Tar sands and oil shales†	78×10^9	0.83	1.2

 * Expressed in tonnes of coal equivalent.
 † These include what are called 'conventional' reserves: that is, those which it is estimated could be extracted at a price below \$(US)30 per barrel in 1978. At a price of \$(US)50 per barrel, it is currently estimated that Q_∞ for tar sands and oil shales would be comparable with that for oil itself.

Table 2.7   Geographical distribution of fossil fuel reserves in t.c.e. (source: UNO, 1973)*

Region	Coal and lignite 10^9 t.c.e.	%	Natural gas 10^9 t.c.e.	%	Oil† 10^9 t.c.e.	%
North America	1300	16.99	129	20.0	9.5	7.0
South America	33	0.43	13	2.0	7.0	5.2
Middle East	1	0.01	107	16.6	86	63.4
Africa	87	1.14	70	10.9	5.2	3.8
Asia (without China)	133	1.74	13	2.0	3.1	2.3
Oceania	46	0.60	6	0.9	0.4	0.3
Western Europe	118	1.54	63	9.8	3	2.2
Eastern bloc‡	5935	77.55	244	37.8	21.4	15.8
Totals	7653	100.0	645	100.0	135.6	100.0

 * 1 tonne oil = 1.5 t.c.e.; 1 tonne lignite = 0.5 t.c.e.
 † Proved reserves.
 ‡ Including China which alone possesses 1000×10^9 tonnes or about 12 per cent of the total world reserves of coal.

that really exist (Q_∞) predict a rather longer time for the useful period of the fuel. Nevertheless, the statistics on exploration (that is, the shape of the real dQ_d/dt curve) and the catalogue of potential oil deposits do not encourage optimism. As time goes on, new deposits are found to occur more and more deeply in sedimentary layers and/or further and further out to sea, while the amount of oil discovered per square kilometre of exploration carried out decreases throughout the world year by year. These factors account for some of the pessimism to be found in statements about the future of oil. Thus, A. Robin, director of Electricité de France, writing in 1971 at a time when large thermal power stations based on solid fuel were being built, said: 'To put all your money on oil in the long term is to assert that 280 000 m tonnes of reserves will be proved between now and the year 2000 . . . this is a huge quantity when you remember that only 15 000 m tonnes of new deposits have been discovered in the last ten years. . . . '

To take an example, the total oil resources of the North Sea, which the media have often presented as a veritable gold mine, are estimated to be 2.5×10^9 t, which is not even 4 years consumption for the whole of Europe at the rate predicted for 1985 (0.7×10^9 t).[1] In the same way, the celebrated oil reserves of Prudhoe Bay in Alaska, exploited at the price of real ecological damage in a region of tundra, amount to hardly a dozen years of consumption in the USA at current rates, and thus much less if growth continues.

The validity of the theory outlined in the previous section on the production cycle of a fuel has been tested in the USA. From the beginning of the 1960s it had been forecast that US production would reach its maximum between 1970 and 1975. This has since been confirmed, as the data on oil production plotted in Figure 2.9 show. It can also be seen in the figure that the maximum rate of discovery (dQ_d/dt) was reached towards 1960, thus preceding by $10\frac{1}{2}$ years the peak of the production rate at the beginning of the 1970s.

Figure 2.10 shows two production cycles for oil in the USA, one corresponding to the predictions of Hubbert (1967) and the other to those of Berg et al. (1974). The difference between the two occurs because Berg et al. have taken into account the Alaskan deposits and also the estimates of the US Geological Survey which includes deep offshore oil in sediments up to 2500 metres in depth.

The length of the useful period for world oil supplies has been calculated by Ryman of Standard Oil, New Jersey (Exxon), basing his estimates on various detailed geological considerations and on statistical data. He fixes the value of Q_∞ as 2100×10^9 barrels (1 barrel = 158.6 litres): in other words, about

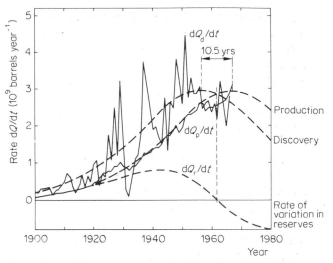

Figure 2.9   Variation in rates of discovery and production of crude oil in the USA between 1900 and 1980. Broken line curves are analytical derivatives, continuous lines are actual yearly data. (From Hubbert, 1969, p. 178)

330×10^9 t or some 15 Q (see footnote on p. 26 for the definition of 1 Q); this includes those offshore deposits located in shallow waters and capable of exploitation by current techniques. It is a somewhat optimistic value in that it assumes a 40 per cent recovery rate for the total oil contained in the original rock, whereas present available technology achieves little more than 30 per cent.

Hubbert (1969) has put forward a lower estimate for Q_∞ of only 1350×10^9 barrels or 215×10^9 t. He points out that the continental reserves currently existing in the USA, where the peak of the production curve is a known factor, have turned out to be 50 per cent

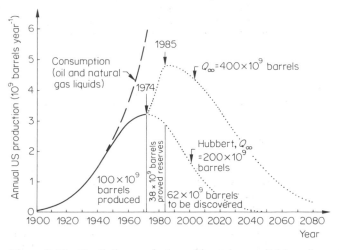

Figure 2.10   Predictions of the ultimately available oil resources (Q_∞) which can be extracted from continental and offshore deposits in the USA. The estimate of 400×10^9 barrels for Q_∞ includes those deposits that have still not been exploited in Alaska, in the continental shelf and in the oceans themselves up to a depth of 2500 metres. (From Berg et al., 1974, p. 332)

1. According to Wilson et al., 1977.

lower than those predicted by Weeks—and Ryman's estimate quoted above is an extrapolation of Weeks's data for the USA to the whole world.

In spite of the disagreement between the two values of Q_∞ for oil, they do provide an effective range within which it is highly probable that the actual total lies. Recently, some experts have put forward a more optimistic estimate for Q_∞ of around 600×10^9 t, a figure that is so different from the other two that it looks unrealistic. In fact, it includes potential offshore deposits located in depths up to 2500 metres. There is, however, no economically acceptable technique of exploration and drilling which would allow these to be exploited in the foreseeable future and, in any case, what of the problems caused by pollution as a result of exploiting such deposits: accidents, for instance, at a drilling platform (e.g. Ekofisk) or, even worse, at a well-head (e.g. the disasters of Santa Barbara in California and Ixtoc 1 in Mexico)?

A value of the useful period for oil can be calculated using data accumulated since 1860 for the initial part of the production curve as well as the Q_∞ estimate of Ryman (2100×10^9 barrels). The result is a period of 64 years, beginning in 1965 and ending in 2029, with a maximum rate of extraction occurring in the year 2000 and then decreasing rapidly. If the same calculation is carried out using the Hubbert value for Q_∞ (1350×10^9 barrels) instead of the Ryman value, the useful period turns out to be 58 years from 1962 to 2020, with a production peak in 1991 (Figure 2.11). Comparison of the two production curves compels us

to recognize an important fact: the discovery of additional reserves amounting to 50 per cent of the original estimate (i.e. the excess of the Ryman figure over that of Hubbert) puts only 6 more years on the useful period, which increases from 58 to 64 years—in other words, an extension of barely 10 per cent. This is in line with the exponential character of the consumption from which such curves are calculated, and it shows clearly that predictions based on oversimplified arithmetical arguments can be quite deceptive.

The curves of Figure 2.11 also show that, even if oil consumption is slowed down or stopped, the changeover to other energy sources should begin now and will demand efforts in research and development on a scale unequalled in the whole history of technology. Humanity has at best about 20 years at its disposal to find replacements for the decline in oil: already, since the beginning of the 1970s, the rate of discovery of oil reserves in the non-communist world has fallen below the rate of production, as is shown in Figure 2.12. Not only that, but recent economic studies (de Montbrial, Lattès and Wilson, 1978) show that a serious gap would open up between supply and demand around 1985, even without any voluntary limitation on production by OPEC (Figure 2.13).

Natural gas resources

Natural gas consists principally of methane (75–95 per cent), the rest being light hydrocarbons (propane, butane and pentane) with traces of nitrogen and

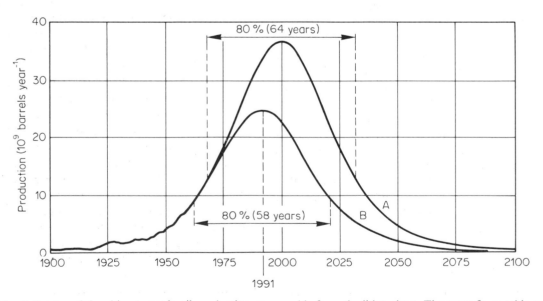

Figure 2.11 Estimates of the ultimate crude oil production recoverable from the lithosphere. The more favourable estimate of Ryman (curve A) is of the order of 2100×10^9 barrels or about 330×10^9 t, while the more pessimistic one of Hubbert (curve B) is 1350×10^9 barrels or 215×10^9 t. Note the limited duration of the useful period on the assumption that the present rate of production is maintained: 64 years in the most favourable case, with peak production occurring at the end of the century and decreasing thereafter. Although these forecasts were made at the end of the 1960s, accumulated data during the last decade have largely confirmed their validity. (From Hubbert, 1969, p. 196)

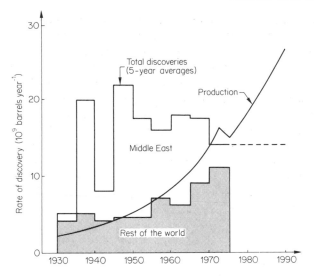

Figure 2.12   Variations in the rate of discovery of oil from 1930. Note the halt or even decline in the discovery of new reserves since the end of the 1960s and, in contrast, how the rate of production has exceeded that of discovery since 1970. (From de Montbrial *et al.*, 1978)

sulphur compounds, and it is thus by far the least polluting of all the fossil fuels. The construction of gas pipelines and later of liquefied-gas tankers for intercontinental transport has led to a rapid development of this resource, which now accounts for about a quarter of world energy consumption.

The Q_{∞} of natural gas is much less certainly known than that of oil, and considerable variations are to be found in its estimation, depending on the author: some would put it as low as 150×10^9 t.c.e. while more optimistic values can be around 1500×10^9 t.c.e. The most probable estimate lies at some 600×10^9 t.c.e. (i.e. about 20 Q) giving an order of magnitude for potential reserves comparable with that for oil. What is different from oil, however, is its geographical distribution, which is much more favourable to Europe with about 10 per cent of world reserves. The deposits at Groningen in Holland have for a long time been the largest in the world with reserves of some 2000×10^9 m³. Even in France, where the proportion of gas in national energy consumption is less than half the world average (Table 2.4), the Lacq field discovered in 1957 was still providing half the gas required by the country in 1974, although its production is bound to decline from 1985 onwards.

Tar sands and oil shales

These resources consist of rocks containing appreciable quantities of hydrocarbons having a high viscosity or even quasi-solid. The best known of these sedimentary deposits are the *tar sands* of Alberta in Canada covering nearly 100 000 km², the largest being that at Athabasca near Fort McMurray. The sand is impregnated with the highly viscous oil so that the fuel cannot be extracted using normal drilling processes. On the other hand, its chemical composition is very close to that of ordinary crude oil and it can thus be refined in the same installations.

No quantitative worldwide survey of tar sands has been carried out but according to most geologists the deposits of Alberta are by far the largest, with at least 47×10^9 t of hydrocarbons recoverable using present techniques:[1] almost exactly half the proved world oil reserves! There are also, in Venezuela, the tar sands of the Orinoco containing some 20×10^9 t of asphalt. Finally, according to some geologists, there are deposits in Madagascar which would yield 150×10^9 t of hydrocarbons, an estimate that surely needs confirmation in view of the enormous energy resources they represent!

Oil shales or *pyroschists* contain a mixture of solid hydrocarbons called kerogen having a composition very different from that of crude oil. There are also various nitrogen compounds and other inorganic impurities present, and all this means that refining cannot be carried out in conventional oil installations.

The largest deposit in the world occurs in the USA, extending over the states of Colorado, Utah and Wyoming and associated with the Green River basin that has given its name to the corresponding geological formation. This was discovered around 1860 by workers constructing the transcontinental railway who were surprised to notice that the broken rocks with which they had enclosed their camp fire had ignited!

Duncan and Swanson (1965) have made the estimates of world oil shale resources given in Table 2.8, from which it can be seen that world reserves recoverable under present economic conditions were as high as 190×10^9 barrels of hydrocarbons in 1965. The same authors estimated that there are some 17×10^{12} barrels contained in shales having between 100 and 400 litres of oil equivalent per tonne of rock; 350×10^{12} barrels in those with between 40 and 100 litres per tonne of rock; and 1750×10^{12} barrels in those with 20 to 40 litres per tonne. They also evaluated proved reserves for the whole world as a total of 720×10^9 barrels, taking account of shales with a level of more than 100 litres of oil equivalent per tonne of rock: that is the concentration above which the reserves can be extracted even today at an acceptable cost of production at around $50 per barrel of kerogen. Thus, in the Green River basin deposits of the

---

1. That is, by mechanical excavation of the sand and treatment with a mixture of boiling water and steam. This separates the hydrocarbons from the inert material in which they were absorbed and they then float on the surface of the water. Only sands with more than 8 per cent of oil are currently extracted, but any technological progress in treatment will bring a corresponding increase in the quantity of available reserves.

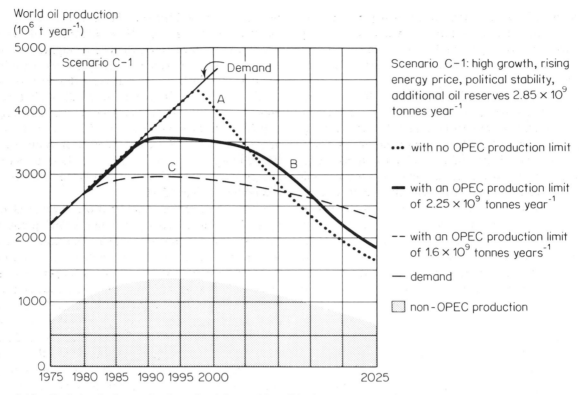

Figure 2.13   Variation in the production of and demand for oil in the non-communist world. With no OPEC production limit (curve A) a sharp deficit in supply is anticipated by 1995. If OPEC production is limited to 2.25 × 10⁹ t year⁻¹ (curve B), the deficit would be first noticed from 1985 and would become serious by the year 2000. With an OPEC limit of 1.6 × 10⁹ t year⁻¹ (curve C), the gap between supply and demand would have begun to appear in 1981. (Reproduced with permission from Wilson *et al.*, *Energy: global prospects 1985–2000*, Report of the workshop on alternative energy strategies, McGraw-Hill, 1977)

USA containing an estimated 1800×10^9 barrels of kerogen, only (!) the 117×10^9 barrels occurring in the layers of oil shale having a thickness greater than 10 metres and a concentration higher than 100 litres per tonne are considered to be recoverable (Metz, 1974). Even if the total estimated energy resources contained in world deposits of oil shales are reasonably accurate in yielding some 2×10^{15} barrels, only about 50×10^9

tonnes of hydrocarbons at the very most are capable of being extracted, and 20×10^9 tonnes of this are contained in the Green River basin.

In France, there are more than 3×10^9 tonnes of kerogen in the oil shales of Toarcien, which lie on an arc around the south-eastern fringe of the Paris basin. However, as in other European countries, the deposits are widely dispersed and only reach an average con-

Table 2.8   *Energy content of oil shales expressed in oil equivalent* (from Duncan and Swanson, 1965)

	Recoverable under present economic conditions (10⁹ barrels of oil equivalent)	Estimate of total resources for oil shales of various grades (in 10⁹ barrels of oil equivalent)		
		100–400 litres per tonne	40–100 litres per tonne	20–40 litres per tonne
Africa	10	4 000	80 000	450 000
Asia	20	5 500	110 000	590 000
Oceania	small	1 000	20 000	100 000
Europe	30	1 400	26 000	140 000
North America	80	3 000	50 000	260 000
South America	50	2 000	40 000	210 000
Totals	190	16 900	326 000	1 750 000

centration of 40–100 litres of kerogen per tonne. In spite of that, the oil shales of Autun were being exploited using pyrolysis until 1954, when the installation which had been producing tens of tonnes per day was closed down.

The method of extracting the shale itself is the same as for coal (open-cast or underground mines), but the mass of rock that has to be handled is many times greater than that of the hydrocarbon produced. For a parent rock containing 100 litres of kerogen per tonne, for instance, the mass is eleven times greater. In addition, the recovery of the kerogen requires great quantities of heat since it separates from the parent rock only above 480 °C. Large amounts of water are also needed: 3 m^3 per tonne of naphtha produced. For Europe at any rate, therefore, rather than exploit shale deposits containing on the average some 100 kg of coal equivalent per tonne, it would appear more logical to produce extra coal from the considerable resources available: coal provides 10 times more energy for the same quantity of rock extracted, leading to lower costs and a smaller environmental impact.

In conclusion, it is clear that the presentation of oil shales as a solution to the problem of oil shortage, as the media and some of the politicians in the West have done, is at the very least unrealistic. In the present economic climate, and with the current state of technology, the hydrocarbon content of oil shales seems destined to replace oil mainly as a raw material in the synthesis of organic chemicals. At most, the exploitation of the high-grade Green River deposits might partially alleviate the problems of oil supply in the USA.

Coal resources

The size and geographical distribution of world coal reserves are known with much greater precision than those of hydrocarbons. Unlike oil and natural gas, coal is not distributed erratically but occurs in stratified deposits over wide areas which are easy to identify and hence to map.

Averitt, of the US Geological Survey, published in 1969 an estimate of the total quantity of technically producible coal present in the lithosphere. He considered as extractable 50 per cent of the coal deposited in seams with thicknesses of 36 cm or more and with depths up to 1200 m. In exceptional cases, he also took into account deposits up to 1800 m, while pointing out that those deeper than 1200 m contain at most 10 per cent of the total coal in the Earth's crust.

Averitt arrived at two estimates: one of 4300×10^9 t for coal resources contained in fields already established by mapping, and one of 7600×10^9 t for ultimate world resources Q_∞. According to his figures, the USSR alone possesses some 4300×10^9 t of coal,

or nearly three-fifths of the ultimate world total. Second in size comes the USA, at 1486×10^9 t, which is more than three times the energy equivalent of the Q_∞ for world oil resources using the most probable estimate of 2100×10^9 barrels. The rest of Asia outside the USSR is credited with 681×10^9 t, which appears relatively small, and more recent data suggest that China alone might possess 1000×10^9 t of exploitable coal. As for Western Europe, coal resources here are estimated to be 377×10^9 t or close to twice the energy equivalent of proved world oil reserves now available (cf. Figure 2.11).

Hubbert (1969) has calculated the world production cycle for coal, assuming that the annual production rate would not double more than three more times to a maximum rate of eight times the then world consumption. Under those conditions, the useful period for coal with Averitt's lower estimate would lie between AD 2000 and 2200 with a production peak in 2110. The same calculation carried out with a Q_∞ of 7600×10^9 t gives a useful period from 2040 to 2390, and a production peak in 2160 (Figure 2.14). With the second hypothesis, world resources would thus satisfy human energy requirements for more than four centuries!

The enormous size of these reserves inevitably implies an increasing reliance on coal as a primary energy source. It also implies a revival of the coal industry, even if such an irreplaceable fuel source was once considered by many European economists and technocrats as out of date and unprofitable, an opinion which led a number of EEC countries to shut down some of their coal mines rather hastily. France is a particularly notable example of this, as can be seen from Table 2.9 showing the decrease in the proportion of the total energy supply provided by coal.

France is the one large industrialized nation with extensive coal deposits which has not conducted any research into new ones over the last 20 years. Discoveries that were made during that time, such as those at Lons-le-Saunier and in the Chateauroux

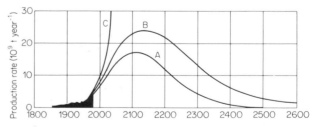

Figure 2.14 Cycles of world coal production. Curve A shows the variation in production rate for total reserves that were considered economically exploitable at the beginning of the last decade (4.3×10^{12} t), and curve B the variation for a Q_∞ of 7.6×10^{12} t. Curve C represents an indefinite growth in production at 3.6 per cent per year. (From Averitt, in Hubbert, 1969)

Table 2.9   Coal production in France

Year	Coal produced (10^6 tonnes)	% of total energy consumption
1960	58	45.4
1970	40	18.0
1973	24.5	9.6
1980	19.7	7.0

region, were quite fortuitous. It would undoubtedly be going too far to advance the coal resources of that country as a means of improving its adverse energy balance by a significant amount. All the same, in spite of official remarks about the alleged absence of coal reserves in France, it is worth recalling that available estimates credit the country with 3×10^9 t of the fuel. About 1.5×10^9 t of it could already have been considered as economically exploitable before the oil price increase of 1979: from 1977, heat energy from coal was costing 2.5 centimes per therm as against 4.5 centimes for fuel of equivalent sulphur content. Today it is even less costly in comparison with oil, which reached a price of 15 centimes per therm at the end of 1979.

A new policy for French coal is all the more essential since nuclear power will not provide enough to compensate for the deficit in oil between now and the end of the century. J. Ricour, president of the French Geological Society, wrote in 1976:

> Conjuring up a parity between heat energy from coal and from nuclear sources belongs to the realm of high fantasy. Even if the nuclear programme is developed as intended, France needs to have available by 1985 electrical energy from non-nuclear sources amounting to 30 million t.c.e. It is true that we could resort to oil to provide this amount, but that doesn't stop the exploitation of French coal deposits appearing, not only desirable, but impossible to avoid if we are to halt the growth in unemployment and the outflow of funds from the country.

Perhaps paying for oil at $80 a barrel, which cannot be long delayed, will help the 'decision-makers' to understand more clearly the difference between the general good and certain short-term 'interests'.

In fact, coal alone exists in sufficient quantities to be capable of replacing the total consumption of oil as a fuel. Although coal is a major source of pollution when it is burnt in the crude state, it can nevertheless be transformed into gaseous or liquid fuels that cause little pollution and this is in great contrast to the derivatives of oil that are in use at present.

Gasification of coal (Figure 2.15) is a rapidly developing technique. It produces hydrogen, carbon monoxide and methane, which can be used directly as substitutes for natural gas or as base materials for the synthesis both of liquid fuels like petrol and methanol and of lubricants. The development of the Fischer-Tropsch gasification process helped Germany to wage the Second World War thanks to (if that is the right phrase) the synthetic petrol it produced. Nowadays the only large synthetic fuel plant in operation is in South Africa: it produces more than 200 000 tonnes of fuel each year at a cost comparable with that of petrol from crude oil at the current price per barrel.

Investigations are also being carried out into the possibility of gasification *in situ* where deposits are too deep or the seams too thin to be exploited in the normal way. If such a process came into general use, it would also eliminate the risks to miners associated with underground shafts and galleries because only opencast mines would be left.

To sum up, there are two central facts that emerge, whatever our view as to the most reliable data on the ultimate reserves of fossil fuels:

(a) Neither our generation nor the next will see the exhaustion of fossil fuels or even a decline in their consumption in view of the enormous coal resources in the world.

(b) A serious shortage of conventional energy resources will arise between now and the end of the next century if our civilization persists in using fossil fuels as the principal source of energy in circumstances of uninterrupted growth.

The average rate of increase in coal production over the period 1860–1965 was 3.6 per cent per year and, if even this relatively modest rate were maintained indefinitely, the resources would be exhausted within two centuries. If instead we assume that coal will become a quasi-unique energy source, then the figure we need to use is the average growth in world consumption of fossil fuels as a whole. During the last two decades, this growth has been slightly higher than 6 per cent per year, giving a doubling time of 11 years. This would lead to a 1000-fold increase in production in little more than a century (i.e. ten doublings of 11 years each, since 2^{10} is about 1000). If we assume that 0.1 per cent of the total fossil fuel resources of the lithosphere have already been extracted, then this 1000-fold increase would lead to a consumption of $0.1\% \times 1000 = 100\%$: in other words, complete exhaustion of existing coal reserves in less than 100 years.

Pursuing the same idea, but assuming optimistically that a gross error has been made in the estimate of total resources, let us take it that we have really only consumed 0.01 per cent (that is, an underestimation of Q_∞ by a factor of 10!). Even in that case, the continual 10- or 11-year doubling of consumption would mean that the period until complete exhaustion would be extended by a mere 30 years or so!

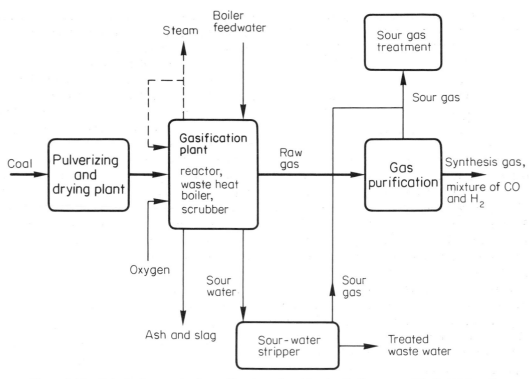

Figure 2.15   Principal stages in the gasification of coal by the Shell process. (Shell, 1978, p. 7)

These considerations not only reveal the futility of the idea of indefinite growth, but show as well that it is absolutely essential that we find one, or several, replacements for fossil fuels. It might appear from the above discussion that the serious problems arising from the exhaustion of coal resources would only concern our descendants and would not appear before the middle of the next century. However, the current energy crisis is in fact far more serious than might be thought.

A study group set up by the Massachusetts Institute of Technology (MIT), under the direction of C. L. Wilson, quite recently published a report on world prospects whose conclusions are very disturbing (*Energy: Global Prospects 1985–2000*, 1977). Because they have failed to make a massive reduction in their energy demands, Western countries are fast approaching a catastrophe of unequalled proportions. According to the MIT study, even if drastic efforts were made to conserve and diversify energy resources in a climate of moderate economic growth, the non-communist world will be short of 750 to 1000 million tonnes of oil per year by the end of the century. This prospect already assumes a doubling of the world oil production rate (which would then exceed 6000 million tonnes per year!) and an increase of 50 per cent in the price of oil in real terms.

If Saudia Arabia had limited its oil production to 500×10^6 tonnes per year from 1978, the deficit would have begun to appear in 1981. In fact, its onset has been delayed by the reduction in world oil consumption caused by the present economic crisis (1982–83). However, this does not invalidate the general conclusions of the MIT study: world demand for oil would still outstrip supply by the end of the 1980s if there were only a modest but sustained revival of industrial activity in developed countries. This would clearly have considerable economic consequences in view of the fact that the price increases from 1973 onwards occurred when potential supply clearly exceeded demand.

Following a similar line of argument, experts in Washington have estimated that in spite of the enormous size of their potential resources in fossil fuels the USA will experience a real economic disaster in 1985 unless strict conservation measures are taken and alternatives to oil and natural gas as energy sources are found.

Thus it can be seen that, even on the assumption that energy growth in industrialized countries will cease, there is an urgent need for other types of source to take over from the declining fossil fuels. Such alternatives are also required to meet the growing demand from the Third World in which the acquisition of a certain minimum standard of living—called for by a respect for human dignity—implies a significant increase in available energy *per capita*.

In a context as dramatic as this, humanity must have recourse not only to what are called 'new' energy sources but also—whether one likes it or not—to nuclear energy. This appears to be inevitable for, as will be demonstrated in later sections, no resource seems capable on its own of entirely replacing fossil fuels.

2.2.2   Potential sources of energy, 2: nuclear fuels

There are two fundamental processes that enable nuclear energy to be produced:

(a) *Fission*, in which the nucleus of one element disintegrates spontaneously into nuclei of other elements with smaller atomic masses. The original nucleus is said to be *fissile*: examples are uranium-233, uranium-235 and plutonium-239.

(b) *Fusion*, in which two nuclei of light elements combine to produce the nucleus of a heavier element.

In both cases, the reactions take place with the release of considerable energy and, at the same time, the final mass of the reaction products is less than the initial mass. The decrease in mass (Δm) is related to the energy released (ΔE) by the famous Einstein equation:

$$\Delta E = (\Delta m)c^2 \qquad (1)$$

where c is the speed of light.

Nuclear fission

Every atom of uranium-235 that disintegrates releases 200 MeV or 3.2×10^{-11} J of energy. A gramme of uranium contains 2.56×10^{21} atoms so that this mass of fissile matter would be equivalent to the energy from the combustion of 2.7 tonnes of coal. Since natural uranium contains 0.71 per cent of uranium-235 (1 atom for every 140 atoms of uranium-238), 1 gramme of the natural element has an energy equivalent of 19 kilogrammes of coal.

Most nuclear reactors which are currently in operation for the production of electricity are known as *burners* since they use only the controlled disintegration of uranium-235. Some, however, are *converters* and in these a proportion of the energy released is used to produce significant quantities of another fissile material, plutonium-239: the British and French gas-cooled graphite-moderated reactors using natural uranium are examples. In these cases, if Q_0 is the initial quantity of fissile material in the reactor and if Q is the quantity remaining after one cycle of the reaction, then $Q/Q_0 < 1$ for converters but $\rightarrow 0$ for burners.

There is a third type of reactor known as a *breeder* (or generally *fast breeder* because it uses fast neutrons to cause fission) in which $Q/Q_0 > 1$: after one cycle of the nuclear reaction there is a greater quantity of fissile material in the reactor than was initially present. Fast breeder reactors use materials which are said to be

fertile: they are not themselves readily fissile but are capable of being converted into fissile elements. The main examples are uranium-238 and thorium-232, both of which are to be found in appreciable quantities in igneous rocks. These fertile elements are placed in a 'blanket' surrounding a core of fissile material (uranium-235 or plutonium-239) which emits neutrons. The uranium-238 or thorium-232 captures the neutrons and is transformed into plutonium-239 or uranium-233 respectively. These are both fissile and produce an energy per unit mass equivalent to that provided by the disintegration of uranium-235. Figure 2.16a shows the breeder cycles for thorium and uranium-238, and Figure 2.16b(C) the reactor structure.

Breeder reactions create the possibility of utilizing the total uranium and thorium contents of geologically exploitable deposits. As soon as a reaction cycle in a fast breeder has been completed, uranium-235 can be dispensed with and the active core of the reactor loaded with the uranium-233 or plutonium-239 produced. Plutonium is particularly important for the fast breeder because its neutron yield is very high.

In theory, the use of fast breeders multiplies the quantity of fissile material available per tonne of natural uranium by a factor of 140. In practice, when the efficiency of the reaction and other energy-absorbing aspects are taken into account, the multiplying factor reduces to about 50: that is, in relation to the direct use of uranium in a classical burner reactor. Finally, fast breeders also enable thorium to be used as a nuclear fuel through its conversion to uranium-233, and the potential reserves of that element in the lithosphere seem to be distinctly greater than those of uranium at equal costs of extraction.

At the present time, the only burner type reactors producing a significant output of nuclear electricity are those that function with light water (that is, ordinary water as opposed to heavy water in which the hydrogen is replaced by deuterium). In this type of reactor, the fuel elements consist of enriched uranium (with the proportion of uranium-235 increased to 3 per cent) encased in a material with suitable neutron and thermal properties (zircalloy = an alloy of zirconium and tin, for instance). The neutrons which maintain the nuclear reaction are moderated (i.e. slowed down) by ordinary demineralized water (with added boron) which also acts as the coolant or carrier of heat. There are two main varieties: the pressurized water reactor (PWR) and the boiling water reactor (BWR).

In the PWR (Figure 2.16b(B)), water circulates in the core at a pressure of 150 atmospheres and a temperature of 320 °C. In normal operation, its fuel elements are subject to various faults which show up as cracks in the metallic casing. Although the rate of incidence of such faults is low (0.1 per cent per year), a

Plate III   Energy and the environment

1   Cooling towers at a nuclear installation. Large quantities of heat discharged at low temperatures by power stations create difficult environmental problems. Cooling by air is tending to become more widespread because it avoids the thermal pollution of continental waters. In addition to their unsightliness, these enormous structures lead to a significant loss of water in the form of vapour, sometimes at a rate as high as several cubic metres per second.

2   A 300 000 tonne oil tanker discharging its cargo at Fos-sur-Mer. World industrial development has been based on oil for more than half a century: unwisely, because it is the least abundant energy source and the most badly distributed.

3   A petro-chemical installation at Lavera in the south of France. The use of heavy fuels rich in sulphur produces a formidable amount of atmospheric pollution whose effects can become apparent a long way from the point of emission.

(Photographs F. Ramade)

Plate IV   Energy and the environment
(continued)

1   A pressurized water reactor (PWR) under construction on the banks of the Rhône at Bugey.

2   High-voltage overhead cables in the neighbourhood of Paris. The development of a sizeable electro-nuclear programme entails the construction of high- and medium-voltage supply lines which will cover tens of thousands of hectares with a continuous network of cables if the policy of burying them underground is rejected.
(Photographs F. Ramade)

PWR with an output of 900 MW(elec.) contains some 40 000 fuel rods (Figure 2.17) in its core, so that there does occur a permanent radioactive contamination of the primary coolant circuit by fission products. Each year, a PWR of this power output discharges into the atmosphere several thousand curies of krypton-85 and a greater quantity of radioactivity through tritium, mainly as tritiated water.

In the BWR, the core coolant is still light water but there is no secondary circuit: the steam formed in the reactor is taken directly to the turbine, which thus also becomes contaminated with radioactivity and raises problems in maintenance. One advantage of the BWR, on the other hand, is that it does not produce tritium.

Experts in nuclear safety refer to the various types of light water reactor as 'dirty'. This is because they are designed to be shut down only once a year for the replacement of fuel elements that have been in the core for 3 years (for example, one third of the total of 90 t of enriched uranium fuel is replaced annually in the core of a 1000 MW(e) PWR). This design even makes removal of faulty elements impossible when the reactor is operational and they thus have plenty of opportunity to produce contamination during the time between shut-downs of the primary circuit.

The second category of nuclear reactor consists of various types of *converter* characterized by a conversion ratio (i.e. the number of plutonium nuclei produced for every uranium-235 nucleus that has undergone fission) greater than that of the burners but still less than 1. One such type is the gas-cooled graphite-moderated reactor using natural uranium which was developed independently and almost simultaneously in France and in Great Britain where it is known as the 'magnox' series of power stations. The use of natural uranium in the fuel elements saves the cost of enrichment plants essential for light water reactors. The fuel is encased in an alloy of magnesium (hence the name 'magnox') having a high resistance to thermal and mechanical deformation. Blocks of graphite placed around a cluster of fuel elements act as the moderator to slow down the neutrons, and the heat developed is carried away by CO_2 gas at a pressure of 40 bar, CO_2 being well known for its low chemical reactivity.

This type of reactor is very safe in operation: an accident involving loss of coolant carries no fear of a catastrophe like the partial melting of the core because of the large thermal inertia. Not only that, but because this type of reactor is designed to be charged and discharged while still operating, any faulty fuel elements (which are rare in any case) can be immediately extracted. The primary circuit thus suffers very little contamination, so that these reactors are considered to be 'clean', unlike the PWRs and BWRs.

The main disadvantage of the type is that they are larger than others for a given power output so that they are more costly to build. Moreover, their conversion factor, even though slightly above 0.5, is not as high as the conversion factor of those to be discussed a little later. Finally, the core becomes 'poisoned' by some of the fission products, particularly xenon-133 and samarium-149, which absorb neutrons and thus remove them from the fission process. All these factors lead to a relatively moderate efficiency in the use of fissile material, amounting to barely 25 per cent.

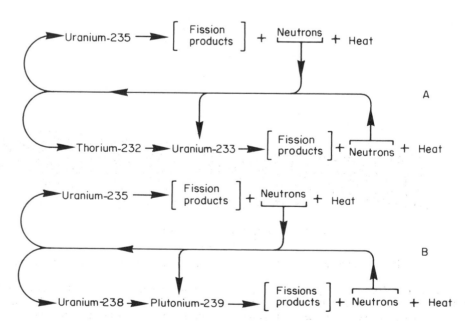

Figure 2.16a   Schematic representations of breeder reactions: A, thorium cycle; B, uranium-238 cycle

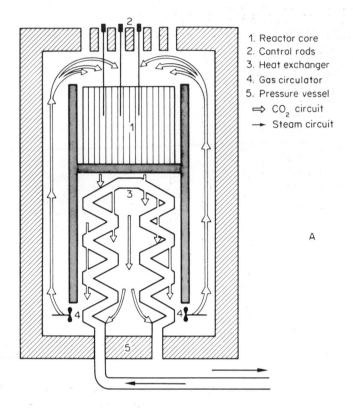

1. Reactor core
2. Control rods
3. Heat exchanger
4. Gas circulator
5. Pressure vessel
⟹ CO_2 circuit
→ Steam circuit

A

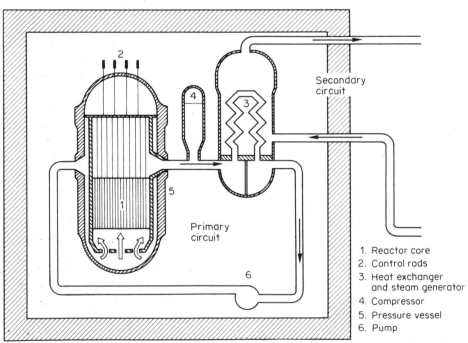

Secondary circuit

Primary circuit

1. Reactor core
2. Control rods
3. Heat exchanger and steam generator
4. Compressor
5. Pressure vessel
6. Pump

B

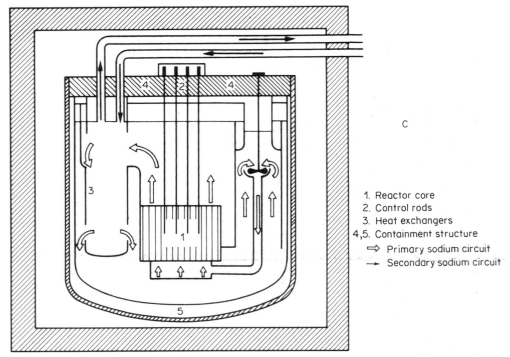

C

1. Reactor core
2. Control rods
3. Heat exchangers
4,5. Containment structure
⇨ Primary sodium circuit
→ Secondary sodium circuit

Figure 2.16b Principal types of nuclear reactor: A, gas-cooled graphite-moderated natural uranium reactor; B, reactor cooled and moderated by light water (PWR); C, fast breeder reactor. (From an Electricité de France document)

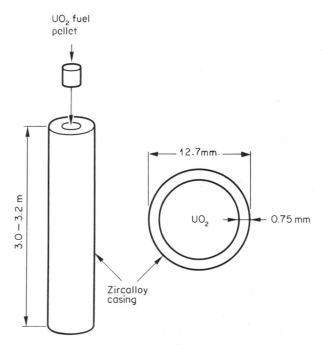

UO$_2$ fuel pellet

12.7mm

UO$_2$ — 0.75 mm

3.0 – 3.2 m

Zircalloy casing

Figure 2.17 Structure of a fuel element in a PWR. Each element is a hollow tube with walls of zircalloy forming a casing containing UO$_2$ pellets in which the uranium-235 is enriched to 3 per cent

Another series of reactors which can also be classified as converters but which have a greater efficiency are the Candu type (Canadian Deuterium Uranium). These have the advantage of using nuclear fuel very economically, with a consumption 30 per cent less than that of a light water reactor of equal power. They use natural uranium with heavy water as a moderator and possess a high conversion ratio (higher than 0.7) while they can also be used after some modification as a type of breeder using slow neutrons by placing fertile elements like thorium-232 round the fuel. Finally, Candu reactors are 'clean' and very safe: immediate evacuation of the fluid moderator allows a rapid shut-down in the event of an accident.

HTRs (high temperature reactors) are still being developed, unlike those previously discussed which have long been commercially established. They form an intermediate class between converters and breeders and seem to offer some advantages. Their fuel elements consist of small particles of fissile material embedded in pyrolytic graphite, a design that allows much higher working temperatures than those of previous types of reactor, thus bringing a significant increase in thermo-dynamic efficiency. Not only that, but HTRs could bring about a real revolution in the field of combustible fuels. Current research is aimed at using the heat generated at temperatures as high as 1000 °C to produce hydrogen by the catalytic pyrolysis of water. This would facilitate the production of fuel in large quantities limited only by the amount of fissile material available.

The construction of HTRs known as 'feed-breed' reactors has also been considered, operating with a conversion ratio greater than or equal to 1. The fuel in these consists of a mixture of particles, some of fissile material (the 'feed') and others of fertile material (for example, thorium-232) to 'breed'. New fissile material (uranium-233) formed in the 'breed' particles and not burnt up during operation can be separated when the irradiated fuel is reprocessed and used to form part of the next 'feed' charge—or the whole of it, if it operates as a fast breeder. The resulting economy in the use of fissile material is quite considerable in view of the high conversion ratio of these reactors.

The final class of reactor we shall consider is that of the *fast breeders*, whose principles of operation have already been indicated. The adjective 'fast' refers to their use of neutrons emerging straight from the fission reaction with high energy: no moderators are used. Such neutrons are capable of transforming fertile elements into fissile ones (Figure 2.16). There is a further important principle involved: if the flux of neutrons in a reactor can be sufficiently increased, there will be a large excess of them over and above those needed to maintain the fission chain reaction, an excess which can be used for other purposes like breeding. Plutonium is a particularly fruitful source of fast neutrons in that it produces on the average 2.4 of them upon disintegration. In principle, only one is needed to sustain the chain reaction, so that there is an excess of 1.4 neutrons per nucleus undergoing fission. It is true that such reactions take place in any nuclear reactor, but in the case of those using fast neutrons a conversion factor greater than 1 is more easily obtained. This means that more fissile material (in the form of plutonium-239 or uranium-233) is produced than is consumed: hence the name 'breeder'.

Fast breeders can in theory extract 50 times more energy than can burner type reactors from the same initial quantity of natural uranium. Moreover, they use uranium more efficiently and also make it possible to use thorium, which is otherwise difficult to employ. All these factors enable us to exploit mineral deposits with quite low concentrations of fissile and fertile materials and this means that we could count on resources in nuclear fuels distinctly greater than those considered economically workable if only light water reactors were being supplied. In fact, the uranium ore deposits that can be used in the types of reactor now in operation are at present much more limited in quantity.

Reserves of nuclear fuel[1]

The known reserves of economically workable fissile material will just satisfy world requirements for about

15 years given the present rate at which the production of nuclear energy is growing. In 1970, proved reserves of uranium oxide available to the Western world at a cost of less than $20 per kg amounted to 840 000 t of U_3O_8. To that should be added another 750 000 t

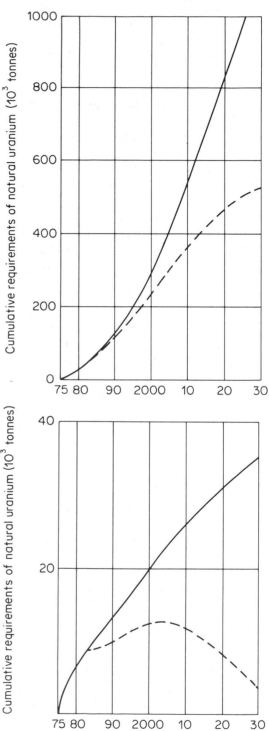

Figure 2.18 Forecasts of the growth in consumption of uranium by France during the next half-century. Continuous curves show the cumulative and annual requirements if the nuclear programme is based strictly on PWRs. Broken lines indicate the effect of a programme in which fast breeders progressively take over. (From Thiriet, 1976, p. 64)

1. Throughout this section, the costs of extraction per kilogram are quoted in US dollars at rates that prevailed generally during the 1970s when the original articles were compiled. The corresponding costs in French francs as originally given by the author can be obtained by using exchange rates characteristic of that period, varying between 4.3 and 4.8 francs to the dollar.

Table 2.10   Annual requirements of natural uranium (tonnes) (from the CEA (Commissariat à l'Énergie Atomique), in Thiriet, 1976)

	1970	1975	1985	1990
EEC	5 000	12 500	32 000–34 000	45 000–50 000
of which West Germany	1 000	4 500	9 000–10 000	12 000–13 000
France	1 500	5 000	9 000–10 000	12 000–13 000
USA	10 000	26 000	40 000–45 000	65 000–75 000
Japan	2 000	4 000	9 000–10 000	15 000–17 000
Total Western World	21 000	55 000	105 000–115 000	105 000–185 000

whose existence was less certain, so that at most a total of only 1 600 000 t was available at low cost. Since then, the projected development of nuclear power in Western Europe and North America during the next decade has implied that another 1 million tonnes of uranium oxide must be discovered and extracted by 1985!

France, although relatively well endowed with uranium ores, is an important example which illustrates the same point. If the national programme of 1973 for the construction of nuclear power stations using PWRs is taken through to completion, it will lead to an installed power of 50 GW(e) by 1985 and of 170 GW(e) by the year 2000. This would mean an annual consumption of 7000 t of natural uranium in 1985 and of 24 000 t in 2000. Proved French reserves stand today at 50 000 t, and the estimated ultimate resources are about 120 000 t, so that national requirements will lead to their exhaustion before the end of the 1990s (Figure 2.18).

World uranium consumption has been showing considerable growth amounting to an increase of some 15 per cent per year, which corresponds to a doubling time of roughly 5 years. Table 2.10 shows the figures of actual consumption for 1970 and 1975 together with predicted requirements in 1985 and 1990. It can be seen that by 1985 the world PWR programme will have exhausted the whole of the proved and probable uranium reserves of about 1 650 000 t which can be extracted at a cost of less than $90 per kg (Table 2.11). It is thus quite clear that unless there is a rapid transition to the use of fast breeder reactors, the scarcity of

high-grade uranium deposits will lead to a serious shortage of fissile material by the end of the century and very probably by 1995 (Figure 2.19).

It seems, therefore, that the present policy of concentrating on light water reactors is in the medium term not an adequate solution to the problem posed by the shortage of fossil fuels. Suppose, for instance, that the world Q_∞ for natural uranium extractable at a cost lower than $70 per kilogram is taken to be 10 million tonnes—an optimistic figure since it represents a total quantity over five times greater than current proved reserves. Even that amount would only lead to an energy equivalent of 5 Q if all its uranium-235 were used directly in PWRs, and 5 Q is barely one third of the Q_∞ for world oil reserves.

Ultimately, if industrialized countries continue to use up the Earth's stock of uranium-235 in light water reactors before fast breeders have been properly developed, it will become almost impossible to achieve such a development later on. In that case, the uranium era would have been even more ephemeral than that of oil. This wastage of yet another non-renewable resource would also be one of the greatest disasters to

Table 2.11   Potential uranium resources in the Western world (10³ tonnes) (adapted from OECD–IAEA data in Thiriet, 1976)

Categories of cost in 1978 $ per kg of uranium	Proved and probable reserves	Estimated additional resources	Totals
<$90 per kg	1650	1510	3160
$90–150 per kg	540	590	1130
Totals <$150 per kg	2190	2100	4290

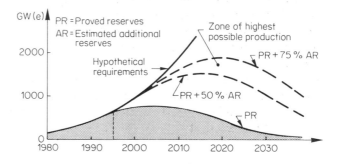

Figure 2.19   Evolution of world nuclear power production based on various projections of available uranium resources. If nuclear installations are built at the rate predicted by the IAEA, reserves that were known at the end of the 1970s would become insufficient to meet demand by 1995. Even if 75 per cent of the more speculative resources are added, supplies would become critical by the beginning of the next century. This study assumes that the installations will be PWRs and that no reprocessing of irradiated fuel is carried out. (From Langlois *et al.*, 1979)

Table 2.12 World reserves of uranium* expressed in energy equivalent (adapted from Felden, 1976)

Type of reactor	Type of ore	Proved or probable reserves	Estimate of ultimate resources (Q_∞)
PWRs	high grade ($\leqslant$\$150/kg)	1 Q†	3–4 Q
	low grade (\$150–500/kg)	15 Q	45–60 Q
Fast breeders	high grade	70 Q	200–300 Q
	low grade	1050 Q	3000–4000 Q

* Excluding communist areas for lack of reliable data.
† 1 Q = 36.6×10^9 t.c.e.

hit industrialized society as regards energy since it seems quite clear that a limited reliance on nuclear energy is inevitable before new types of source can take over.

A transition to fast breeders would, on the other hand, produce a considerable increase in the amount of fissile material available for three reasons:

(a) because theoretically it would multiply the amount of energy available from natural uranium by a factor of 20,
(b) because, given such a factor, it would enable the exploitation of low-grade ores unsuitable for PWRs to be undertaken,
(c) because only fast breeders enable thorium to be used, an element that is much more abundant than uranium in the lithosphere for a given cost of extraction.

World reserves of fissile materials expressed in terms of their energy equivalent are shown in Table 2.12. The figures clearly demonstrate the considerable increase in available energy which could be made possible by the development of fast breeder reactors, which would enable low-grade ores to be exploited. This point is illustrated by the example of the uranium-bearing schists of Chattanooga in the Appalachians which contain 60 g of natural uranium per tonne in a deposit covering an area of 750 km² and with a thickness of 5 m. The amount of nuclear fuel here would be the equivalent of the total coal reserves of the USA. Again, the thorium-bearing granites of Conway, New Hampshire, of a similar area and with a depth of 100 m, contain some 20 times the energy equivalent of the total American fossil fuel reserves. As Table 2.13 demonstrates, the quantity of natural uranium available increases rapidly as ores of lower and lower grades are worked.

The fact that fast breeders allow the use of thorium brings another advantage over uranium in that its breeder cycle does not produce plutonium with all its attendant problems of proliferation and toxicity. Moreover, although exploration for thorium deposits is

Table 2.13 Uranium reserves of the USA in terms of the concentration of U_3O_8 in the deposits (from Rose, 1974b)

Concentration of U_3O_8 (p.p.m.)	U_3O_8		Equivalent electrical energy (GW/year)	
	Cost of extraction (\$/kg)	Available quantities (10^3 t)	PWRs	Fast breeders
>1600	<22	1 127	6 600	880 000
>1000	22–33	1 630	9 500	1 270 000
>200	33–66	2 400	14 000	1 860 000
>60	66–110	8 400	49 000	6 500 000
>25	110–220	17 400	102 000	13 500 000
3*	several thousand	10^6–10^7	—	—

* Mean concentration of U_3O_8 in the Earth's crust.

being undertaken on a far smaller scale than for uranium, its abundance in the lithosphere is four times greater: the average concentration of thorium in surface rocks is 12 p.p.m. as against 3 p.p.m. for uranium.

The probable US reserves of thorium (cost < $90/kg) have been estimated at 3.5×10^6 t in comparison with 500 000 t of uranium (cost < $150/kg). The ratio of these two quantities can presumably be extrapolated to world deposits, and this would mean that the lithosphere contains ten times more thorium than uranium available at a given cost of extraction by current and foreseeable techniques. In the long-term perspective, therefore, say over a period of 100 years, the emergence of fast breeder technology ought to mean that no energy crisis would occur owing to a lack of resources.

It will be quite different in the medium term, however, since the only reactors operating in large numbers at present are PWRs. The worrying question of uranium supplies for this type of reactor in 15 years or so has already been raised: the uranium needed for the operation of the reactors under construction and those already completed still remains to be discovered! Given the long lead times necessary for the development of any new system of energy production, it is difficult to see how there can be enough time to construct the number of fast breeders needed to take over from the PWRs.

This last point is particularly relevant to the use of the thorium cycle, since there is no foreseeable development of it and thus no prediction is possible of a date by which it would become available for large industrial use. Considerations of this sort are even more relevant to nuclear fusion, which even today is incapable of being realized in the laboratory. We now turn to this aspect of nuclear power.

Nuclear fusion

The nucleus of the most abundant form of hydrogen is written as ^1_1H, where the upper figure gives the mass number (the total number of neutrons and protons that it contains) and the lower figure is the atomic number (the number of protons). Two other isotopes of the element exist, one known as deuterium (^2_1H or ^2_1D) and the other as tritium (^3_1H or ^3_1T).

Controlled nuclear fusion involves two of the possible reactions between these nuclei:

$$^2_1D + ^2_1D \rightarrow ^3_2He + \text{neutron} + 3.2\text{ MeV} \quad (1)$$

$$^3_1T + ^2_1D \rightarrow ^4_2He + \text{neutron} + 17.6\text{ MeV} \quad (2)$$

Tritium does not occur naturally and the only practical way of obtaining it is to bombard the alkaline-earth element lithium with neutrons, producing:

$$^6_3Li + \text{neutron} \rightarrow ^4_2He + ^3_1T + 4.8\text{ MeV} \quad (3)$$

Combining this with (2) yields the overall result:

$$^6_3Li + ^2_1D + \text{neutron} \rightarrow 2\,^4_2He + \text{neutron} + 22.4\text{ MeV} \quad (4)$$

Work on controlled nuclear fusion is still at the stage of fundamental research and is likely to remain there for an indefinite period of time. Indeed, it has not yet been achieved experimentally because it runs up against two obstacles: one is the temperature that has to be reached in the plasma and the other is the length of time that the reactants must be confined. Methods using lasers have succeeded in getting close to the threshold temperature (about 5×10^7 K for the D–T reaction), while others have achieved a sufficiently long confinement time, but none has succeeded in satisfying both conditions simultaneously. This state of affairs gives us some idea of the time that is likely to elapse before fusion energy is available for industrial purposes. As a matter of interest, it is useful to recall that 30 years passed between the success of the first atomic pile and the opening of the first nuclear power station with an output comparable with that of the oil-fired or coal-fired ones. . . .

However, in spite of these reservations, the potentialities opened up by nuclear fusion are theoretically very considerable indeed. Until now, all the research has been based on the deuterium–tritium reaction, which has a lower threshold temperature than the deuterium–deuterium one. However, the problem of finite resources will still be encountered because the limiting factor is the scarcity of lithium-6, the element indispensable to the preparation of tritium.

Natural lithium contains 7.4 per cent of lithium-6, the main isotope being lithium-7. The element itself is not very abundant in the Earth's crust and its principal crystalline form is spodumene, a mineral constituent of pegmatites. Lithium also occurs in significant concentrations in efflorescent brines of salt-water lakes that are subject to high evaporation: the Great Salt Lake of Utah, for example, or the chotts of North Africa.

At present, the known reserves of lithium amount to 9×10^6 t, which would yield 66 500 t of lithium-6. Since the fusion energy produced per lithium-6 atom is 3.19×10^{-12} J, the reserves are equivalent to 215×10^{22} J of available energy or the equivalent of that contained in the whole of the estimated fossil fuel reserves. This is an enormous quantity, certainly, but it is limited nevertheless.

It is therefore not possible to claim that the energy put at our disposal by nuclear fusion would resolve the resource problem as long as the basis is the deuterium–tritium reaction. Only the D–D reaction (1) is capable of producing a fundamental change in future prospects and that is because each cubic metre of water contains 34.4 g of the isotope. Moreover, the D–D reaction would have the advantage of not discharging toxic tritium into the environment. Unfortunately, this reaction is even more difficult to bring about than the D–T reaction so that we are forced to the conclusion that any calculations of nuclear energy from fusion must remain speculative and that its availability will be subject to delay for an unforeseeable period of time.

2.2.3 Natural sources of energy, 1: general discussion; geothermal sources

There are three types of natural energy source in the ecosphere: solar, geothermal and gravitational, in decreasing order of size. Solar radiation is by far the largest, with a total power of $179\,000 \times 10^9$ kW, while geothermal sources account for 32×10^9 kW and gravitation in the form of tidal energy provides 3×10^9 kW.

Figure 2.20 shows what happens to the incoming solar radiation: 33 per cent of it is lost by direct reflection back into space; nearly a half is transformed directly into heat; and 23 per cent is consumed by evaporation, precipitation and storage in the hydrosphere where it acts as the driving force behind the water cycle. Only 0.2 per cent of the total radiation goes to·the production of winds and phenomena associated with them, like ocean currents, convection, etc. Finally, a mere 0.023 per cent is converted into biochemical energy through photosynthesis, the process on which all terrestrial life depends.

Only solar radiation and tidal energy are really inexhaustible sources on the human scale (and even

they are ultimately limited in cosmic terms). Although they are sometimes called sources of 'free' energy, their use is still limited, but by the flux available rather than by the total stock as is the case with the 'exhaustible' sources already discussed. Geothermal energy is difficult to classify, but is ultimately stock-limited since it is really a kind of nuclear fossil source. It arises from heat released in the Earth's interior by the radioactive decay of various unstable elements originating from the time when the solar system condensed. Some of the energy may also be the remnants of the heat generated by that very condensation itself.

Possibilities and limitations of geothermal energy

The 32×10^9 kW of geothermal power represents an energy equivalent of 1 Q per year dispersed as heat: in other words, it is about four times the rate of consumption of primary energy by humans. However, the energy flux per unit surface area is too small, and its temperature too low, except in active volcanic regions,[1] to lead to its general industrial use. Note also that it involves an amount of energy about 3700 times less than the solar flux at the Earth's surface.

On the other hand, there do exist large local concentrations of geothermal heat in volcanic areas or, more generally, in the regions of contact between tectonic plates. These concentrations arise from recent intrusions of magma from the Earth's mantle into the upper parts of the lithosphere. The most spectacular examples are provided by hot springs and, more exceptionally, by geysers.

On the average the temperature rises by 1 °C for every 33 m increase in depth (the *geothermal gradient*), so that it is possible to find hot plutonic rocks anywhere by going deep enough. Concentrations of heat sufficiently near the surface to be usable are, however, distributed very unevenly and the great geothermal regions obviously correlate with active volcanic zones, containing rock masses with temperatures between 200 and 400 °C. There are some especially favourable localities where the rate of increase of temperature with depth is ten times greater than the average geothermal gradient (that is, it increases 1 °C every 3 m!). In these cases, rocks at a depth of about 1 km reach a temperature of 300 °C.

The total amount of recoverable energy in the various known geothermal 'deposits' would be considerably greater than 100 Q, but very little surveying has been done and this quantity could well be revised upwards with future advances in technology.

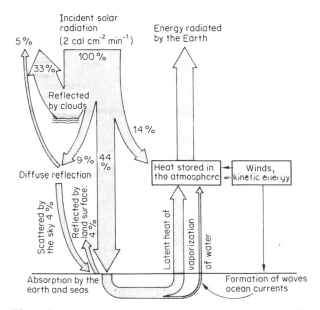

Figure 2.20   Diagram showing what happens to solar radiation incident on the Earth's surface. (Ramade, 1978a, p. 81)

---

1. The maleo fowl, a bird native to Sulawesi in Indonesia, uses geothermal heat for incubation. It does not make a nest, but simply deposits its eggs in volcanic sands having a high enough temperature.

Plate V   Natural energy sources

1   Geysers in the Norris Basin in the Yellowstone National Park, Wyoming, USA: the most spectacular form of geothermal energy. While it is true that sources producing steam dry enough to drive turbo-generators are quite rare, the potential resources of low-temperature geothermal energy (for central heating of dwellings, for example) are very considerable indeed.

2   Water pumps for irrigation driven by windmills on the Lassithi Plateau in Crete. It is wrong to think that the use of wind energy is comparatively recent: in fact, it is the oldest source of energy used by humans for mechanical work. The use of modern wind-driven generators could produce the equivalent of several million tonnes of oil per year over EEC territory as a whole.
(Photographs F. Ramade)

Plate VI   The environmental impact of mineral extraction

1   A bauxite quarry near Brignolles (Var) in France. After being abandoned when the mineral deposits are exhausted, such a quarry cannot be recolonized by vegetation.

2   An open-cast coal mine abandoned several years ago in the Cévennes Basin, France. Reafforestation of mining zones of this sort is a slow and tedious process if the land is not rehabilitated by filling-in after excavation has ceased. Here, a few pines just about manage to grow on the sterile slopes away from the old mine face.

(Photographs F. Ramade)

There are two types of process for the exploitation of geothermal sources, one classified as 'wet' and the other as 'dry'.

The *wet* techniques, using natural hot water or steam, were the first to be developed and they separate into high- and low-temperature methods.

The *high-temperature method* using dry steam (without water droplets) formed in some volcanic areas was developed at the beginning of the century at Lardarello in Tuscany, where the first geothermal power station was built in 1904: its present power output is 400 MW(e). The largest installation of this kind every built is the power station at Geysers in California with an output of 900 MW(e). The steam here is produced with a temperature of 200 °C under a pressure of 6 atmospheres, and it costs 20 times less than that obtained in the same condition from oil-fired stations (1979 prices). The total capacity of the Geysers field is estimated at 4000 MW(e). Finally, there is an installation in Japan which was started in 1960 and still has a very low output of only 20 MW(e).

The *low-temperature method* uses hot water. In very favourable conditions, such as those at Wairakei in New Zealand where the water is under pressure at 120 °C, 'wet' steam can be produced. This is then dried and used to drive a turbo-alternator. The power output at Wairakei is 200 MW(e). Other such sources occur in Iceland, Japan, the USSR and Mexico.

In practice, only 1 per cent of 'wet' geothermal sources can be used to produce electricity. The remaining 99 per cent consist of expanses of water at 60 to 90 °C which are suitable for local heating or other uses not requiring temperatures above 100 °C. In France, for instance, the hot waters of the Dogger near Melun, with temperatures of around 80 °C, have enabled more than 3000 houses to be heated since 1971: by the end of the 1980s, several hundred thousand could be heated in this way. It has been estimated that the use of geothermal energy for space heating could allow France to save the equivalent of at least 10^7 t.c.e. annually.

Urban heating using natural hot water is not a recent innovation. Reykjavik in Iceland has been using it for some 60 years, thanks to the wealth of hot springs and wells in a region of high volcanic activity.

The use of low-temperature geothermal energy is also being developed in the USA (Oregon, Idaho, Nevada), in Hungary (600 MW were being used in urban heating in 1973)—and in the USSR, where research has been undertaken to try and produce electricity from the heat in underground water at relatively low temperature. The method uses an organic liquid (butane, for example) with a boiling point below 90 °C, but the technique does not look very promising.

The *dry* techniques have a distinctly greater

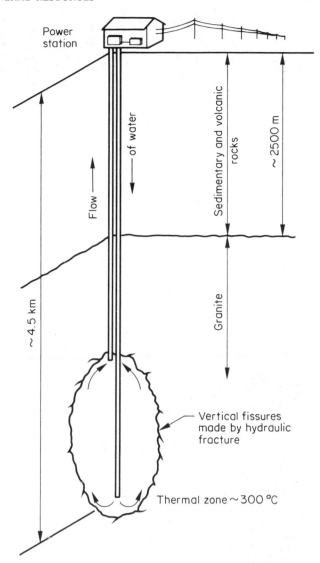

Figure 2.21 Proposed method of extracting heat from a 'dry' geothermal source. (From Smith, Los Alamos Scientific Laboratories, in Hammond *et al.*, 1973)

potentiality than the wet ones, but they have not yet been developed on an industrial scale. The method uses explosives or some other fracturing technique to hollow out one or more cavities in magmatic or plutonic rocks (e.g. granite) with a temperature higher than 300 °C. Water is injected into these cavities and the heat is then extracted from it by pumping and circulation around the network of underground reservoirs which act as a collection of geothermal boilers (Figure 2.21).

The dry geothermal sources have a potential output at least ten times that of the wet ones. In France, for example, the Massif Central is particularly rich in pockets of plutonic rock capable of being used industrially. In a single *département* of the Auvergne, some 100 km³ of rock at depths between 3 and 5 km could alone provide enough energy for several years' national consumption. In other parts of the world there exist pockets of hot magmatic rock quite close to the surface: in the USA a dry geothermal deposit has recently been discovered in Montana having a diameter of 8 km, a temperature between 500 and 800 °C, and with its upper limit no deeper than 1 km.

To conclude this brief survey of geothermal sources, it should be noted that they are not without disadvantages as a resource. Like others we have studied, this resource is essentially exhaustible and is not without environmental impact: production of electricity is accompanied by sizable emissions of hydrogen sulphide and other atmospheric pollutants derived from sulphur. Thus at Geysers in California the quantity of sulphurated gas released is comparable with that produced by a conventional power station of the same output using an industrial fuel with a low sulphur content!

In addition, the majority of the sources of natural hot water are rich in dissolved salts and have a salinity that is as great as, and sometimes greater than, that of sea water. The thermal waters of Imperial Valley by the Salton Sea in California are capable of providing an electrical output of 20 000 MW by wet geothermal techniques. Unfortunately, there are 200 g of dissolved salts per litre, which is six times the concentration in sea water. Hot water with that degree of salinity cannot be discharged after use into the surface waters of the region and have to be reinjected into deep geological strata.

2.2.4   Natural sources of energy, 2: tidal energy

Gravitational energy, although essentially inexhaustible, is not capable of providing large power outputs. Its capacity is limited to the equivalent of 1 per cent of the world hydroelectric supplies at the very most, and in addition there are only a few favourable sites where construction of power stations dependent on tidal energy would be worthwhile. The height of the tide must be greater than 5 m and the necessary barrage must not involve prohibitive amounts of civil engineering work. These conditions are in practice only encountered in one or two narrow bays or estuaries such as those of the Rance and of Mont-St-Michel in France and at Fundy Bay in Canada.

The French electricity authority EDF (Electricité de France) achieved the world's first generator of this kind with its 250 MW(e) installation on the River Rance. This was inaugurated in 1966 and has since been operating very satisfactorily in spite of the novel technology involved. Eventually, the barrage of the Iles Chausey, designed to harness the tidal energy of the bay of Mont-St-Michel, should be capable of producing more than 15 000 MW(e).[1]

Apart from the aesthetic considerations with such well-known sites, installations like these could well have other environmental effects and such matters as the consequences for sedimentation need very careful evaluation.

In the end, however, it seems unlikely that there will ever be a significant contribution to global energy supplies from tidal sources, in spite of a few recent successful installations.

2.2.5   Natural sources of energy, 3: solar energy

Solar radiation is by far the largest source of 'free' energy, with an annual flux over the whole ecosphere equivalent to 3800 Q, or 13 500 times the total human energy consumption in 1980—a colossal amount! Astrophysicists estimate that the sun will continue to provide a flux density equal to its present value for some 5×10^9 years to come, so that on a human scale it is effectively an infinite source.

Not only that, but solar energy is the only source that is completely and rapidly transformed into heat at the ambient temperature. This means that any extraction of energy from it by humans as it passes has no overall ecological effect, provided that large concentrations of devices for trapping it are avoided: in particular, it does not create thermal pollution. Finally, in addition to those thermodynamic advantages, the use of solar energy cannot lead to the emission of gaseous and liquid effluents, so that it is essentially non-polluting on all counts.

Characteristics of solar radiation

Solar resources possess two fundamental properties which condition the way it is used: it is unevenly and irregularly distributed; and it has a low energy density. These are major disadvantages and they account for the derisory amount of research and development devoted until recently to its industrial or domestic use. Its irregularity in time gives it the unfortunate property of being non-existent at night (by definition!) and smaller in winter: in other words, it is reduced at the very times when there is most need of extra energy. In addition, its surface density is very low and barely reaches 1.35 kW per square metre of area normal to the radiation at the upper limit of the atmosphere. By comparison, any domestic gas burner produces 100 times more energy per unit area per unit time!

As the solar radiation passes through the atmosphere, there are further reductions in intensity produced by reflection back into space and by scattering due to particles. Furthermore, as it has a spectrum similar to that emitted by a black body at 5800 K, there is selective absorption of certain wavelengths by various gases like CO_2, ozone and water vapour.

---

1. According to G. Gibrat, private communication, 1975.

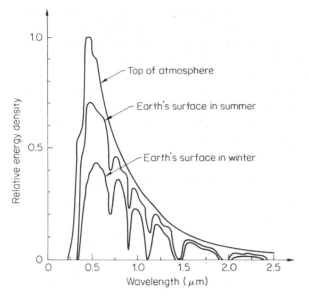

Figure 2.22 Energy density of solar radiation over the whole spectrum at the top of the atmosphere and at ground level at latitude 52° N. (From Brinkworth, 1975)

Figure 2.22 shows the effect of all these losses, which occur mainly in the infra-red and to a lesser extent in the ultra-violet (no ultra-violet radiation with a wavelength less than 0.325 μm reaches the Earth's surface). In the end, the solar flux that reaches ground level rarely exceeds 1 kW per square metre and even that figure only occurs in the most favoured situations.

The influence of latitude Solar flux obviously decreases along a meridian from the Equator towards the polar regions. This arises from the Earth's spherical shape and from the fact that radiation from the sun is normal to the surface at the Equator, so that it becomes more and more oblique on moving away from it. The continuous decrease of solar flux with latitude at noon on midsummer's day is shown along the vertical axis of Figure 2.23a, which also illustrates the daily variation in intensity. At the Equator, the intensity at midday with a clear sky is more than 1 kW m^{-2}, while the maximum value at the polar circles (latitudes 66.5°) is only 0.77 kW m^{-2}. As the latitude increases, however, the day lengthens and this partly compensates for the lower flux: in the end, the energy supply from the sun over a whole day of 24 hours shows unexpectedly little variation with latitude at the height of summer. This total supply amounts to about 30 MJ m^{-2} on a horizontal surface: this is the theoretical daily maximum and corresponds to an average flux over the whole 24-hour cycle of 0.347 kW/m^2. Figure 2.23b shows the distribution of mean solar flux over the entire biosphere.

In practice, distinctly lower intensities are found because of the seasonal cycle and variations in the weather. The inclination of Earth's equatorial plane to the plane of the ecliptic is 23°27′. A place situated at latitude L thus has a maximum angular height H of the sun in the sky given by $H = 90° - L + 23°27′$: at Paris, for instance (latitude 48°50′), the midday sun at midsummer has an elevation of 64°37′. Similarly, in midwinter the maximum angular height is given by $h = 90° - L - 23°27′$ and is thus about 17°44′ at Paris.

The length of the period of daylight also decreases rapidly in winter with increasing latitude, and beyond the polar circles there is continuous night-time for at least some of the year. The combined effect of longer nights and lower elevation of the sun produces a far smaller solar flux in winter than in summer. As an example, the solar intensity on a horizontal surface in the Paris region is nine times weaker in midwinter than in midsummer. The same applies to the total daily energy supply: as Figure 2.24 shows, this is more than seven times greater in July than in January around Paris. Note, however, that the flux received by a vertical south-facing surface is greater in winter than that incident on a horizontal surface, although the opposite is true in summer (Figure 2.25). This is another result of the relative directions of the rays from the sun and the Earth's horizontal surface at the point.

In spite of the large daily and seasonal variations in the direction and elevation of the sun, they are at least absolutely regular at any given place and can be accurately predicted for the purpose of designing devices to harness solar energy. The same is not true of variations connected with climatic factors affecting the amount of atmospheric shielding: cloud cover, fogs and so on. These are much more uncertain and difficult to anticipate, but they can strongly affect the intensity at ground level.

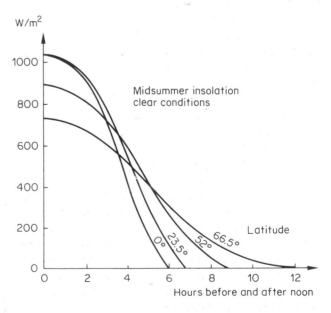

Figure 2.23a Variation of solar energy flux on horizontal surface (insolation) with time of day for various latitudes. (From Brinkworth, 1975)

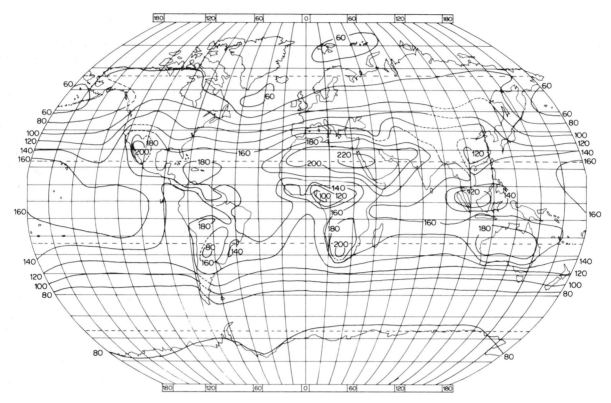

Figure 2.23b World distribution of the mean solar flux at ground level expressed in kcal cm^{-2} year^{-1} (1 kcal cm^{-2} year^{-1} = 1.32 W m^{-2}). (From Budyko, in Sellers, 1965, p. 25, and Ramade, 1984)

As an example of the consequences of all these effects, the combination of different latitudes and climates over France produces variations in hours of sunshine over the country ranging from 1750 hours to over 3000 hours per year. The resultant energies received per square metre per year on a horizontal surface vary from 1100 kWh to 1900 kWh, and the flux of energy ranges from 0.125 to 0.220 kW m^{-2}.

Over the whole ecosphere, the solar flux received ranges from 0.09 kW m^{-2} in polar regions to 0.29 kW m^2 in tropical deserts, which are clearly most favoured in hours of sunshine and latitude.

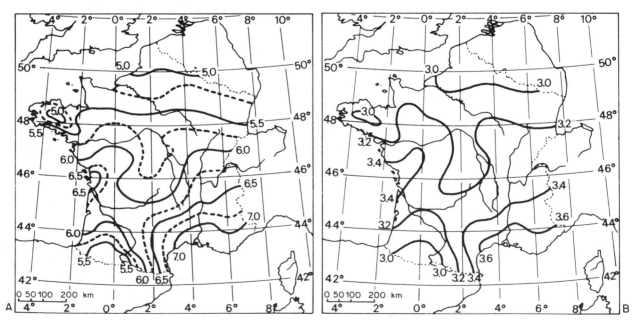

Figure 2.24   Distribution of the solar flux over France incident on a horizontal surface: A, daily flux in kWh m^{-2} in July; B, the same in January. (Météorologie Nationale de France)

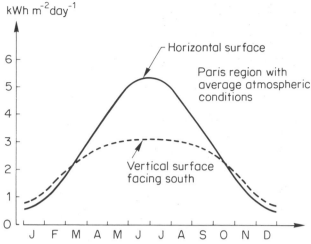

Figure 2.25   Seasonal variation of the solar flux received by a horizontal or vertical surface in the Paris region. (Météorologie Nationale de France)

Techniques for harnessing solar radiation

These can be divided into two broad categories. The first consists of *direct methods* of using solar energy by converting it either to thermal energy or electrical energy (with or without an intermediate thermal stage). The second category covers the use of various types of energy originating from solar radiation: hydro-electricity, winds and waves, photosynthesis, for example. These can be called *indirect methods*.

Direct utilization by thermal conversion

Direct conversion into heat is perhaps the simplest and quickest way of using solar energy and is currently a very satisfactory method of providing space-heating and air-conditioning for individual dwellings. Solar water heaters have, in fact, already passed beyond the prototype stage and so are ready for rapid commercial development.

It is also possible to envisage the use of direct thermal conversion for air conditioning by absorption refrigerators.[1] Every square metre of horizontal surface at temperate latitudes receives 5000 kcal of energy per day and this would be enough to produce 4.5 kg of ice from water at 30 °C (assuming an efficiency of 10 per cent) or of an equivalent amount of cold. Another possible application of direct thermal conversion is in the desalination of sea water, something which could be developed on a very large scale in arid regions of the Earth.

---

1. It seems absurd that the maximum demand for electricity in the USA now occurs in the summer because of air-conditioning, whereas the high proportion of individual dwellings in the country makes it very suitable for the installation of solar air conditioners.

Direct utilization by thermo-mechanical conversion

Thermo-mechanical conversion is another technique that ought rapidly to become more popular in tropical regions. It is based on the idea of the solar pump, a modern form of the process due originally to Georges Claude.

A prime example of the principle is provided by the solar pump of Masson and Girardier illustrated in Figure 2.26. Water circulates through the solar collectors and acts as a heat-carrying fluid. Its accumulated heat energy is transferred to a low-boiling-point liquid like butane in a heat-exchanger or boiler, and the pressure produced by the hot evaporated gas is sufficient to drive a pump. The gas is then cooled in a condenser, being liquefied by cold water lifted from a well by the pump. The liquefied gas is reinjected into the evaporator and starts a new cycle.

Such devices have a low thermodynamic efficiency, but in spite of that they have the advantages of great reliability and little need for maintenance. Moreover, this is all at a cost per cubic metre of pumped water considerably less than that of electric motors or diesel engines in bushland or steppe far removed from conventional energy sources.

Direct utilization by helio-electric conversion

There are two kinds of process that might be described in this way. The first, discussed in this section, uses the heat from solar radiation to produce a vapour under pressure which then drives a turbo-generator. The second kind is based on the photovoltaic effect and converts solar energy directly into electric current. This is discussed later.

Helio-electric conversion has recently been brought into use with the establishment of the French solar power plant at Odeillo in the Pyrenees. In this installation, the 63 plane mirrors belonging to the boiler of the solar energy research station of the CNRS (Centre National de la Recherche Scientifique) are used to reflect sunlight on to a parabolic mirror producing a temperature of 3800 °C at its focus. The electric power produced is quite modest at 60 kW with an overall efficiency of 6 per cent relative to the incident energy (Figure 2.27).

One of the first projects involving a 'solar tower' was that suggested at the end of the 1960s by Hildebrandt and Gregory of the University of Houston. They proposed to reflect solar radiation from mirrors covering an area of 2.6 km² on to a solar furnace at the top of a 500 m tower and then convert the heat at 2000 K to electricity using magneto-hydrodynamic (MHD) methods. Some of the electrical energy would be used for the production of hydrogen and oxygen, which would then be reconverted overnight to electricity using fuel cells. The net power of such an installation

52 ECOLOGY OF NATURAL RESOURCES

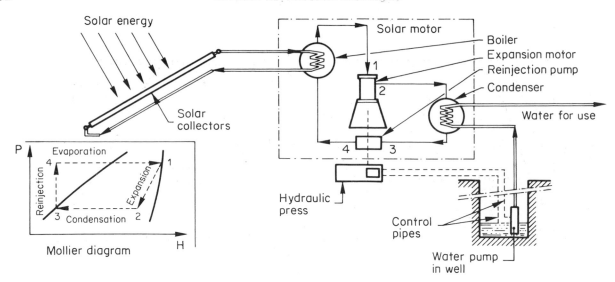

Figure 2.26   Principle of a pump driven by a solar motor (patented process). (From Sofretes, 1979)

Figure 2.27   The solar boiler at Font-Romeu-Odeillo-Via in the Pyrenees constructed by the CNRS in 1969. A power plant was added in 1976. A more powerful 2.5 MW(e) unit, the THEMIS solar power station, is coming into operation in June 1983 on a neighbouring site. (Photograph F. Ramade)

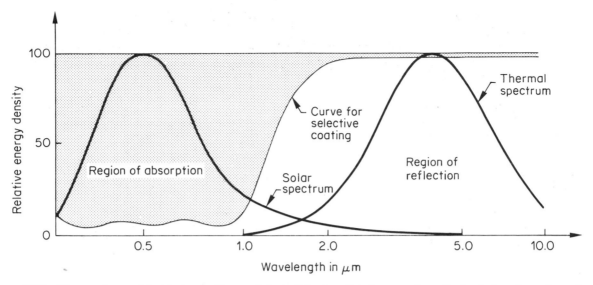

Figure 2.28   The spectrum of incident solar flux and that of the reradiated energy from the heated surface of a collector. Solar energy is captured because of a special coating on the collector having strong absorption in the visible region and low emissivity in the infra-red. (From Meinel, in Hammond *et al.*, 1973)

would be about 500 MW(e) and its efficiency 20 per cent.

A. and M. Meinel (University of Tucson, Arizona) have proposed a less futuristic project based on the use of simple planar collectors and avoiding the construction of the much more elaborate and costly optical devices like plane and parabolic mirrors. The surface of these collectors is coated by vacuum evaporation with a thin film of silica or other material that does not reflect appreciably over the wavelengths contained in the solar spectrum. Such a surface absorbs the solar flux but transmits infra-red radiation, because it also has a low emissivity in the latter part of the spectrum. Thermal energy is thus trapped in the collector by a sort of greenhouse effect (Figure 2.28), and heats a fused K–Na mixture to about 540 °C. The energy is then carried via a fused NaCl–KCl mixture to a heat exchanger where steam is generated to drive a turbo-alternator. Assuming an overall conversion efficiency in the production of electricity from the heat received by the circulating fluid, the authors estimate that 3.6 km^2 of the Arizona desert would be enough for a power plant with an average annual output of 100 MW(e). Collectors of the type just described would have the advantage of harnessing diffuse solar energy and thus of being capable of functioning even in overcast weather. This is in contrast to installations using optical systems involving reflection and focusing, where direct radiation (and thus a clear sky) is essential.

However, all these projects belong more to the realms of fiction than reality at the present time. There are technological problems associated with them which, while not insurmountable, have not yet had allocated to them the resources for research and

development that they warrant. Rather more modest in its objectives but much more realistic is the THEMIS project of the French CNRS currently being completed in the south of France. The plan involves the building of a solar tower fed by some 5000 mirrors each with a reflecting area of 50 m^2. The collecting surface would occupy about 64 hectares of land (or a square of side 800 m) but, because the mirrors must be well dispersed so as not to cast mutual shadows, they would in fact cover only about a third of the terrain: the rest would be available for agricultural use and would not be completely sterile. The power output of this unit will be 2.5 MW(e). Theoretical calculations on the optics of such systems show that it would hardly be possible to place the reflecting mirrors further away from the tower supporting the solar boiler than 1000 m. This would limit the power output of such installations to 100 MW(e).

Direct utilization by photovoltaic conversion

The photovoltaic cell was the first type of photoelectric device to be used for the production of electrical power because of the development of space technology from the end of the 1950s. The circumstances were such that cost was not a determining factor in the production of the cells, and other considerations had priority, such as weight, efficiency and above all absolute reliability. Because of that, the cost per kW of electricity from silicon or cadmium sulphide photocells was around $100 000 as compared with the few thousand dollars per conventional kW(e).

Photovoltaic converters do, however, possess the advantages of a high overall efficiency of conversion to electricity (currently achieving 10 per cent as against a theoretical upper limit of 23 per cent for silicon cells), and of needing no maintenance once installed.

Present research is aimed at significantly lowering the price per kW since it is still a factor of at least 10 above the threshold figure which would make it competitive enough to be available for its enormous number of potential applications. A cost of $10 000 per kW can be conceived in a few years, but the breakthrough figure of $1000 or $2000 per kW seems a very distant prospect, and one that assumes either great improvement in the properties of silicon or the discovery of a more economic converter.

Indirect utilization through water power

We now turn to natural forms of energy which are originally derived from solar radiation. One of these is water power, first used in Roman times and thus one of the earliest of the non-biological human energy sources. It represents the largest concentration of solar energy produced in the ecosphere by a natural process and because of this it enables power stations to be constructed with an output exceeding 2000 MW!

The potential world hydroelectric power was estimated at the end of the 1960s to be about 3000 GW, with a distribution as shown in Table 2.14. One remarkable feature is that South America and tropical Africa, whose relative poverty in fossil fuels is notorious, possess the greatest potential water power, but it is clearly in Western Europe and North America that the main development has occurred, with 30 and 20 per cent respectively of the available capacity already established.

The potential world resources in hydroelectricity would be enough to satisfy demand if consumption stabilized at its present level. However, in spite of the unused capacity, it would be very risky to rely entirely on water power as a means of coping with the majority of human energy requirements. This is because

hydroelectric installations are not everlasting: the capacity decreases with time owing to erosion and sedimentation which cause the lakes behind the barrages to fill up year by year. Within a century, or two at most, the majority of the large schemes at present operating will become unusable. This problem would not be overcome by the construction of numerous 'microstations' in the smaller waterways, such as are now envisaged.

There are, too, some unfavourable ecological consequences which accompany the diversion of rivers and even their drying up over some stretches when water levels are low. In fact, hydroelectric installations often produce considerable effects on the environment: the slower rate of flow in the waterways upsets sedimentation and alters the character of lakes; they are a potential source for eutrophication of the whole hydrographic network in industrialized countries through pollution by organic matter; and finally, such a proliferation of power stations dries up waterfalls, engulfs fertile valleys so that they are permanently lost to agriculture, and disfigures many a well-known landscape with huge dams.

In spite of these adverse effects on the environment, however, hydroelectricity is the only primary energy source which at the same time produces no pollution, is renewable on a time scale of a century or two, and which has already for several decades yielded sizeable quantities of power.

Indirect utilization through wind power

Wind power, like water power, has its ultimate origin in the solar energy received by the Earth. For more than half a century it has been totally neglected by the majority of industrialized countries, and even one or two applications like the pumping of water which persisted for a long time have now been turned over to machines driven by internal combustion engines.

Wind power is the oldest of the 'new' energy sources

Table 2.14   World water-power capacity (from Hubbert, 1969, p. 209, data of Adams, 1961)

Region	Potential (10³ MW)	% of world potential	Power already developed (10³ MW)	Power already developed as % of potential
North America	313	11	59	19
South America	577	20	5	0.9
Western Europe	158	6	47	30
Africa	780	27	2	
Middle East	21	1	—	
S.E. Asia	455	16	2	
Far East	42	1.5	19	
Australasia	45	1.5	2	
USSR, China and satellites	466	16	16	3
Totals	2857	100	152	

and was being used as long ago as 3000 BC by the Parthians and the Chinese. It was still playing a far from negligible role in the supply of energy at the beginning of the industrial era, not to mention the fact that it provided practically the only motive power for travel by sea until the middle of the last century.

The total wind power potential over the whole ecosphere is very large indeed, amounting to 180×10^9 kW (or 1.6×10^{16} kWh of energy per year!). Out of this total, 100×10^9 kW is provided by winds in the lower layers of the troposphere. Some idea of what this means can be gained from an example: the harnessing of only 1 per cent of the wind power near ground level in France would be enough to satisfy half the present national demand for electricity.

As with all other free sources, wind power has the advantages of being non-polluting, inexhaustible, and costing practically nothing to exploit after the initial installation except for maintenance of the equipment. Its disadvantages stem from its uneven distribution in space, its irregularity in time and its low power density. However, at any given place, statistics show that winds blow less capriciously than might be thought: as a general rule, their occurrence can be predicted with a margin of error no greater than 15 per cent. There is, moreover, a considerable advantage in that wind power, unlike solar radiation, is greater in winter than in summer in temperate latitudes. In Brittany and the Mediterranean regions, for example, it can be as much as three times greater in winter. On the other hand, a

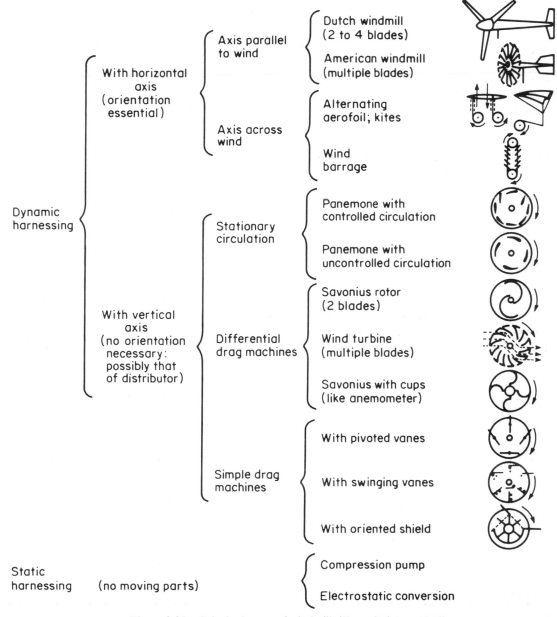

Figure 2.29   Principal types of windmill. (From Sofretes, 1979)

careful survey of wind forces is essential since it varies so much even within a microclimate. A difference of 10 m in height between two points in the same locality can reduce the wind power by half.

The principal types of windmill construction are illustrated in Figure 2.29. There are two main categories: those with horizontal axes and those with vertical ones. Only the former type have until now been used practically: some of them, built over 40 years ago, have already achieved an output of more than 1 MW(e).

Contrary to what is often thought, even the very large windmills can be almost silent. For that to be the case they have to be free of vibration, of course, but that is necessary in any event to avoid the destruction of the various rotating parts. With the development of helicopters, the technology of rotors and blades has made great strides in the last few decades, and the construction of windmills generating 2000 kW now raises no major difficulties, while 100 kW can be regarded as routine. At the end of 1978, for example, the General Electric Company began operating a wind-generator near Boone in North Carolina with an output of

2 MW(e), and Boeing is constructing a 2.5 MW(e) machine in collaboration with NASA (Goethals, 1980).

The average efficiency of windmills with horizontal axes is between 25 and 40 per cent in relation to the kinetic energy of the winds: this means that it is at best no more than 60 per cent of the Betz theoretical limit for most installations. However, a triple wind-generator with a power output of 800 kW(e) was built in 1958 by Electricité de France at Nogent-le-Roi, and this represents a particularly good performance since it reaches 85 per cent of the Betz limit.

Machines like these can either be linked to the national grid or be used as independent units supplying relatively isolated sites. In the latter case, a storage system needs to be added so as to compensate for the irregularity of the winds (Figure 2.30).

In view of the potential wind power, some authors have suggested the installation in France of some 2000 windmills, each of 2 MW, to provide 10 per cent of the current demand for electricity. However, it would appear that such generators are better adapted to the provision of decentralized and independent supplies by a multiplicity of small units than to the construction of large installations of high power feeding the electrical supply network. Given the low power density (0.4 kW m^{-2} at ground level in the most favourable sites), achieving 1000 MW would mean the creation of gigantic structures. Take as an example an area where the average wind power density is 0.2 kW m^{-2}. Assuming that conversion to electricity can be carried out with an efficiency of 40 per cent, which is quite high, the production of 1 MW would need a windmill fitted with rotating blades of 125 m diameter, and a thousand of these would need to be assembled to produce 1000 MW!

In the same way, to satisfy the present European demand for electricity, a windmill of 30 m diameter would be needed in every 4 km^2 of land, assuming an average wind speed of 30 km per hr and an efficiency of 30 per cent. The enormous amount of land taken up by such a scheme seems to make its use rather utopian except in countries with large desert areas. However, even if wind power seems little adapted to the construction of large power stations, that does not prevent its use in numerous cases where it can replace internal combustion engines consuming hydrocarbon fuel nor its use in supplying electricity to small villages, farms and other isolated communities. Furthermore, the great progress recently made in the fields of synthetic fibres and metallurgy now means that sailing ships or other types of vessel propelled by wind can be built with performances that would eclipse those of the famous 'clippers' of the last century.

A more sustained effort in research and development would eventually enable wind power to satisfy 2 per

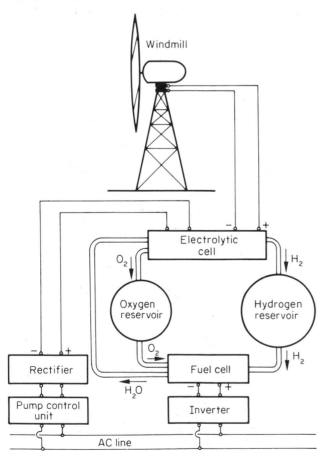

Figure 2.30 A windmill coupled to a fuel-cell storage system forming an electric generator which is capable of functioning independently (From Summers, in Ramade, 1978a)

cent of the energy requirements of a country like France by the end of the century. That appears a somewhat limited contribution, but the economic effect would in fact be quite considerable. It would mean, for example, a saving of 3.8 million tonnes of oil per year if the rate of consumption stabilizes at its present value. The effect of this on the balance of payments of an importing country would be very large in view of the likely cost of hydrocarbon fuels by that time.

Finally, there is another form of natural energy[1] which originates from the winds, and that is wave power. Its world potential has been estimated at 0.08 Q per year, which is clearly somewhat limited in comparison with the 3 Q per year available from low-level winds. In addition, the harnessing of wave power presents various technical and practical difficulties such as the disturbance of coastal traffic by the installations. For the moment its utilization seems somewhat unrealistic.

2.2.6 Ecological limitations and the energy crisis

Even if we accept that improved technology might make inexhaustible energy resources available to mankind, we should still have to acknowledge that other limitations, no less restricting, would put a definite term to the growth of energy consumption. These limitations are of an ecological nature and fall into three distinct categories:

(a) various forms of pollution,
(b) unavoidable thermodynamic constraints because of the second law,
(c) the amount of available space.

These ecological constraints will limit the continuing growth of world energy consumption long before civilization faces the problem of the complete exhaustion of resources.

The severe public health problems and the degradation of natural resources, both of which are consequences of the various forms of pollution that invariably accompany all artificial production of energy, ought to provide food for thought for those who oppose a policy of zero growth in the use of energy.

---

1. There are two other sources of natural energy that we shall not discuss: the use of thermal gradients in the oceans, and what is known as helio-hydroelectricity.

The first uses the difference in temperature that exists between the ocean depths and the warmer surface waters, which can be as much as 25 °C in tropical seas.

The second is to do with vast continental basins below sea level which could be connected by a channel or pipeline to the nearest sea. The resultant water flow would drive a turbo-generator and the difference in levels would be maintained by the high rate of evaporation from the basin, particularly in tropical areas.

Neither method appears at present to be practicable.

Atmospheric pollution

Atmospheric pollution is one example which shows that there is a limit to the number of times that fossil fuel consumption can be doubled in the future. Beyond a certain point, the degree of pollution would become unacceptable, even if that already reached in some large cities and industrial regions can be considered as tolerable.

A reduction by a factor of 10 in the concentration of air pollutants produced by combustion assumes technological advances and financial investment on a scale that no industrial country has yet achieved. Meanwhile, any sudden improvement of this size in the fight against atmospheric pollution would be completely neutralized if fossil fuel consumption doubled just over three more times ($10 \sim 2^{3.3}$), and the level maintained even then would only be one that is already found intolerable in many of today's industrial regions. It is hardly necessary to point out that Lave and Seskin were suggesting in 1970 that air pollution in cities, due to various forms of combustion, would reduce the expectation of life of the population by 3 years.

Considerable atmospheric pollution is caused by the combustion of the various fossil forms of carbon. Sulphur dioxide, produced mainly by the use of fuel oil

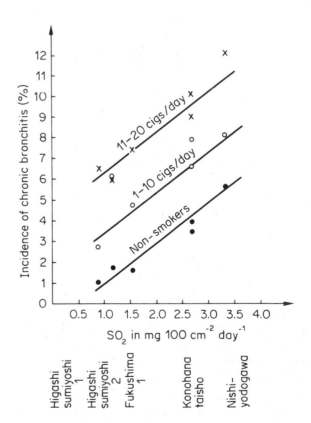

Figure 2.31 Synergic effect of the pollution of air by SO_2 and of nicotinism on the incidence of chronic bronchitis in various Japanese cities. (From Nishishwaki *et al.*, in Ramade, 1979a)

but also by sulphur-rich coal, is by far the largest pollutant by volume, but there are a good many other gaseous materials with appreciable toxicological effects on the environment: examples are the oxides of nitrogen, peroxyacetyl nitrate (PAN), and the products of incomplete combustion of some fuels, the fearsome polycyclic hydrocarbons like benzopyrene, for example.

Numerous epidemiological studies have shown the high correlation between the concentration of such pollutants in the air and the incidence of sickness through pulmonary infection, as in Figure 2.31. In addition, recent investigations have shown that many combustion products like PAN and the oxides of nitrogen can induce mutations.

Air pollution also has a disastrous effect on renewable energy resources. Plants are particularly sensitive to SO_2, to oxides of nitrogen, to PAN, and so on. Thus, hundreds of thousands of hectares of forest have already been destroyed in industrialized countries, and there are still larger areas that suffer reduced productivity. Considerable losses in agricultural production occur because of atmospheric pollution: estimated, for example, at more than $100 million per year for the state of California alone!

However, it is not only the combustion of fossil fuels that causes so much pollution: the stages at which they are extracted and transported contribute to it as well.

Pollution of the seas by oil

The principal cause of marine pollution is contamination by oil spillages of various types, and the degree of such pollution that could be generated by the oil-producing industries is potentially quite fantastic. This point is well illustrated both by accidents occurring at offshore wells like Santa Barbara in California or more recently Ixtoc 1 in Mexico (where more than 500 000 tonnes of oil were discharged into the sea between June and November 1979), and by accidents to supertankers with catastrophic consequences like those from the running aground of the *Amoco Cadiz* in March 1978. In total, the combined activities of extraction, transport and utilization of hydrocarbons are accompanied by the discharge each year of 2.5×10^6 tonnes of crude oil into the marine environment!

Radioactive pollution

This is an unavoidable consequence of the development of nuclear power and could be a major obstacle to its expansion if considerable technological efforts are not made from now on to control it more effectively. The main problems of contamination connected with peaceful uses of nuclear energy occur at reprocessing plants for irradiated fuel, where all the waste products from the nuclear fuel cycle are to be found. The total production of electricity from nuclear reactors involves the reprocessing of large amounts of material: 5 100 tonnes per year for the French programme alone by the turn of the century, which by itself would produce a quantity of radioactive waste equivalent to that from several tens of thousands of Hiroshima-type bombs!

Reprocessing plants yield various types of liquid and gaseous waste. Liquids of low radioactivity are discharged into inland waters or the sea, while radioactive inert gases and tritium are passed into the atmosphere. So long as this problem of radioactive waste is not satisfactorily resolved, it must be considered a burden that will weigh heavily against any sizeable development of the nuclear industry.

Figure 2.32 illustrates the fuel cycle for light water reactors and its attendant risks: there are 'potentially catastrophic' risks of pollution at the majority of the stages which are currently developed. These also exist for the stored waste products, given the present methods of dealing with them. However, the figure does not include methods for the isolation of the wastes from the biosphere in a form which would remain recoverable, and these new techniques (involving vitrification), without being in my opinion a final and permanent solution, would at least avoid the difficulties connected with the storage of large volumes of liquid.

The problem of nuclear waste disposal could not be avoided even if the cumulative quantities were quite small. The only clear-cut solution would be to recycle the actinides in reactors and to perfect a method of getting rid of the radioactive elements of lower atomic mass (like strontium-90 or caesium-137, for which a storage time of more than five centuries must now be contemplated!).

Personally, I consider certain statements by authorities in the electro-nuclear industry to be dangerous, and even factually open to question, in playing down the matter of discharging low-level wastes into surface waters and into the sea at concentrations said to be 'acceptable'. It may be true that higher aquatic organisms constitute more of an eliminating filter for actinides and fission products than anything else, but nevertheless there remains the possibility that some of the important radioactive isotopes could reach high biological concentrations in food chains leading to humans. We do not know enough to be sure that such cases of concentration in food webs are exceptional, as is sometimes said. Amiard-Triquet and Amiard (1976 and subsequently) have analysed the various ways that radiocontaminants behave in aquatic organisms and biocoenoses. They stress the role played by the physico-chemical form of the radioactive elements in water and emphasize cases where some of these elements can become concentrated in food webs.

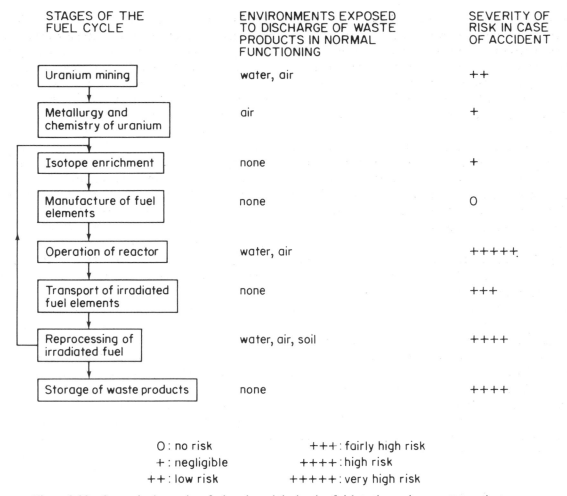

STAGES OF THE FUEL CYCLE	ENVIRONMENTS EXPOSED TO DISCHARGE OF WASTE PRODUCTS IN NORMAL FUNCTIONING	SEVERITY OF RISK IN CASE OF ACCIDENT
Uranium mining	water, air	++
Metallurgy and chemistry of uranium	air	+
Isotope enrichment	none	+
Manufacture of fuel elements	none	O
Operation of reactor	water, air	+++++
Transport of irradiated fuel elements	none	+++
Reprocessing of irradiated fuel	water, air, soil	++++
Storage of waste products	none	++++

O : no risk   +++ : fairly high risk
+ : negligible   ++++ : high risk
++ : low risk   +++++ : very high risk

Figure 2.32   Stages in the nuclear fuel cycle and the levels of risk to the environment at each stage

However, they also point out our imperfect knowledge of the behaviour of actinides, especially the transuranic elements, in ecological systems.

In fact, a whole series of radioisotopes is known, each of which is capable of becoming highly concentrated in food chains. These are isotopes either of elements that themselves occur in living matter (^{32}P, ^{131}I, ^{14}C) or of elements chemically similar to the first group and thus capable of replacing them (^{90}Sr, chemically similar to calcium; and ^{137}Cs, chemically similar to potassium). Moreover, our relative lack of understanding of the behaviour of the heavier radioactive elements in ecological systems should again be emphasized: it was recently reported that the discharge of sterile waste from uranium mines into Czechoslovakian waterways produced a significant contamination of lake water by such elements, of which radium was one. Trout and other fish eaten by humans were particularly affected. Even if the doses have turned out to be lower than the maximum permitted exposure, it is nevertheless true that the injection of actinides into the food web of freshwater teleostei was not something that had been foreseen. The current standards governing the discharge of low-level radioactive waste into water do not take into account this phenomenon of concentration in food webs.

It should not be necessary to remind the authorities in the nuclear power industry that improvements in decontamination coefficients, even in some cases to the stage of 'zero release', are now technically possible. Even if putting such procedures into practice would add an extra 10 per cent, or at most 20 per cent, to the price per kWh, that would seem a lesser evil than the risk of environmental pollution which would be difficult to control. Similar experiences with contamination by various chemical products show that, in practice, the standards for the maximum tolerable discharge have nearly always had to be revised downwards following the discovery of previously unknown toxic effects. It is therefore surprising to see that, in contrast to the chemical example, the competent authorities in radioactive protection have recently raised quite significantly the maximum threshold of the 'tolerable' amounts of tritium that can be discharged into the environment by nuclear reactors! Yet it would be quite possible with current techniques to recover all the waste products in the various stages of the fuel cycle. It was estimated at the end of the 1960s that such a total recovery of waste products from the United States nuclear industry would add at the very most 2 per cent to its overall cost. Nobody is capable of weighing such a marginal increase against the risks inherent in possible concentrations of radionuclides in food chains or in the effect of accumulation of them (e.g. ^{85}Kr) in the atmosphere.

The disposal of radioactive waste should be based on the following principles:

(a) All radioactive material is biologically harmful. It must be isolated from the environment for a period of time equal to 20 half-lives.[1] Thus, radioactive strontium and caesium must be isolated for 600 years, and plutonium-239 for 500 000 years! (See Figure 2.33).

(b) The rate of production of waste is strictly proportional to the rate of consumption of fissile material. At the present time this is doubling every 5 years.

(c) The practice of discharging waste with low levels of activity into the atmosphere and into waters anywhere whatsoever must not be allowed to occur without the certainty that it will not present a hazard even when the rate of discharge has grown by several orders of magnitude. Such practices must be forced to stop as soon as techniques for confinement become available.

(d) Safety in nuclear matters should not be compromised for the sake of economic profitability.

Occupation of space

The harnessing and production of all forms of energy take up space and reduce the amount of this valuable natural resource that is freely available.

The extraction of fossil fuels, especially of coal by open-cast mines, has in the past caused the devastation of considerable areas of land. Extensive wooded regions in the Appalachians of the eastern USA, for example, have been destroyed by such mines, which leave bare rock at the bottom of huge excavations and completely disrupt the hydrological network. Recent laws enacted there have made it almost obligatory to restore the landscape in line with the practice in Germany: this entails using sterile materials to fill in the excavations, adding a superficial fertile layer and reestablishing grassland or woodland cover, according to the local ecological conditions.

Even the mining of nuclear fuels can lead to a considerable occupation of space since exploitation of low-grade materials requires the extraction of large quantities of rock. This means that the high density of energy potentially contained in the uranium and the thorium is counterbalanced by their low concentration in the parent rocks from which they are extracted. At the present time (the end of 1979), ores capable of producing uranium at a cost price lower than $150 per kg have a concentration of 500 p.p.m. The energy equivalent in coal of this uranium being known, a calculation shows that extraction of nuclear fuel from open-cast mines requires five times less rock to be handled than for a volume of coal giving comparable amounts of heat energy. Although this constitutes one of the advantages of using uranium, the consumption of space would still be significant in view of the large requirements of a PWR programme. Fast breeders, on the other hand, use nuclear fuel much more efficiently and greatly reduce the extraction needed for a given energy output.

There are other constraints connected with the amount of space used up which militate against an indefinite growth in energy production. The hallowed rule that this production doubles every 10 years would lead to a 100-fold increase in electricity generation by the middle of the next century. If we see what this entails for a country like France, we take the power installed in 1973 (27 GW), and realize that by the year 2040, 2700 GW would be needed: in other words, some 2000 reactors of 1300 MW(e) each. Even if gases were used to help out as refrigerants, the cooling of such a vast assemblage would convert into steam too high a proportion of the inland water supply, so that the units would need to be sited near the coast. It can be said at once that France's 3000 km of coast would not be sufficient for her needs since there would need to be a nuclear power station every 5 km. Even if grouped in 'parks', to use the technocratic euphemism for these gigantic systems, they would certainly constitute a futuristic-looking wall along the Atlantic and Mediterranean coastlines.

We are also entitled to ask about the long-term future of nuclear power stations. Since they are planned to operate for about 30 years, what will be done with them when they go out of service? At the moment, the only scheme envisaged is the abandonment of the downgraded reactors on site after removal of the fuel elements. In that case, shall we see the land increasingly studded with their remains? All that can be said about relics of that sort is that they hardly have any chance of being registered by future generations in lists of historic sites!

The absurd aspects of these various arguments show the futility of an uninterrupted growth in energy consumption supported by a kind of 'nuclear monoculture' taking over from the declining 'oil monoculture'.

Nevertheless, it must not be thought that changing

---

1. The laws of radioactive decay mean that a time equal to 20 half-lives is necessary for an initial mass of any radioactive element whatsoever to decrease by a factor of 1 million (since $2^{20} \sim 10^6$).

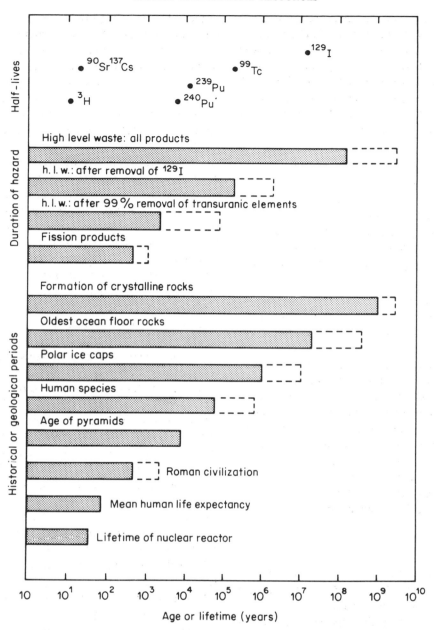

Figure 2.33   Duration of storage needed for various nuclear wastes to decay to a low level of activity, together with the half-lives of some of the more significant constituents (at top). Note that for many radionuclides the times are comparable with those of geological periods. (From Rochlin, 1977, p. 25)

to what are called the 'new' energy sources would remove the constraints imposed by the amount of space taken up. On the contrary, such constraints are even greater than in the case of non-renewable resources like fossil and nuclear fuels, which at least have the advantage of a high energy density. With renewable resources, the low fluxes of free energy would mean the occupation of very large areas in total: the high concentration of windmills needed to produce great quantities of electrical power has already been mentioned. Solar energy is similar, and it can no longer be counted on to maintain the sacrosanct energy growth advocated with such surprising stubbornness

by technocrats, economists and politicians of many ideologies! If photovoltaic conversion were to be used on an industrial scale, two-thirds of the territory of the United States would be covered by solar collectors by the year 2070, assuming that the growth rate for energy consumption remains at its level during the early 1970s.

Again, take the case of France in a hypothetical situation in which all current energy needs (about 280×10^6 t.c.e. per year) are to be satisfied by solar energy through photovoltaic conversion. The average solar flux at ground level in France is 0.19 kW m^{-2} or 1662 kWh m^{-2} year^{-1}. If a conversion efficiency of about 10 per cent is assumed, more than 15 000 km^2 would have to be covered[1] by solar collectors to satisfy the energy requirements. The immobilization of such an area (amounting to that of three *départements*) can certainly be contemplated in order to ensure the supply of energy, but it could hardly be claimed that solar energy collected on such a scale would be without environmental effects. The considerable space occupied by collecting panels prevents solar energy from replacing more than a fraction of France's current energy consumption.

Thermal pollution

Thermodynamic constraints also impose a strict time limit on energy growth if it relies on the use of fossil fuels or nuclear fission. The second law, as we have seen. on p. 5, stipulates that heat engines can only do work with loss of heat to the surroundings:

$$\Delta G = \Delta H - T\Delta S$$

where ΔG is the change in usable energy, ΔH the change in enthalpy (that is, the heat exchanged between the system and the surroundings), T the absolute temperature, and ΔS the change in entropy of the system.

The heat lost at low temperatures and not converted to useful energy is discharged into the atmosphere and into nearby water systems. As a result, rivers that cross highly industrialized regions and waters at certain points on coastlines are particularly affected and have already reached worrying levels of thermal pollution.

In the long term, however, the climatic consequences of releasing so much heat at low temperatures into the ecosphere are much more disquieting because there is the possibility of upsetting the terrestrial energy balance: the prospect of a global climatic change is not merely a subject for science fiction. This means that there are two consequences of the use of fossil fuels occurring consecutively which can combine to produce meteorological changes on a planetary scale: the first is the discharge of CO_2 as a result of combustion; the second, in the longer term, is the discharge of heat at low temperatures into the environment. The climatic effects of these little understood aspects of the energy crisis are examined in detail in the next chapter.

2.2.7 Economic aspects of energy and growth

The way in which energy is used in our present civilization would seem to call for some comments at a socio-economic level.

It is quite clear, as we have already emphasized, that the constraints produced by increasing cost will slow down energy growth,[1] or even stop it altogether, long before the problem of limited resources is reached. The economic limitations imposed by growing expense in a system based on a gross squandering of energy will show up a long time before resources in fossil fuels are exhausted, and an even longer time before nuclear supplies give out—assuming that nuclear energy turns out to be capable of replacing the declining oil resources completely at some stage (which still remains to be seen).

In fact, the growing economic difficulties that Western countries in general, and not only Europe, will find themselves facing (for the 'crisis' has really only just begun) are not yet the result of any absolute shortage of primary energy resources. Although energy is not yet in short supply, we are already beginning to lack cheap oil and gas. The funds needed to pay for a mammoth system involving incessant growth in energy supplies are becoming scarcer. In addition, we are beginning to come up against environmental problems connected with the discharge of effluents and waste products (not only nuclear) resulting from the production of energy, and this also entails extra cost to the community as a whole: pollution, whether combated or not, is never free, and so we now face the increasing social cost of unrestrained energy growth. A final point is that the time necessary for the adaptation of our industrial civilization to these new economic realities appears to be more and more a matter for concern with every day that passes. Twenty years ago, humanity had a certain amount of room for manoeuvre: today, this freedom is a lot more restricted because of the long 'lead time' needed to develop a new strategy based on a levelling-off of consumption, first by industrialized countries and then by the rest of the world. A massive changeover to free natural energy, plus a total reorganization of systems of production and consumption designed to cut out waste and to optimize the use of primary energy sources, will involve even longer delays and in any case cannot give immediate results. In this field at least, any attitude of irresolution and 'wait and see' is clearly a cardinal error since time is so short.

Quite clearly, then, current problems do not arise from a lack of resources but rather from a combination of two socio-economic factors. First, there is the enormous increase in the price of oil and the virtual absence during the last 40 years of research into and development of alternative energy sources, save that of nuclear power. In addition, at the risk of repeating myself, the

---

1. Because of the need to move between and around the collectors and because of other technical restrictions. it is more realistic to double the area.

---

1. The connection between gross national product and energy consumption *per capita* that is often proposed by economists appears to me to involve more dogma than rationality.

major role played in the energy crisis of the West by the never-ending growth in the 'need' for energy (or rather, in its greedy consumption) should not be forgotten.

It is thus essential to begin immediately the setting up of new socio-economic structures adapted to a complete and permanent halt in energy growth. In my view, adequate planning of intermediate stages on the way to that goal would enable a transition to the new form of civilization to take place smoothly and continuously: a form in which the primary energy consumed *per capita* would stay constant or even decrease to a level of perhaps half the present one. This development is entirely possible without the standard of living of the population in industrialized countries being seriously affected—it might even improve some of the more positive aspects through the application of technology to everyday life.

Such an operation implies first of all a systematic diversification of energy supplies. Second, it implies a more efficient use of fossil fuels, which goes along with the fight against atmospheric pollution. Finally, it forces us to resort to the various natural forms of energy, and to geothermal sources, as rapidly and extensively as possible, so as to replace oil wherever its use is not absolutely essential.

A definite time scale should also be adopted for the elimination of energy wastage in our 'consumer society'. All that has been done until now is derisory in comparison with what not only could, but must be, achieved. Limiting the speed of vehicles and lowering the temperature of centrally heated dwellings to a physiologically healthy level: these are not useless, but are merely the tip of the iceberg in terms of energy conservation. In fact, it is only by a searching analysis and a total recasting both of industrial production and the methods of manufacture that optimum use of energy can be achieved.

There are many examples to illustrate this. One is in the area of systems of transport, where developments in the last two decades reveal a systematic tendency to favour methods that waste energy. In this connection, the technocratic and damaging way in which the transport problem has been tackled in France during the last few years must be criticized. The method adopted has been to break down the economic analysis of the communication network into its various sectors: railways, roads, canals, air transport—a method in contradiction with the very nature of the subject. Such a restricted form of economic thinking, which produces only incomplete, and even tendentious, estimates of cost, has consistently led European and American political leaders to favour road over rail. Yet a railway consumes 5.7 times less energy per tonne–km than roads, and 7 times less when the railway is electrified (Tables 2.15 and 2.16).

So what are we to think of the growth in the stock of noisier and more polluting diesel locomotives at a rate of 5.9 per cent per year in France between 1969 and 1973 while the increase in numbers of electrically powered ones over the same period was only 2.2 per cent per year? What are we to say about plans currently being carried out to give France an additional 5000 kilometres of motorways between 1973 and 1985? How are some of the official declarations to be interpreted when they seem to rejoice in the fact that the amount of freight carried more than 100 kilometres by road has grown by 10 per cent per year since 1970, while that carried by the French railways has remained the same? In 1965, the SNCF carried 68 per cent of the long-distance freight and road haulage accounted for 32 per cent: at present, only 35 per cent goes by rail, and 65 per cent by road.

It is exactly the same with passenger transport, and Table 2.16 shows the crushing superiority of rail over other forms of travel in terms of energy.

Railways have a number of additional advantages over other means of transport. Not only are they less noisy and non-polluting when electrified, but for the same amount of traffic they take up an area of land that is ten times smaller than that needed for roads. Thus, a railway track 12 m wide can carry as much merchandise and more passengers than a motorway taking up a width of 120 m. The impact on the environment of an additional 5000 km of motorway is also very considerable. It means a total loss of about 50 000 hectares of land, either directly or indirectly (through gravel pits, etc.), and this still does not take into

Table 2.15 *Relative efficiency of the main methods of transporting goods in terms of energy consumption* (from Hirst, Oak Ridge National Laboratory, in Hammond *et al.*, 1973)

Type of transport	kWh per tonne–km	Ratio to railways
Pipeline	0.96	0.68
Heavy lorry	8.07	5.68
Railway (diesel)	1.42	1.00
Barge	1.44	1.01
Aircraft	89.20	62.82

Table 2.16   Relative efficiency of the main methods of transporting passengers in terms of energy consumption (from Watt, 1973, p. 155)

Vehicle	Distance covered per litre of fuel (km)	Number of passengers	Passenger–km per litre of fuel
Motor car (US)	6	1.3*	7.8
		4†	24
Bus	2.28	5	11.4
		40	91.1
Jet aircraft	0.1	54‡	5.4
Electric train	0.8	600	480

* Mean occupancy in urban areas, light loading.
† Mean occupancy, heavy loading.
‡ Mean number of passengers carried worldwide by aircraft companies, 1967.

account the damage to farms that are cut by roads that run across them: the regrouping that is necessary all along the road network brings great disadvantages with it. Moreover, motorways constructed on embankments and in cuttings have much more devastating effects on the biotopes they cross than railways which have to run over viaducts and through tunnels with small gradients. 'Development' in the form of motorways causes quite serious disturbance of the water systems in valleys sealed off by embankments and also creates pollution of the underlying water table from oil and other hydrocarbons lost from vehicles and oozing through the surface layers. Lastly, road networks tend to encroach on areas devoted to national and regional forests because they are easier to take over by compulsory purchase. This leads to their fragmentation and to a significant deforestation, partly directly because of the roads themselves and partly unintentionally owing, for instance, to attacks by disease on trees along the roadside.

Government authorities cling stubbornly to a policy of providing the country with a road network that will eventually be superfluous because of the unavoidable rise in petrol prices. Yet at the same time the development of railways is held back in the name of profitability, in the narrow sense. Now that half the municipal revenues in France—and other countries—go directly or indirectly into highway maintenance so as to 'adapt the city to the motor car', it might be as well to point out that an individual car consumes between 12 and 60 times more energy per person transported than underground railways (the exact figure depends on the type of vehicle and the number of occupants).

The breaking down of economic analysis into sectors leads to many other anomalies in Western countries and gives rise to as much energy wastage in other areas of activity as in that of transport. For example, at a time when 26 per cent of the oil imported into France is used for domestic and commercial heating, it might well be wondered what arguments there are against linking urban heating to the cooling systems of power stations. Not only would this save a large amount of energy but it would dramatically reduce the thermal pollution of the waters into which the heat is normally discharged and hence completely lost. Power stations are often situated near large conurbations which consume the normal electrical output in local heating. It would be more logical to use the heat discharged at low temperatures (which cannot be used to produce work) for space and water heating and to reserve the electricity for operations where it is irreplaceable. In other words, rather than resort to all-electric heating with its lower overall efficiency, it would seem preferable to link dwellings to the cooling systems of power stations, at least in regions where the density of housing justifies the investment needed for the construction of an underground network of hot-water pipes. We must certainly not fail to recognize the technical problems inherent in such schemes, but whatever is said about them, they would seem to be surmountable provided the temperature of the cooling system is raised by a few tens of degrees. The loss of efficiency entailed by this would be largely compensated for by the saving of millions of tonnes of fuel made possible by the arrangement.

The sector of activity that is a consumer of energy *par excellence* is that of industrial production; its wastage of energy is widespread and chronic, even if less apparent than in other cases. A careful examination of the manufacturing processes in areas as different as the metallurgical and chemical industries shows that, with rare exceptions, neither the design nor the use of production plant are concerned with energy-saving ideas. It is, of course, true that fuel prices in general until 1973 did not constitute a large item in the costs of production. However, as an example of the conservation of energy that is possible, take the steel industry: the replacement of oil-fired by electric induction furnaces would reduce the energy consumed by a factor of 2.5!

It might also be mentioned that older equipment,

which was relatively economic in its use of energy, has been systematically replaced during the last two decades by other types that consume a good deal more. In addition, the increasing obsolescence of modern consumer goods means their faster renewal and again leads to a considerable waste of primary energy. The motor car is a glaring example of this. Whereas before the Second World War it had an average life of about 30 years, it now has one of less than 10 years, yet more than the equivalent of 2 tonnes of oil is needed to produce a vehicle whose basic weight is 1 tonne.

Turning to another area, the replacement of glass bottles by plastic containers has also led to an unfavourable energy balance. In France, it has been estimated that it takes the equivalent of some 300 000 tonnes of oil per year in various oil products to make the plastic bottles used for liquid nutrients (mineral waters, milk, cooking oil, etc.) involving around 15×10^9 containers annually. To turn from glass bottles, which can be completely recycled several dozen times[1], to plastic containers again results in increasing energy consumption: the containers are either burnt in incinerators after use, thus also adding to atmospheric pollution, or they are thrown away and even abandoned in the countryside, contributing to the environmental problems there. Comparison of energy usage shows that the replacement of all plastic containers for liquid nutrients by glass bottles recycled on average 10 times would save more than 200 000 tonnes of oil per year in France alone. With a similar idea in mind, it has been calculated that the recycling of the aluminium in cans used for beer, etc., in the USA would reduce the energy needed in the production of this metal for food containers by a factor of 20.

In conclusion, when ecologists consider the economics of energy production and consumption, they are led to a radical reassessment of current industrial strategy in the civilized world. A complete rethinking of production methods, proper insulation of buildings, the production of more robust articles in general (for example, more solid and hence more durable cars that will be changed less frequently), the total elimination of the recent growth in obsolescent and unnecessary gadgets—all these objectives assume considerable changes. Establishing a new industrial structure with the aim of producing goods involving lower energy consumption both in their manufacture and use; redirecting industrial strategy towards the production of 'quality' goods having a large input from the skills and 'grey matter' of humans and a relatively small one of energy and raw materials: all this implies a complete overturn of present economic policies which are based instead on a continuous urge towards the frequent replacement of consumer goods and even of capital goods.

The unquestioning search for short-term profit, with economic efficiency and incessant attempts to lower costs as the only criteria, is leading civilization towards an evolutional cul-de-sac. When shall we begin to consider, in evaluating production costs, the relative merits of manufacturing processes that are known to be 'profitable' in terms of the environment and of our natural resources?

These are the various considerations that make ecologists very sceptical about the capacity of our society with its present structures to find rational solutions to energy problems.

2.3 Mineral Resources

The situation regarding raw materials is at present much less disquieting on a world scale than that concerning the fossil hydrocarbons. Nevertheless, the quantity of non-ferrous metals and other minerals of strategic importance that will be used during the next decade will exceed the total amount that has been exploited since the beginning of civilization. The rates of increase in the use of these materials, in many cases equal to or greater than 10 per cent per year, mean a doubling time of less than 7 years in their annual consumption.

Unfortunately, the infinite resources demanded by incessant growth do not exist for metals and technically important minerals any more than they do for fossil fuels. Even with an effectively constant population, the annual consumption of metals in Western countries has continued to increase over the last 10 years. When it is realized that at present less than 20 per cent of the world population uses practically the whole of the output of raw mineral materials, it is not difficult to see what the demand will become with the industrialization of the Third World.

The majority of high-grade metal-bearing deposits are already exhausted and it has become necessary to exploit ores with lower and lower concentrations of the elements involved. As an example, the average concentration of copper in the deposits worked at the beginning of the century was 2 per cent, whereas today it barely reaches 0.5 per cent. In Canada, only seven of the metal-bearing deposits out of 1000 discovered during the last 10 years have turned out to be workable, and the effort needed to extract the important elements has to go on increasing merely to maintain the same level of production. This means that the amount of energy needed per unit of production continues to increase as recourse is made to deposits of ever lower grades situated ever deeper in the Earth's crust (Figure 2.34).

---

1. In Great Britain, it has been estimated that milk bottles make on the average 25 trips between the dairy and the consumer and 50 trips in the southern counties. This trippage is, however, declining.

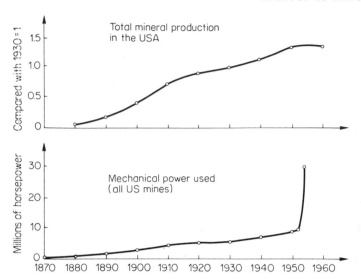

Figure 2.34 Changes in the mechanical power installed in US mines compared with the annual quantity of minerals extracted. (From Lovering, in National Academy of Sciences, 1969, p. 122)

There are other factors which make the supply of mineral elements a more acute problem than appears at first sight. One of these is the extreme non-uniformity of their distribution over the Earth's surface. Even with iron, whose production provides a good indication of a country's degree of industrialization, the workable deposits are so unevenly distributed that some nations with a strong steel industry (Japan, Italy) are forced to import practically the whole of the necessary ores! In Europe, the first region of the world to become industrialized, the majority of the high-grade metal-bearing deposits are exhausted so that the fulfilment of requirements for non-ferrous metals depends to a large extent upon imports.

Many of the various mineral resources only exist as workable deposits in a few regions of the globe: a good proportion of the world reserves of nickel occur in New Caledonia, and of cobalt in Shaba (a province of Zaire which is also a dominant producer of industrial diamonds), while Bolivia and Malaysia possess a near monopoly in the world production of tin. With the possible exception of the USSR, there is in fact no country that is self-sufficient in minerals, no matter how large its territory: even the USA has to import a whole series of strategically important metals.

Another factor that accentuates the problem of mineral supplies is the technological development that has made modern industry dependent on certain scarce elements. It is not many years since rare-earth metals (lanthanides) were regarded as laboratory curiosities in courses on mineral chemistry. Today they have many applications: in metallurgy, in electronics, and in the preparation of optical glass with a high refractive index. Again, what was gallium used for before the

advent of semiconductors? Or zirconium before the development of the nuclear industry?

For several reasons, therefore, the growth in consumption of mineral resources is beginning to present serious problems, and the population explosion in the Third World, coupled with its industrialization and rising standards of living, is going to produce a dramatic growth in demand in the years to come. To see the effect of this, suppose the entire population of the world adopted the same standards of consumption as the USA. Total production of zinc and iron would have to increase 7.5 times, that of copper 100 times, and that of tin 250 times! Nobody can claim that such production is ecologically feasible: even if we ignore the resultant pollution, mineral extraction on that scale, turning as of necessity towards lower-grade deposits, would devastate a vast area of agricultural land through open-cast mining and would seriously affect the food supply of the existing population (cf. chapter 5, p. 148).

In reality, the Third World will not be able to achieve the level of consumption of industrialized countries in a few decades. Nevertheless, the whole world is already facing a scarcity of raw mineral materials at rising prices. In 1979 the US Bureau of Mines estimated the number of years that available reserves of various inorganic elements and minerals could satisfy demand, making two different assumptions about the growth in consumption (Figure 2.35). For a number of important materials it is clear that available reserves are only enough for a few decades. Kessler (1976) has also made an estimate of mineral reserves from various geological data (Table 2.17). This shows that supplies of a number of metals and minerals would become critical in at most a quarter of a century, but that there should be no major problems for iron, aluminium and other indispensable metals for some time to come. In fact, the situation is less happy than is apparent even for these elements because of the never-ending growth in demand which the table does not take into account. Moreover, the working of lower and lower grade reserves becomes more difficult as the large amounts of energy needed become more expensive.

The only remedy for these problems is a continual and systematic development of recycling programmes, which produce considerable savings in energy and a reduction in pollution as well as postponing or even preventing the exhaustion of the principal deposits. Unfortunately, we are still a long way from achieving a high level of recycling, as Table 2.18 shows. This is true even for precious metals such as silver, where recycling is far from 100 per cent effective. In view of the technological importance of silver in photography, etc., the supplies of this metal are particularly worrying. The world has been consuming more silver than it

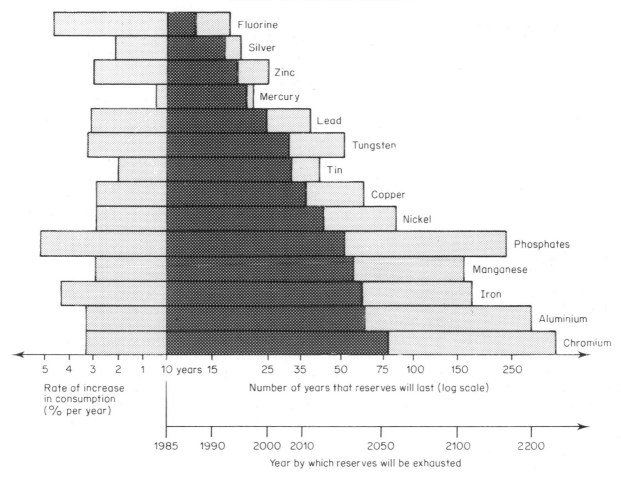

Figure 2.35 Available reserves of the 14 principal minerals that are indispensable to modern industrial processes. The light grey bars on the right of the figure give the lengths of time that the reserves will last and the dates of their exhaustion, assuming that the demand stabilizes around their 1975 values. The dark grey portions give the times and dates if consumption continues to grow at the rates observed at the end of the 1970s (given at the left of the figure). (Adapted from US Bureau of Mines, *Mineral trends and forecasts*, 1979, in Ramade, 1982)

Table 2.17 *Estimates of world reserves of the principal mineral resources* (from various publications, especially of the US Geological Survey in Kesler, 1976)

Element or compound	Estimated reserves (tonnes)	Annual consumption in 1975 (tonnes)	Lifetime of reserves* (years)
Iron	160×10^9	500×10^6	320
Aluminium	5×10^9	20×10^6	250
Manganese	2×10^9	10×10^6	200
Chromium	0.5×10^9	2×10^6	250
Nickel	100×10^6	0.5×10^6	200
Tungsten	2×10^6	30×10^3	66
Copper	300×10^6	7.5×10^6	40
Lead	140×10^6	3×10^6	46
Zinc	250×10^6	6×10^6	42
Mercury	250×10^3	10×10^3	25
Tin	3×10^6	200×10^3	15
Fluorine	30×10^6	2.5×10^6	12
Potassium	$>20 \times 10^9$	25×10^6	>800
Phosphates	30×10^9	110×10^6	270

* Assuming a levelling off of consumption at the 1975 rate. If the growth in demand is taken into account, some of these figures should be reduced by a factor of 5 or 10.

Table 2.18 Recycling rates for the principal metals used in the USA (from the US Bureau of Mines mineral yearbook)

Metal	% recycled
Aluminium	16
Nickel	19
Zinc	22
Copper	22
Tin	26
Mercury	32
Lead	42
Iron	52
Antimony	60

has extracted for 20 years and only 20 per cent of the deficit is made up by recycling. The rest of the deficit of around 7000 tonnes is provided essentially by stocks of the metal made available by its demonetization: for example, the stocks of coins in India amount to some 16 000 tonnes. There remains a lot to be done, therefore, even if only photography is being con-sidered: this alone consumes nearly a third of the 17 000 tonnes of silver used annually in the West and not even half of that is recycled at present.

In conclusion, we should remember that the various mineral resources are interdependent and it does not follow that they will continue to be produced in increasing amounts just because the deposits are far from exhausted. Thus, although the great abundance of bauxite seems to mean that there will be few problems in producing aluminium, the situation could be com-plicated by a shortage of fluorine which is indispens-able for the preparation of cryolite. In the same way, silver is a by-product of the metallurgical extraction of lead and zinc and there would not necessarily be a will-ingness to produce excess quantities of those metals just for the silver obtained.

Finally, where the use of a material involves its widespread dispersal (as with phosphates) recycling is virtually impossible. However, in some cases there is the possibility of solving the problem by substitution: for example, the elimination of lead from petrol by using added methanol instead.

Chapter 3

The Atmosphere and Climates

Contemporary technological civilization uses the atmosphere—as it does the oceans—both as a resource and as a medium for the discharge of its waste products. Such an ambivalent and contradictory attitude results in physico-chemical changes in the air around us that have harmful biological and economic consequences. The chemical composition of the atmosphere is being modified by the injection into it of particles, gases and volatile substances, all more or less toxic to living organisms. This hinders the renewal of certain natural resources, upsets the normal functioning of the atmosphere in varying degrees, and runs the risk of inducing irreversible ecological changes on a worldwide scale.

At the local level, atmospheric pollution in urban and/or industrial regions is generating serious public health problems and is causing grave danger to cultivated plants and forests. At the level of the entire ecosphere, it is upsetting the principal biogeochemical cycles and producing climatic effects of uncertain magnitude.

3.1 Perturbation of the Principal Biogeochemical Cycles

At the present time, we are witnessing undoubted human interference in the main biogeochemical cycles that pass through a gaseous phase. This includes the cycles of all the principal biogenic elements with the exception of phosphorus: carbon, oxygen, nitrogen, hydrogen and sulphur all pass through the atmosphere in transit from one stage to another of their cycles.

Our industrial civilization has already disturbed the carbon and sulphur cycles quite considerably, while its effect on the nitrogen cycle can no longer be regarded as negligible. There is, too, an increasingly worrying interference with the ozone cycle, which is an offshoot of that of oxygen.

To take an example: the increasing use of fossil fuels has meant the discharge of great quantities of CO_2 gas into the atmosphere, amounting now to the equivalent

of more than 6×10^9 tonnes of carbon per year. That should be compared with the 65×10^9 tonnes of equivalent carbon per year injected by the respiration of all terrestrial communities.

In a similar way, natural processes produce some 100 million tonnes of CO per year, whereas fuel combustion, especially in petrol engines, discharges 300 million tonnes annually. Again, natural phenomena introduce the equivalent of 84×10^6 tonnes of sulphur into the atmosphere every year, while the production of SO_2 by domestic and industrial combustion of fuel oil and coal puts out nearly 100×10^6 tonnes.

For some of the biogenic elements, therefore, the quantities generated by human technology are of an order of magnitude comparable with those produced by natural processes.

3.1.1 Perturbation of the carbon cycle

The most perfect of the biogeochemical cycles is undoubtedly that of carbon because of the speed of its passage between its various inorganic stages and of its exchange between those and the biomass. It is a predominantly gaseous cycle since CO_2 is the main vehicle of transfer between the atmosphere, the hydrosphere and the various biocoenoses.

Carbon is the fundamental biogenic element since it occurs in all organic molecules, but its preponderant forms in the biosphere are inorganic: gaseous CO_2 in the atmosphere, and various dissolved carbonates in the hydrosphere. The atmosphere and hydrosphere are the two principal pools of carbon for autotrophic organisms but, in fact, they fulfil their roles in the carbon cycle relatively independently: green plants and atmospheric CO_2 are associated on the one hand, and photosynthetic marine organisms and carbonates dissolved in the hydrosphere on the other. These two sub-cycles are nevertheless interconnected by exchanges of CO_2 between the oceans and the atmosphere according to the following set of equations which regulates the concentration of CO_2 in the air:

$$\text{atmospheric } CO_2 \leftrightarrows CO_2 \text{ dissolved in waters}$$
$$\leftrightarrows CO_2 + H_2O$$
$$\leftrightarrows H_2CO_3 \text{ carbonic acid}$$
$$\leftrightarrows HCO_3^- \text{ carbonic anion}$$

In addition, the CO_2 in freshwater environments such as rivers and lakes dissolves the calcium in surface rocks and the water cycle then takes the carbon to the oceans in the form of bicarbonates:

$$CO_2 + H_2O + CaCO_3 \leftrightarrows CaH_2(CO_3)^2$$
$$\text{(soluble calcium bicarbonate)}$$

In spite of a number of uncertainties that still exist in connection with the carbon cycle, it remains the best known of all the biogeochemical cycles.

The atmospheric pool of some 700×10^9 tonnes of equivalent carbon in the form of CO_2 is much smaller than the hydrospheric (see Figure 3.1). The latter has the equivalent of about 560×10^9 tonnes of carbon as bicarbonates in surface waters, but the total quantity occurring as CO_2, mostly in the ocean depths, is far greater and has been estimated at $38\,000 \times 10^9$ tonnes of carbon.

The third pool of carbon in the biosphere consists of all its organic matter in three main categories. On land, there is first the plant biomass containing some 827×10^9 tonnes of carbon, and secondly the dead organic matter in the soil (humus) estimated to contain between 1000 and 3000×10^9 tonnes of carbon in all. Thirdly, although the biomass in marine environments

is negligible, there is a pool of considerable size in the organic matter 'dissolved' in the water, also estimated at 1000 to 3000×10^9 tonnes of carbon.

Sedimentary rocks, especially the fossil fuels in them, contain a much larger quantity of carbon, but in the absence of human intervention it would remain quite separate in the biosphere, isolated in the Earth's crust by geological processes. These deposits would hardly be involved at all in the carbon cycle, except through the release of CO_2 from lithospheric carbon during volcanic activity. Such contributions are, however, very small and probably less than 50×10^6 tonnes of carbon per year. In contrast to that, sizeable quantities of CO_2 are discharged into the atmosphere as a result of the increasing use of fossil fuels by our industrial civilization.

The role of biological processes in the carbon cycle

Photosynthesis and respiration are the two opposing processes which form the fundamental driving forces of the carbon cycle. The first of these processes extracts CO_2 from the atmosphere and the oceans, transforming it into biochemical substances, while respiration (and fermentation) decompose organic compounds into CO_2 with the production of cellular energy (Figure 3.2). Only organisms containing chlorophyll (that is, the autotrophic organisms: green plants and phytoplankton) are capable of effecting photosynthesis and this is the origin of their classification as primary producers. Respiratory activity, on the other hand, is as much a function of heterotrophic organisms (animals and decomposing micro-organisms) as of autotrophic ones.

Changes in the atmospheric concentration of CO_2

The competition between photosynthesis and respiration, and the exchange of CO_2 between atmosphere and ocean, together bring about a nearly perfect equilibrium in the carbon cycle. There are many indications now available which point to an almost completely steady concentration of atmospheric CO_2 for at least the whole of the Quaternary period up to 1850, when industrial development began to consume the fossil fuel stock at a significant rate.

Systematic observations of atmospheric CO_2 concentration made since 1958 at the Mauna Loa observatory in Hawaii by Keeling *et al.* and at other analysing stations show that it has experienced a continuous and uninterrupted rise. In the middle of the last century it was estimated at 268 p.p.m. (Stuiver, 1978) while now it is 335 p.p.m.

The short-term fluctuations in the concentration vary with the latitude and altitude of the observatory and also with the time of year (Figure 3.3). The

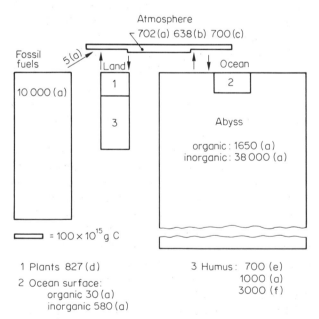

Figure 3.1   The world carbon balance. The relative magnitudes of the various pools and flows are indicated by the sizes of the boxes and arrows. The figures give the quantities of equivalent carbon in units of 10^9 tonnes, and the labels (a) to (f) indicate different sources of the estimates. (From Woodwell *et al.*, 1978, p. 143)

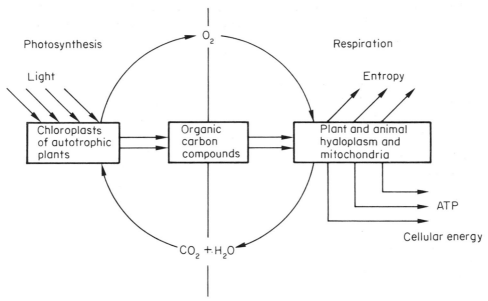

Figure 3.2   The carbon cycle and energy flow at the cellular level. (From Ramade, 1978a)

amplitude of annual oscillations is a maximum at sea level and decreases with increasing altitude. It is also a minimum at the Equator, increases with latitude, and is larger in the Northern than the Southern Hemisphere. The cyclic variations are due to changes in photosynthetic activity: the minimum concentrations of CO_2 occur at the end of spring in the Northern Hemisphere when take-up by chlorophyll is high. The maxima occur in winter when plants are dormant.

The average annual increase is between 0.5 and 1.5 p.p.m. according to latitude, with a mean value for the whole globe estimated at 1 p.p.m. per year in 1980.

Increase in CO_2 concentration from the use of fossil fuels

The release of CO_2 into the atmosphere from the consumption of fossil fuels is undoubtedly a basic cause of perturbations in the carbon cycle. In the middle of the 1970s, it was estimated that such discharges were injecting the equivalent of some 5×10^9 tonnes of carbon into the atmosphere *per year*, whereas the *total* amount released between 1850 and 1950 was similarly estimated at 60×10^9 tonnes. The increases in quantities of CO_2 produced by industry since 1950 is illustrated by Table 3.1.

Increase in CO_2 concentration from biological processes

Human activity clearly produces changes in the pool of organic carbon and contributes to the increases of concentration in atmospheric CO_2. However, recent studies by several authors have shown that the changes

cannot be entirely accounted for by the consumption of fossil fuels alone. Deforestation, another factor of human origin, is capable of playing just as large a role as the combustion of oil and coal (Adams *et al.*, 1977).

Over the last 1000 years, but especially since the middle of the last century, the combined effects of the population explosion and industrial growth have brought about a continual clearance of many afforested regions. Thus in North America, the huge forest which used to cover the whole of eastern USA from the Mississippi to the Atlantic coast has had 90 per cent of its original area turned over to agricultural land. In Canada, the great northern coniferous forest has similarly been reduced to at most a third of its original biomass.

In the case of tropical forests, there has been reckless exploitation carried out at such an excessive rate compared with the speed of their restoration that they face total annihilation by the middle of the next century. In Venezuela, for instance, the area covered by rain forests decreased by 32.5 per cent between 1950 and 1975. Again, whereas 60 per cent of the state of São Paulo in Brazil was covered by forest in 1910, only 20 per cent was covered in 1950! Madagascar is another example: in the middle of the last century, three-quarters of its land mass was covered with varied types of forest; today, the island is a semi-desert over two-thirds of its territory.

The total biomass of world forests is estimated to contain 743×10^9 tonnes of carbon. The average annual rate at which they are being cut down can be taken as somewhere between 1 and 2 per cent of their total area. If two-thirds of this mass is assumed to be released into the atmosphere in the form of CO_2

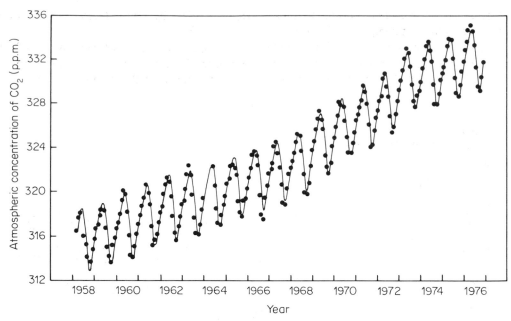

Figure 3.3   Fluctuations in the atmospheric concentration of CO_2 between 1958 and 1975 at the Mauna Loa observatory. (From Keeling, in Woodwell *et al.*, 1978)

because of its exploitation in various ways, a calculation shows that deforestation injects between 5×10^9 and 10×10^9 tonnes of carbon into the atmosphere every year, of which from 3 to 6×10^9 is from tropical forests alone. In addition, if the land is put to agricultural use after clearing it, then the resulting oxidation of the forest humus stored in it will release a further 2×10^9 tonnes of carbon per year.

In total, therefore, keeping in mind the intrinsic uncertainties in such estimates, we can say that deforestation leads to the discharge of a quantity of CO_2 into the atmosphere equivalent to some 8×10^9 tonnes of carbon per year, the actual figure probably lying within the range from 2×10^9 to 18×10^9 tonnes.

Isotopic methods for measuring the flow of CO_2 between the biomass, the atmosphere and the oceans

The use of isotopic methods has recently made it possible to attempt indirect estimation of how much CO_2 has been released into the atmosphere by various human activities and of its flow between different segments of the biosphere.

Table 3.1   *CO_2 produced by various fossil fuels* (in 10^9 tonnes of CO_2 per year)

Year	Coal	Lignite	Oil	Natural gas	Total
1950	3.7	0.9	1.4	0.4	6.4
1960	5.0	1.4	3.1	1.0	10.5
1970	6.7		7.8	3.0	17.5
1980	11.1		10.8	4.0	25.9

There are three isotopes of carbon: carbon-12, ^{12}C, which is by far the most abundant in the universe; a heavier isotope, ^{13}C, which is also stable; and ^{14}C, which is radioactive with a half-life of 5370 years and which is formed in the upper atmosphere by neutron activation of nitrogen induced by cosmic rays.

Carbon-14 occurs in living organisms in approximately the same ratio to ^{12}C as it does in atmospheric CO_2. On the other hand, the $^{14}C/^{12}C$ ratio starts to decrease as soon as the organism dies since the take-up of carbon ceases *ipso facto*: this decrease is used in archaeological dating. Fossil fuels, whose ages can be anything up to tens or even hundreds of millions of years and thus far greater than the half-life of ^{14}C, contain practically no radioactive carbon.

The other heavy isotope ^{13}C, unlike ^{14}C, does not occur in the same proportions in living organisms as in atmospheric CO_2: plants 'prefer' ^{12}C during photosynthesis, so that organic matter is less rich in ^{13}C than the atmosphere. They thus have a smaller ratio $R = {}^{13}C/{}^{12}C$. Variations in R found in any organic or mineral sample are related to the ratio in a reference standard called R_{ref} by defining a quantity $\delta^{13}C$ as follows:

$$\delta^{13}C = \frac{R_{\text{sample}} - R_{\text{ref}}}{R_{\text{ref}}} \times 1000$$

Clearly, $\delta^{13}C$ expresses the variation in parts per 1000 (i.e. as 0.1 of 1 per cent) usually written 'per mil'.

Fossil fuels are found to have a $\delta^{13}C$ ranging from -22 to -28 per mil, while atmospheric CO_2 has a $\delta^{13}C$ of -7 per mil. The CO_2 produced by living organisms through respiration and fermentation has a value lying between −21 and −26 per mil, but a difficulty arises from the fact that plants taking the C_3 route in photosynthesis have a different $\delta^{13}C$ from those taking the C_4 route. This means that aerobic or non-aerobic degradation from C_3 plants to C_4, or vice versa, will be accompanied by the release of CO_2 with different values of $\delta^{13}C$.

Changes in the isotopic ratios of carbon in atmospheric CO_2 and in the biomass have inevitably been produced by the combustion of fossil fuels and

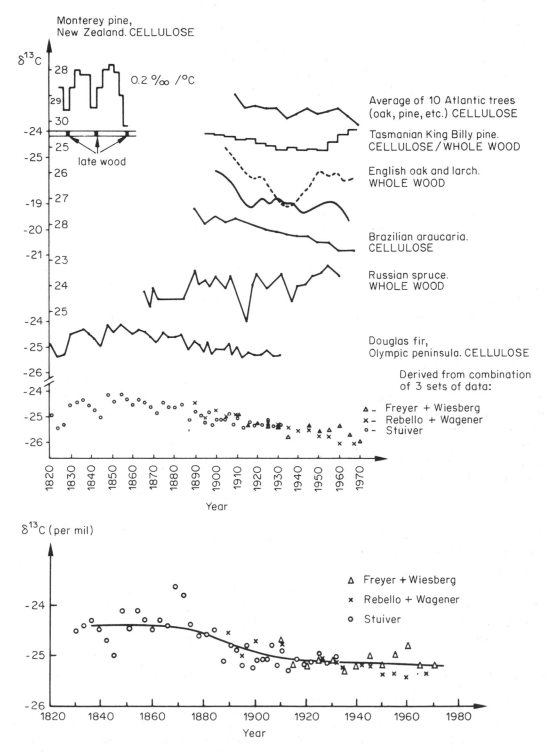

Figure 3.4 Upper diagram: variation of $\delta^{13}C$ in the wood of trees from various regions. Lower diagram: variations of $\delta^{13}C$ for atmospheric CO_2 arising from decomposition of biomass after deforestation (that is, the total $\delta^{13}C$ less that contributed by fossil fuel combustion). (From Stuiver, 1978, p. 255)

the increasing clearance of forests. Fossil fuel CO_2 in the atmosphere reduces the $^{14}C/^{12}C$ ratio since it involves carbon that almost completely lacks ^{14}C. However, both the use of fossil fuels and deforestation reduce the amount of ^{13}C in the atmosphere since they both involve the release of CO_2 from material of biological origin having a lower R value.

Measurement of the concentrations of ^{13}C and ^{14}C in the annual growth rings of trees has made it possible to calculate variations in atmospheric CO_2 over the last few centuries. The rings themselves yield an absolute determination of the year in which the wood was formed, so that cellulose or lignin can be extracted and analysed to give the $^{13}C/^{12}C$ and $^{14}C/^{12}C$ ratios corresponding to that year.

In that way, Suess (1955) was able to show that the relative concentration of ^{14}C dropped by 2 per cent between 1850 and 1950 (the Suess effect): in other words, by the middle of the 1950s it was 2 per cent lower than its theoretical value because of the enormous quantities of CO_2 poor in ^{14}C that had been released by the combustion of fossil fuels over the preceding 100 years.

Stuiver (1978) combined the measurement of the $^{13}C/^{12}C$ ratio in the growth rings of Douglas firs between 1820 and 1970 with the data from the Suess effect, and in that way he succeeded in estimating the relative importance of the releases of CO_2 arising from deforestation and from the use of fossil fuels. The only human factor capable of modifying the ^{14}C concentration in atmospheric CO_2, at least before the aerial explosion of nuclear devices, was the combustion of fossil fuels: deforestation does not affect it because, as has been pointed out, the $^{14}C/^{12}C$ ratio is the same for the biomass as for the atmosphere: the only change would be due to the radioactive decay of ^{14}C, producing a reduction of only 1 per cent every 80 years.

The Suess effect, which relates only to the ^{14}C concentration, is thus entirely due to the use of fossil fuels, but it can also be used to deduce the contribution of these fuels to changes in $\delta^{13}C$. The carbon in fossil fuels contains 18 per mil less ^{13}C than atmospheric carbon, but 100 per cent less ^{14}C (i.e. none). The change in $\delta^{13}C$ *due to fossil fuels alone* is thus 18/100 of the ^{14}C change, or:

$$\Delta(\delta^{13}C) = 0.18S$$

where S is the Suess effect in percentage. In addition, Stuiver and others have measured the *total* $\delta^{13}C$ in the growth rings of trees over the past 150 years. The difference between this total $\delta^{13}C$ and that derived from the Suess effect gives an estimate of how much deforestation has contributed to atmospheric CO_2, and thus allows a comparison of the effects of fossil fuels and the destruction of forests (Figure 3.4).

In this way, Stuiver showed that fossil fuel combustion had reduced $\delta^{13}C$ by 0.7 per mil from the beginning of the industrial era to 1970 and this was the result of injecting some 120×10^9 tonnes of carbon into the atmosphere in the form of CO_2. The same author similarly estimated the same reduction of 0.7 per mil in $\delta^{13}C$ from CO_2 of biological origin between 1850 and 1950, again the result of 120×10^9 tonnes of carbon.

In all, between 1850 and 1950, Stuiver showed that the effect of human activity on the carbon cycle manifested itself in the discharge of 60×10^9 tonnes of carbon into the atmosphere due to the combustion of fossil fuels and 120×10^9 tonnes due to the destruction of forests. Thus a total of 180×10^9 tonnes of carbon were added to the atmosphere as CO_2 during that period, and we now turn to a consideration of what has happened to such a large quantity of the gas.

Transfer of CO_2 from the atmosphere to the oceans

The quantity of carbon currently injected into the atmosphere in the form of CO_2 by human activity is at least 8×10^9 tonnes per year and could even exceed 10×10^9 tonnes (Woodwell *et al.*, 1978). On the other hand, it is calculated that only 2.3×10^9 tonnes of this remains in the atmosphere so that there must be a reservoir in the ecosphere which removes a large proportion of this CO_2.

The size of such a reservoir relative to the whole atmosphere can be estimated from the Suess effect. Stuiver has estimated, as indicated above, that 180×10^9 tonnes of anthropogenic carbon were discharged into the atmosphere between 1850 and 1950, and if this whole amount had remained there the Suess effect would have been 180/700 or 26 per cent (taking Bolin's 1979 estimate of 700×10^9 tonnes as the total size of the atmospheric carbon reservoir). Since the effect during that period has been only 2 per cent, there must be a reservoir in the ecosphere 13 times as large as that of the atmosphere.

In fact, there are two sections of the biosphere capable of accumulating CO_2—the biomass and the oceans—and many arguments suggest that only the oceans are capable of extracting excess CO_2 from the atmosphere effectively enough. The large quantities of $^{14}CO_2$ produced by nuclear explosions during the 1950s have made it possible to confirm the important role played by the oceans in the absorption process. The variation with depth of ^{14}C concentration has been studied in the Atlantic (Figure 3.5) and Indian Oceans and the results have been used to measure the rate of exchange between surface waters and the deeper zones. By following the changes in the distribution as time goes on, the rate at which atmospheric CO_2 is absorbed by the hydrosphere can be deduced. Every year,

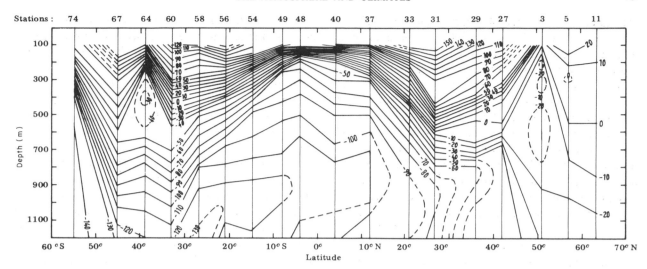

Figure 3.5   Curves of equal concentration of ^{14}C in the Western Atlantic, expressed as per mil deviations from a standard. (From Stuiver, 1978, p. 256)

descending convection currents take 0.5 per cent of the cumulative excess of atmospheric CO_2 into the deep oceans, though this does not occur uniformly but is concentrated in regions where the surface waters sink into the abysses (Figure 3.6). There is, moreover, a clear distinction to be made between the surface layers (approximately the first 300 metres), where the excess CO_2 is quite quickly dissolved, and the deeper regions into which diffusion of the CO_2 occurs much more slowly.

Overall, it has been estimated that by 1950 64 per cent of the excess atmospheric CO_2 had accumulated in the surface layers of the oceans and 24 per cent had been taken to the bathyal and abyssal zones. Subsequently, a calculation by other authors has similarly shown that 49 ± 12 per cent of the CO_2 produced by combustion of fossil fuels between 1959 and 1969 has remained in the atmosphere and the rest has gone into the oceans.

Possible transfer of CO_2 from the atmosphere to the biomass

The biomass is another potential reservoir for excess atmospheric CO_2. According to all the evidence, however, the higher rates of photosynthesis stimulated by any increase in atmospheric concentration of CO_2 were overestimated in studies of the carbon cycle made up to the end of the 1970s (see, for example, the report of the SCEP group of MIT). Bolin *et al.* (1979) even contest that it can play any significant role in the regulation of atmospheric CO_2.

An elementary calculation will confirm that an increase in primary production stimulated by the

increase in atmospheric CO_2 due to human activity is far from being capable of compensating for this increase by incorporating it into the biomass. Let us assume that the net primary production in terrestrial ecosystems is proportional to the atmospheric concentration of CO_2. In that case, the annual storage of carbon in the biomass induced by the accompanying increase in photosynthetic activity will be given by the product:

$$\frac{\text{annual increase in concentration of atmospheric } CO_2 \text{ (1 p.p.m.)}}{\text{total concentration of } CO_2 \text{ in the atmosphere (330 p.p.m.)}}$$

$$\times \text{ world net primary production } (78 \times 10^9 \text{ t})$$

This gives a figure of 0.24×10^9 tonnes of carbon per year for the maximum annual storage in the biomass and it is thus far less (by a factor of at least 10) than the current injection of CO_2 into the atmosphere resulting from forest clearance.

The only way of increasing the storage of carbon in the biomass and of reabsorbing a significant proportion of the excess CO_2 is by putting an immediate stop to the destruction of forests and accompanying that by intensive reafforestation on a continental scale. Yet the present tendency is still towards greater quantities of CO_2 released into the atmosphere both through increasing use of fossil fuels and the continuing destruction of tropical, and even temperate, forests all over the planet. With such a situation, there is no prospect of restabilizing the carbon cycle: on the contrary, a continuation in the rise of the concentration of

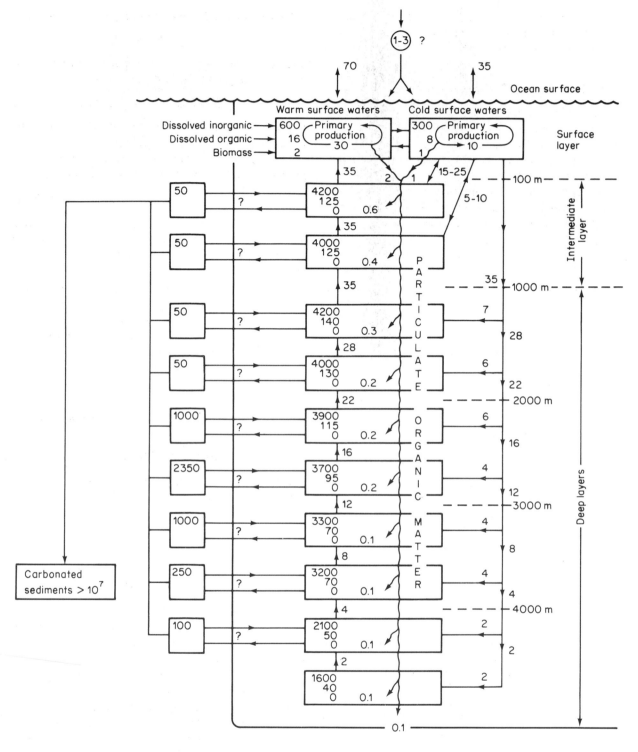

Figure 3.6   Representation of the carbon cycle in the oceans based on a model of the vertical circulation of the water. Flows are given in 10^9 tonnes of carbon per year and stocks in 10^9 tonnes of carbon. (From Bolin *et al.*, 1979; reproduced by permission of SCOPE (Scientific Committee on the Problems of the Environment))

atmospheric CO_2 appears quite inevitable, with consequences for the climate that we shall examine later in the chapter.

Nevertheless, improbable as it is, stopping the discharge of anthropogenic CO_2 into the atmosphere immediately would bring about a rapid decrease in its concentration to the value it had before the beginning of the industrial era. Calculations based on models of

CO_2 flow between the various parts of the ecosphere have shown that half the excess CO_2 present in the atmosphere would be reabsorbed in 21 years.

3.1.2 Perturbation of the ozone cycle

Ozone forms part of the oxygen cycle but has its own distinctive circulation which is entirely dependent on physico-chemical processes: its high toxicity for all living organisms means that they cannot be involved in its geochemical cycle.

Although ozone is a normal constituent of the atmosphere, its concentration at ground level varies considerably depending on the intensity of ultra-violet radiation: it can go up to 5 p.p.b. (parts per 10^9) in regions with much cloud cover but can exceed 15 p.p.b. in some desert areas with high sunshine levels. The concentration increases with altitude and is already as much as 150 p.p.b. at 3 km. The term 'stratospheric ozone layer' is thus inappropriate because ozone occurs at all heights in the atmosphere. It does, however, have a maximum concentration between 20 and 30 km (Figure 3.7) because ultra-violet radiation of wavelength less than 280 nm does not penetrate to lower levels and it is these shorter wavelengths that are involved in its production.

Ozone is formed from molecular oxygen, which is dissociated by ultra-violet radiation of wavelength

between 100 and 245 nm according to the reaction:

$$O_2 + h\nu \rightarrow 2\,O \tag{1}$$

and the resultant atomic oxygen combines with molecular oxygen to form the ozone by:

$$2\,O + 2\,O_2 \rightarrow 2\,O_3 + 2.2\,eV \tag{2}$$

Reactions (1) and (2) are sometimes named after Chapman, who first discovered them.

In passing, the formation of atomic oxygen by the photochemical dissociation of NO_2 should be noted:

$$NO_2 + h\nu(<400\,nm\ or\ >3.1\,eV) \rightarrow NO + O \tag{3}$$

because the atomic oxygen formed in this way reacts as in (2) and provides an explanation for the appearance of ozone in smogs such as those occurring in Los Angeles.

The ozone formed in the upper atmosphere is partially destroyed by opposing reactions involving the oxides of nitrogen, so that its apparently constant concentration is the result of a dynamic equilibrium. One cycle of destructive reaction begins with a retransformation of atomic O to O_2:

$$NO_2 + O \rightarrow NO + O_2 \tag{4}$$

and after that the NO reacts with ozone and re-forms NO_2 and O_2:

$$NO + O_3 \rightarrow NO_2 + O_2 \tag{5}$$

Other authors have suggested a destructive cycle that starts with the photodissociation of water:

$$H_2O \rightarrow OH^- + H^+ \tag{6}$$

with the H^+ subsequently reacting with ozone in the upper atmosphere:

$$2\,H^+ + O_3 \rightarrow H_2O + O_2 \tag{7}$$

However, most experts in atmospheric chemistry now believe that the oxides of nitrogen play the dominant role in the dynamic equilibrium of stratospheric ozone: the CIAP report[1] is just one example.

Natural sources of nitrogen are both biological and physico-chemical in nature. In the biosphere, N_2O is produced by denitrifying bacteria and forms the main source of the oxides of nitrogen in the atmosphere. Some of the N_2O is oxidized to NO and this is capable of initiating the catalytic destruction of ozone by (5). In addition, another Chapman reaction occurs in the mesosphere at around 100 km which transforms nitrogen to NO by

$$N + O_2 + h\nu \rightarrow NO + O \tag{8}$$

and NO is also produced by another type of reaction

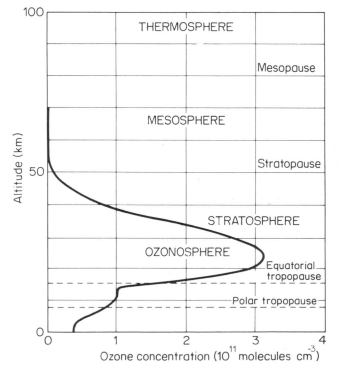

Figure 3.7 Variation of atmospheric ozone concentration with altitude. (From Cieslik, 1976, p. 514)

1. Climatic Impact Assessment Programme of the US Department of Transportation.

between molecular nitrogen and atomic oxygen:

$$N_2 + 2O \rightarrow 2NO \qquad (9)$$

In fact, the efficiency of (5) in destroying ozone is greatly reduced by another reaction which leads to the formation of nitric acid from the products of (5) and (6) in the presence of a catalytic impurity:

$$NO_2 + OH + M \rightarrow HNO_3 + M$$

Thus, as the left-hand side of Figure 3.8 shows, the dynamic equilibrium (stationary state) of atmospheric ozone is maintained in the absence of human intervention by a circulation of natural oxides of nitrogen coupled to the ozone cycle. At the same time, the nitrogen introduced into the atmosphere as N_2O by denitrifying bacteria is returned to the biosphere in the form of nitric acid.

The vertical distribution of ozone in the atmosphere varies with time and with latitude. The existence of the dynamic equilibrium can explain these variations: for instance, why ozone is more abundant in low altitudes in polar regions but has a maximum at about 35 km in Equatorial zones (Figure 3.7).

So far, we have neglected the possible effects of human activity on the ozone concentration. Among the various processes which are capable of interfering with the ozone cycle, we shall be dealing with the explosion of nuclear devices in the atmosphere, combustion which produces oxides of nitrogen (especially that arising from jet aircraft flying in the lower stratosphere), and, lastly, the use of fluorocarbon gas propellants in aerosol sprays. There is a preliminary point that should be emphasized: the atmospheric ozone does not correspond to a stock but rather to a stationary state, so that all human intervention is reversible. It has been estimated that if all the ozone were suddenly removed from the atmosphere it would take times varying from less than 24 hours (in the mesosphere) to over a year (in the lower stratosphere) for the original concentrations to be restored through the Chapman reactions.

One of the main events capable of producing effects that might attack the atmospheric ozone is the testing of nuclear devices. At the temperatures reached in the fireball of a 1 megatonne nuclear bomb, the quantity of NO produced has been calculated at 10^{32} molecules,

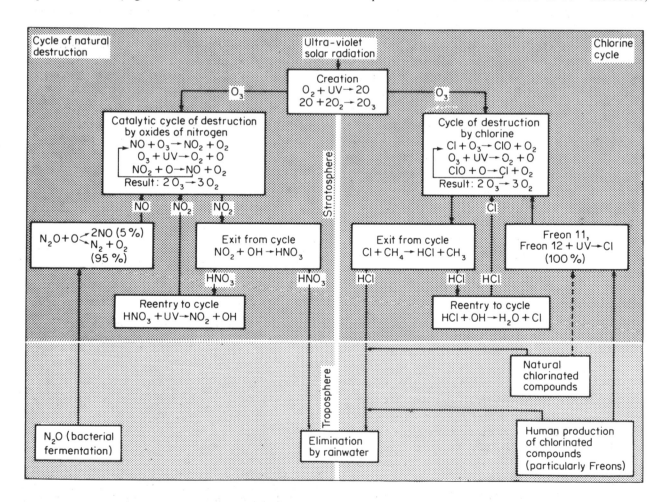

Figure 3.8   Representation of the atmospheric ozone cycle. Natural causes of the destruction of ozone are on the left-hand side of the diagram and human causes on the right. (From Dupas, 1975, p. 971)

Table 3.2   Principal contaminants injected into the stratosphere by supersonic aircraft (from *Man's Impact on the Global Environment*, MIT, 1970)

	Pollutant	Mean concentration introduced into the stratosphere	Maximum foreseeable in the Northern Hemisphere	Concentration measured at cruising height of SST	
				Before Agung eruption	After Agung eruption
Gaseous	H_2O	0.2 p.p.m. (0.3)*	2 p.p.m. (3)		
	CO	6.8 p.p.b. (7.3)	68 p.p.b.		
	NO	6.8 p.p.b. (7.3)	68 p.p.b.		
	SO_2	0.16 p.p.b. (0.2)	1.6 p.p.b. (2.0)		
Aerosols	SO_4	0.24 p.p.b.	2.4 p.p.b.	1.2 p.p.b.	36 p.p.b.
	Hydrocarbons + soot	0.16 p.p.b.	16 p.p.b.		
Total particles		0.34 p.p.b.†	3.4 p.p.b.	1.2 p.p.b.	36 p.p.b.

* The numbers in parentheses give the total quantity of pollutant injected into the stratosphere by military aircraft and SSTs.

† These figures are less than the sum of those in the preceding two rows because 'hydrocarbons' includes liquid compounds that cannot strictly be counted as particles.

compared with the estimated 4×10^{34} molecules of nitrogen oxides contained in the whole atmosphere. As a result of the International Geophysical Year, permanent monitoring of ozone concentration was undertaken from 1958 onwards, and this drew attention to the low concentrations that were observed at some stations in the Northern Hemisphere during 1961 and 1962, when Soviet and US nuclear testing was at its height.

In contrast to that, studies made as part of the CIAP have shown that ecological protest movements have markedly exaggerated the effects of supersonic aircraft that might possibly reach the stratospheric ozone. These studies looked at two effects connected with the flight of SSTs (supersonic transports), and indeed with that of all aircraft, in the lower stratosphere: the injection of water vapour and the introduction of oxides of nitrogen.

Table 3.2 shows the stratospheric pollution which would be produced by a fleet of 500 SSTs each flying 2500 hours per year. The assumptions underlying the CIAP studies are more modest than those of Table 3.2 and are based on the existence of a world total of 100 SSTs each flying 7 hours per day at an altitude of 20 km. These latter studies concluded that the ozone concentration would be reduced by at most 0.5 per cent in the Northern Hemisphere, a figure which should be compared with natural variations of up to 10 per cent in some latitudes according to the season! (See Figure 3.9.)

In spite of the uncertainties which exist over the accuracy of the CIAP model, the inference to be made is that SSTs have a negligible effect on atmospheric

ozone, a conclusion supported by other arguments. For instance, several hundred supersonic military aircraft have been flying every day for several decades in the ozone 'layer' without any detectable decrease in concentration which could be related to flights at those altitudes.

Another possible source of stratospheric pollution by SSTs arises from the discharge of water vapour. The influence of this on the greenhouse effect through

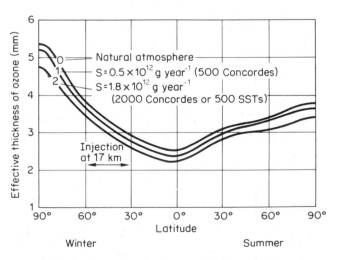

Figure 3.9 Influence of a fleet of SSTs on atmospheric ozone concentration. Curve 0 shows the variation with latitude in the effective thickness of atmospheric ozone; curve 1 shows the same with a fleet of 500 Concordes; curve 2 the same with 2000 Concordes or 500 SSTs. (From Brasseur and Bertin, in Cieslik, 1976)

absorption of infra-red radiation can probably be taken as insignificant, but there are still possible repercussions from condensation. At these altitudes, the water vapour discharged by jet engines produces condensation trails ('contrails') which turn into thin sheets of cirrus cloud. It is thus quite possible for SSTs to produce such formations at their cruising height of 20 km. Since the freezing point at this height for a water vapour concentration of 3 p.p.m. is −87 °C, the necessary conditions could be achieved several days each year on the most frequently used Western air lanes. It should be pointed out, however, that the conditions required for the formation of contrails are much more often met in the upper troposphere where subsonic aircraft fly.

Finally, we turn to the use of what are known as chlorofluorocarbons or Freons, of general formula $C_nCl_mF_p$, compounds which are used as refrigerants but most of all as propellant gases in aerosol sprays. Their domestic use has increased enormously over the past 10 years or so and nearly 2 million tonnes of the substances are released into the atmosphere each year: undoubtedly the form of human activity posing the greatest danger to atmospheric ozone.

Sooner or later, the released Freon gases reach the stratosphere, where they are dissociated photochemically with the production of atomic chlorine. This in its turn reacts with ozone to produce chlorine oxide and molecular oxygen (Figures 3.8 and 3.10). Several theoretical studies based on models of the ozone cycle agree that the use of Freons could reduce the ozone concentration quite significantly by the end of the century. A calculation based on the assumption that the increase in release of Freons will continue at the rate of 10 per cent per year until 1995 and then stop abruptly shows that the atmospheric concentration of ozone would be lowered by 7 per cent. Similarly, a rate of increase of 22 per cent per year until 1987 followed by no further production would diminish the concentration by 11 per cent (Figure 3.11). There are, however, several reservations to be made about such predictions since the amount of chlorine that finds its way into the atmosphere through natural processes compared with that injected by human activity is not known with any precision. In addition, there are biogenic sources of chlorinated organic compounds. Some authors have suggested that several million tonnes of methyl chloride (CH_3Cl) are released each year through two processes: the incomplete combustion of plants (in forest fires, for example) and the reaction between methyl iodide, produced by the metabolism of certain marine algae, and the chlorine ion in sea water (Dupas, 1975).

Nevertheless, in spite of these reservations, it would seem essential to be careful regarding uncontrolled release of Freons into the environment.

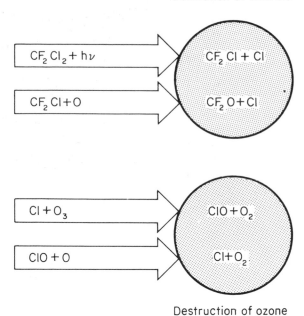

Figure 3.10   Interactions involving Freons and atmospheric ozone: atomic chlorine is formed and decomposes the ozone into oxygen and ClO

3.1.3   Perturbation of the sulphur cycle

The global sulphur cycle would presumably be in equilibrium in the absence of human activity: in other words, there would be a balance between the amounts dissolved in rivers and discharged into the atmosphere on the one hand and those which return to the surface of the land and which become sediments in the ocean depths, etc., on the other. In the real ecosphere, of all the biogenic elements sulphur is the one whose cycle has suffered most from human interference (Figure 3.12).

The sulphur cycle under natural conditions

The atmosphere receives sulphur in various forms from three natural biogeochemical processes:

(a) the formation of sea spray;
(b) anaerobic respiration in marshy areas, in the silt and sludge of river beds and in the sedimentary oozes of the ocean floor;
(c) volcanic activity.

The spray that is formed above rough seas carries sizeable quantities of sulphates into the atmosphere contained in very small water droplets. After the water has evaporated, these sulphates remain as tiny crystals:

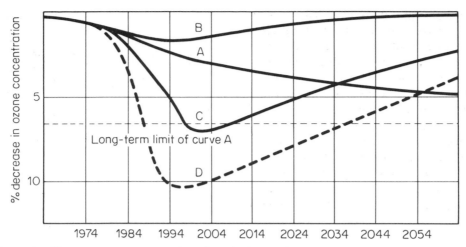

Figure 3.11   Curve A: effect on ozone concentration of continuing production of Freons at the 1972 level. Curves B and C: the same but assuming that the production increases by 10 per cent per year until 1978 and 1985 respectively and then stops. Curve D: the same but assuming that production increases at 22 per cent per year until 1987 and then stops. (From Wofsy *et al.*, 1975, p. 535)

the smaller they are the longer they stay suspended in the lower atmosphere. In all, some 44×10^6 tonnes of sulphur per year are transferred from the oceans to the atmosphere in this way.

The second source of atmospheric sulphur is the anaerobic respiration and fermentation of organic matter which can develop in wetlands, and especially in the muds at the bottom of fresh-water, lagoon and oceanic ecosystems. In these environments, sulphates are reduced by various anaerobic bacteria with the emission of dimethyl sulphide and, more abundantly, hydrogen sulphide (H_2S). Until recently, only two types of sulphur bacteria were known: *Desulphovibrio* and *Desulphomaculum*. In fact, there exists a great variety of micro-organisms which can act anaerobically not only to reduce sulphates but to oxidize all the series of organic acids from formic (C_1) to myristic (C_{14}) and even aromatic compounds like benzoates.

Sulphates can also be reduced by enzymes in a quite different type of mechanism involving intermediate substances. In one chain leading from sulphate to sulphide, for instance, five proteins are involved in the

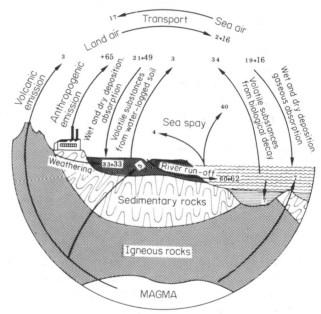

Figure 3.12   The sulphur cycle with rates of transfer expressed in 10^6 tonnes per year. The smaller figures are the transfers which take place in the natural cycle and the larger figures are those resulting from human activity. (From Granat *et al.*, 1976, p. 94)

Table 3.3   *Estimates of H_2S discharged into the atmosphere* (10^6 tonnes of equivalent sulphur per year)

	Terrestrial sources (water-logged land)	Marine sources (inc. tidal flats)	Total
Robinson and Robbins (1970)	68	30	98
Junge (1963)	70	160	230
Granat *et al.* (1976) (Figure 3.12)	3	34	37

reduction which can be simply expressed by:

$$SO_4^{--} + ATP \rightarrow APS^{--} \rightarrow SO_3^{--}$$
$$+ AMP \rightarrow S_3O_6^{--} \rightarrow S_2O_3^{--} \rightarrow S^{--}$$

In deep waters, H_2S is formed and reacts with iron to produce a precipitate of black ferrous sulphide.

In inland surface waters and tidal flats, on the other hand, the H_2S is discharged into the atmosphere but the global quantities emitted are still not known with any accuracy as the estimates in Table 3.3 show, the figures being quoted in 10^6 tonnes of equivalent sulphur per year.

Volcanic activity plays a more modest part and is estimated to discharge only some 3×10^6 tonnes of sulphur per year into the atmosphere in the form of various gaseous or particulate compounds.

Within the atmosphere itself, there is an exchange of air masses between continents and oceans which results in the transfer of 2×10^6 tonnes of sulphur per year from inland air towards the seas, and of 21 tonnes per year from above the seas towards the land, of which 4×10^6 tonnes is due to the formation of spray.

Turning from discharges of sulphur compounds *into* the atmosphere, we now consider the second component of the cycle: the removal of the compounds *from* the atmosphere through their return to inland and ocean surfaces. First of all, oxidation of atmospheric sulphur compounds transforms them back into sulphates which are then taken to land and sea surfaces by rain, snow, etc., either dissolved in the water droplets or as particles consisting of microcrystals of gypsum and so on. These processes, coupled with settling of the densest particles in a dry state and a certain amount of direct solution of gases in the oceans, lead to the return of some 21×10^6 tonnes of sulphur per year to continental surfaces and 19×10^6 tonnes to the oceans.

Another important component of the natural sulphur cycle is its transfer from the lithosphere to the oceans through erosion, leaching and run-off. However, the dissolving of sulphates out of surface rocks accounts for only about half of the concentration found in rivers, the rest being provided by the amounts returned from the atmosphere already mentioned.

Finally, a proportion of the inorganic sulphur contained in the hydrosphere, both in inland waters and in the oceans, is removed from the cycle in the form of insoluble sulphides, either temporarily or permanently through sedimentation. In the oozes and slimes of the ocean floors, the hydrogen sulphide formed anaerobically reacts with iron and other metallic ions to give sulphides which are responsible for the colours of the sea-beds, such as the black of ferrous sulphide. The total quantity of sulphur removed from the biosphere each year by marine sedimentation is still not known.

Perturbation of the sulphur cycle by human activity

Fossil fuel combustion has had a marked effect on the sulphur cycle. Coal frequently contains 1 per cent of the element but its concentration in some deposits can be as high as 5 per cent; heavy fuel oils used in industry can quite legally contain 2 per cent, but there may be more than 6 per cent if the sulphur is not removed. Such levels of impurity led to an estimated 65×10^6 tonnes of sulphur being discharged into the atmosphere as SO_2 in 1965 and by 1979 the figure had reached 100×10^6 tonnes annually.

The SO_2 comes into contact with atmospheric oxygen and water vapour, and is transformed first into sulphuric acid and then to sulphates, both of which are taken back to the Earth's surface by rain, snow, etc. The original SO_2 remains in the atmosphere for a mere 2–4 hours, a time which indicates how rapidly the oxidation proceeds. To understand the magnitude of the disturbance to the sulphur cycle produced by human activity, the 75×10^6 tonnes of sulphur from fossil fuel combustion that reach land surfaces annually as sulphates in rain and snow should be compared with the 42×10^6 tonnes carried by natural processes. A whole series of biogeochemical indices exist to prove the same point.

The most dramatic example of such an index is the great increase in the acidity of rainwater over the industrialized regions of the world. Whereas the rainfall over Europe and North America had a pH value close to neutrality[1] two centuries ago, its value is now consistently below 5.6. As early as 1968, Oden was attributing to SO_2 the continuous fall in the pH value of rainwater over the whole of Western Europe since the mid-1950s, and the same effect was observed over north-eastern USA (Likens and Bormann, 1974): in the mid-1970s, the pH value of rain and snow falling over New Hampshire was as low as 4.0, and that was in a forest zone nearly 1000 km from the highly industrialized East Coast areas. In New York State, the same authors found rainwater with a pH value of 3.0 and a record low of 2.1 was reached in November 1964. In Europe, rainwater with a pH of 2.4 was experienced in Scotland in April 1974: the lowest annual average pH in Europe (3.78) was reached at the end of the 1960s at De Bilt in Holland. Finally, in certain remote districts of Scandinavia the concentration of H^+ ions in rainfall increased by 200 times between 1956 and 1967.

There is no doubt whatsoever that these increases in acidity are related to the combustion of greater and greater quantities of sulphur-rich oil and coal: analysis carried out both in the USA and in Europe show that

---

1. The pH of water obtained from ice formed 180 years ago in the interior of Greenland varies from 6.0 to 7.6 according to the exact point from which it was taken.

SO_4^- ions often account for 50 per cent of the total anion concentration in acid rainfall. In fact, the connection has been recognizable since the middle of the century: the concentration of sulphates in rainfall over Japan at the end of the 1950s was already four times greater than that in 1946: over the same period, industrial emission of SO_2 had increased precisely 4.5 times.

Observations made between 1960 and 1965 in the Yellowstone National Park showed a ten-fold increase in the density of Aitken nuclei: these are particles less than 0.1 μm in size which take a very long time to settle and consist mainly of microcrystals of ammonium sulphate, the principal form in which SO_2 precipitates out from the upper atmosphere.

The increased acidity of rainwater in remote areas of North America and Europe has arisen from the transfer of masses of SO_2-polluted air from the industrialized regions of the respective continents. Oden estimated at the end of the 1960s that 70 per cent of the sulphur present in the atmosphere of southern Sweden came from the distant industrial regions of Great Britain and the Ruhr. Ironically, the technocratic approach to pollution, which consists in diluting or dispersing the contaminant to areas far from the point of emission, has transformed a local problem into a general one. The only method that has been tried for the reduction of pollution around industrial installations using sulphur-rich fuels has been the construction of higher and higher chimneys. The fumes are then injected into the air at such a height that they are carried away by general atmospheric circulation. Before 1955, only two chimneys in the whole of the USA were over 150 m in height: today, every chimney built for industrial use reaches or surpasses that figure. Over the whole world, some 15 chimneys over 300 m in height are built each year. Likens et al. (1979) quote the case of a pyritic cupro-nickel processing plant in Ontario where a chimney more than 400 m in height is responsible for 1 per cent of world SO_2 discharge! This alone has injected more sulphur into the atmosphere over the last decade than all the active volcanoes in the world!

What are the ecological consequences of disturbing the sulphur cycle in this way?[1] Among the immediate and most noticeable of them are the toxic effects on plant and animal physiology of the atmospheric pollution. No plant can grow in air that permanently contains more than 100 p.p.b. of SO_2, and great damage is done by chronic SO_2 pollution to cultivated areas and forests. Conifers and lichens are particularly sensitive: the limit above which the growth of pines is affected has been put by some Swedish authors at 20 p.p.b. of

SO_2 (which incidentally makes them excellent biological indicators of atmospheric pollution).

Sulphur dioxide in the air also affects the health of domestic animals and humans. Its harmful effects on the respiratory channels and pulmonary alveoli make it partially responsible for the disquieting increase in chronic bronchitis through synergism with tobacco smoke (see Figure 2.31, p. 57).

Other consequences show up as disturbances of entire ecosystems. This is particularly the case with fresh-water environments established over substrates of ancient acidic rocks such as granites or quartzites. These rocks confer a natural acidity on the inland waters they support and the resultant poverty in alkaline elements means that there is no buffer against increases in acidity produced by rainfall. There is a rapid lowering of the pH value and a considerable increase in the amount of sulphates in areas of acid rainfall. For example, the mean concentration of SO_4^- ions in the inland waters of Connecticut increased from 0.25 mg per litre in 1937 to 7.0 mg per litre in 1963. Again, a continual decrease in pH value has been observed since the mid-1960s in European and US lakes established over substrates of acidic rocks. A large number of Canadian lakes situated under prevailing winds from metallurgical plant emitting SO_2 have experienced a 100-fold increase in their H^+ ion concentration over the last decade: more than 30 lakes of the Canadian east had a pH lower than 4.5, a value below which there is serious ecological damage to the communities populating such oligotrophic waters.

The lowering of the pH value in these inland waters diminishes the diversity and the primary productivity of phytoplankton. The resulting scarcity of food supplies for the consumers in the food web and the toxic effects of the acidity on invertebrates and fish combine to reduce the secondary productivity. Thousands of lakes in the south of the Scandinavian peninsula, occupying more than 50 000 km² of territory, have already suffered a great depletion in their fish stocks. In the coastal hill torrents of Norway, too, great reductions in the salmon population have been reported during springtime when the melted snow arrives loaded with sulphate ions.

There is a fear that this already serious problem of atmospheric pollution by SO_2 will get worse if heavy, and even light, fuels are increasingly used without the removal of sulphur and if industrialized countries turn massively to sulphur-rich coal as a replacement for increasingly expensive oil. The whole problem is a prime illustration of the short-term attitude towards environmental protection taken by developed countries. Techniques already exist for both the efficient removal of sulphur from oil and coal as well as the production of clean fuel from coal to eliminate its direct combustion in furnaces and fireplaces (Squire,

1. For more details on this topic, see Ramade, 1979a, pp. 134ff.

1972). Even if the use of such techniques were to increase the price of fuels a little, the ecological advantages that would follow from a reduction of pollution would amply compensate for the small extra cost of manufacture.

3.2   Climatic Changes

The socio-economic and political consequences of any climatic changes would have great significance for the whole of humanity. It was no accident that the CIAP, at the request of the US State Department, produced a report in 1977 entitled *Trends in World Population, Food Production, and Climate* whose publication aroused considerable concern in international circles. It was, after all, during the 1970s that various unexpected climatic fluctuations had occurred: droughts in regions bordering the Sahara and then in Western Europe (1976), and the 'cold spell of the century' in eastern and central North America (1977). These had made it clear how much not only agricultural production but industrial activity, too, could be seriously affected by anomalous climatic behaviour, even if quite limited in duration and extent on a planetary scale, as the events of the 1970s had been.

The prediction of future climatic changes has thus become of prime importance because climatic parameters, in their role as limiting ecological factors (Table 1.1), control the equilibrium of the biosphere. However, making such predictions needs a profound appreciation of meteorological processes, a knowledge of past variations in terrestrial temperatures and of climates over the ages ('palaeoclimates') and, finally, a complete understanding of the way both natural factors (solar and volcanic activity) and artificial ones (human activities) influence the Earth's surface temperatures. A mere statement of these requirements is enough to show how imprecise this area of scientific activity must still be: until they can be satisfactorily resolved, all conclusions about the future evolution of climates will be subject to large uncertainties.

3.2.1   The evolution of terrestrial climates

Large variations in terrestrial climates have occurred right from the beginning of the Tertiary period until the present day. Geologists long ago demonstrated the existence of a worldwide tropical climate during the Eocene period, when palm trees grew in Alaska and banana trees in Provence, while the great glaciations of the Quaternary were revealed during the eighteenth century. However, it was not until recently that isotopic techniques provided a means of estimating relative, or in some cases absolute, values for the temperature during past geological periods.

The isotopic 'thermometer'

A method of estimating temperatures during past geological periods ('palaeotemperatures') was developed during the 1960s. It is based on the way oxygen isotopes are distributed at various stages of the water cycle: between water itself and carbonates deposited from it in one variation of the method, and between water vapour and snow in another. Several excellent descriptions of the principles of these techniques and of the results obtained from them have been published (see, for example, Lorius and Duplessy, 1977), and I shall restrict myself to a summary of the basis of the method.

Among the isotopes of oxygen, ^{16}O is by far the most abundant (99.76 per cent) with ^{18}O a long way behind (0.2 per cent), but the exact proportions (obtained by mass spectrometry) may vary a little from one substance to another. In particular, the $^{18}O/^{16}O$ ratio in specimens of fossil carbonate and samples of ice, snow and water have been found to vary with temperature.

The carbonate that precipitates in the shell around a marine organism does not have the same $^{18}O/^{16}O$ ratio as the water from which it came, and the difference is found to vary linearly with the temperature at the time of deposition as shown in Figure 3.13, curve A. Over a restricted range of about 20 °C, the following relationship can be established experimentally:

$$t = 16.9 - 4.2(\delta^{18}O_{carbonate} - \delta^{18}O_{water}) \qquad (1)$$

where

$$\delta^{18}O = \left[\frac{(^{18}O/^{16}O)_{sample}}{(^{18}O/^{16}O)_{reference}} - 1 \right] \times 1000 \qquad (2)$$

and where t is the 'palaeotemperature' in °C giving an estimate of the mean oceanic temperature at the time when the fossil was originally formed.

In a similar way, it is found that the $\delta^{18}O$ of snow increases linearly with the temperature of the Earth's surface at the time of its precipitation, as shown in Figure 3.13, curve B.

Climatic variations since the beginning of the Tertiary period

With the aid of the 'carbonate-water' and 'snow' isotopic thermometers it has proved possible to trace the variations in the Earth's climate from the beginning of the Tertiary period until the present day with a high degree of precision. The results have been obtained by a study of fossils contained in core samples taken from the depths of various oceans (for example, during the *Glomar Challenger* expedition) and of core samples of ice more than a kilometre in length taken from Greenland and antarctic glaciers.

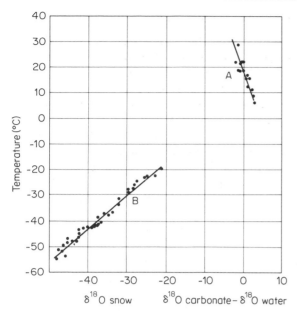

Figure 3.13 The relationship between ¹⁸O/¹⁶O ratios and the temperature of the surroundings. For temperatures above 0 °C, the index used is the difference between the isotopic composition of a carbonate and the water from which it was precipitated (curve A). For temperatures below 0 °C, the δ¹⁸O of snow is used (curve B). In both cases, the index varies linearly with the temperature at the time of deposition or precipitation. (From Lorius and Duplessy, 1977, p. 951)

Such studies have shown that the mean temperature of the Earth decreased almost continuously during the Tertiary era. To take one example: surface layers of the seas to the south of New Zealand were approximately at 20 °C at the beginning of the Tertiary, or nearly 10 degrees higher than at present! From that time, the global temperature began slowly to decrease until about 38 million years ago when the first abrupt climatic change occurred, marked by the appearance of sea ice round the shores of Antarctica (Figure 3.14) as a result of the sharp fall in temperature in those regions. A slight warming up then ensued until the Middle Miocene period, when another abrupt fall in temperature took place and led to formation of the Antarctic ice field some 14 million years ago. Finally, another fall at the end of the Pliocene period about 2.5 million years ago caused the formation of the ice fields of the Northern Hemisphere.

The last 2 million years (in other words, the whole of the Quaternary period) have been covered by studying other sedimentary core samples, this time containing the pelagic foraminifera *Globigerinoides sacculifer*, taken from the Equatorial regions of the Pacific. This is a species which lived in the upper layers of the ocean, to a depth of 50 m, and flourished in tropical waters, but since the isotopic curves of ¹⁸O concentration for

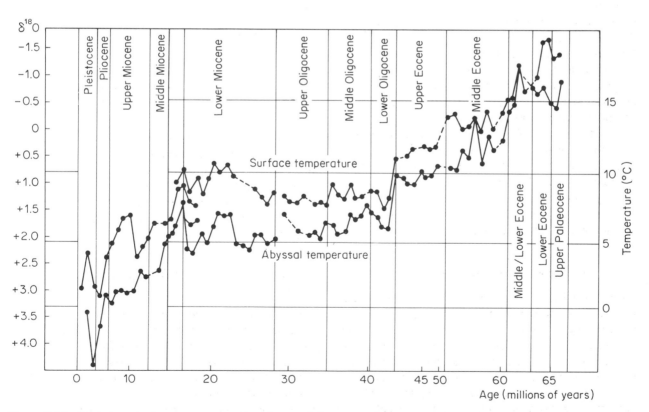

Figure 3.14 Variations in the temperature of surface and deep oceanic waters to the south of New Zealand during the Tertiary period, obtained from measurements of the δ¹⁸O of pelagic and benthic foraminifera shells. These results have made it possible to reconstruct the climatic history of the Earth during the last 70 million years. (From Shackleton *et al.*, in Lorius and Duplessy, 1977)

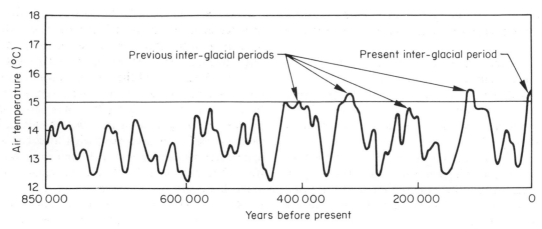

Figure 3.15   Fluctuations in the mean temperature of the Earth's surface during the last million years (Pleistocene). Some 15 well-marked glacial periods can be seen, together with several cool periods of less importance. Four inter-glacial periods comparable with the present one have occurred over the past 500 000 years. (From various sources in Roberts and Lansford, 1979)

the various oceanic regions are quite close to each other this enables global variations in temperature to be correlated. In this way, more than 20 glacial stages have been identified over the past 2 million years, including the four most recent: the Gunz, Mindel, Riss and Würm stages.

The same technique using deep core samples from Greenland (Camp Century) and Antarctica (Byrd and Vostock) has led to a precise evaluation of the climate over the past 100 000 years. These samples have a stratification related to annual cycles so that exact dating is possible. This has enabled the validity of isotopic methods to be confirmed, particularly through the remarkable agreement between results from the Arctic and the Antarctic.

This work has shown that the Würm stage lasted approximately 50 000 years, reached maximum glaciation some 18 000 years ago (a fact already deduced from stratigraphy), and ended abruptly 12 000 years

ago (Figure 3.15). Since then, the climate in the polar regions has been quite stable, but about 5000 years ago there was a modest peak in the temperatures at middle and low latitudes when they were between 1 and 2 °C higher than today. Smaller fluctuations have occurred from that time, but in general the tendency has been that of a continuous but slow cooling. The last 2000 years have been characterized first of all by a period of warmer temperatures that reached its climax between AD 800 and 1200, during which time the Norsemen discovered Greenland (hence its name!). This was followed by a colder period lasting until the seventeenth century and clearly shown in the isotopic profiles from Camp Century. Subsequently, a distinctly warmer period occurred in the nineteenth century and this lasted until it reached its maximum between 1935 and 1945. After that, temperatures have had a tendency to decrease slightly. Broecker (1975) showed that the $\delta^{18}O$ fluctuations over the last 1000 years

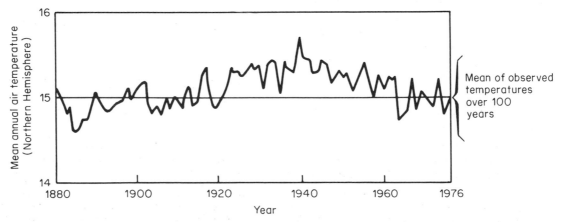

Figure 3.16   Fluctuations in the mean temperature at the Earth's surface in the Northern Hemisphere over the last 100 years. (From National Center for Atmospheric Research and US National Academy of Sciences, in Roberts and Lansford, 1979)

observed at Camp Century could be fitted to a combination of two sinusoidal variations having periods of 80 and 150 years.

Meteorological records show that between 1880 and 1940 the mean annual temperature of the atmosphere between 0 and 60°N rose slightly but significantly by a total of 0.6 °C (Figure 3.16). It then fell by 0.25 °C until 1970 but at the moment the climate appears to be growing slightly milder again.

Damon and Kunen (1976) show that there has been a very clear tendency for temperatures to rise over the Southern Hemisphere since the beginning of the 1960s and they see this as the forerunner of a general rise over the whole globe.

3.2.2 Causes of climatic fluctuations

Long-term climatic changes can be accounted for by cosmic factors like solar activity and the Earth's orientation and location in space, whereas short-term fluctuations of small amplitude can be explained by geophysical phenomena that are either natural (like volcanic activity) or of human origin (like pollution).

Cosmic factors

We turn first to long-term changes in which the intensity of the solar radiation reaching the upper atmosphere is the dominant factor. This is determined both by the Earth's position and orientation in space and by intrinsic solar activity. In connection with the former factors, Hays *et al.* (1976) studied climatic variations over the last 500 000 years by analysing three core-sample profiles: (a) the oxygen isotopic composition $\delta^{18}O$ of planktonic foraminifera (which varies with the disappearance and reappearance of the arctic ice cap); (b) a statistical analysis of radiolarian assemblages (indicators of sea-surface temperatures in the Southern Hemisphere); (c) the relative abundance of a radiolarian species *Cycladophora davisiana* (providing an index of sea-surface structure). As a result of these investigations, the authors showed that climatic changes were in the main a combination of three cyclic variations with periods of 23 000 years, 42 000 years and 100 000 years. These correspond very closely with the periods of (a) the precession of the equinoxes, (b) the obliquity of the Earth's axis to the plane of the ecliptic, and (c) the eccentricity of the Earth's orbit (that is, its degree of ellipticity). It was concluded that these phenomena are the fundamental reasons for the whole succession of glaciations during the Quaternary period: that it is the resultant changes in the Earth's spatial position and orientation that affect the intensity of the solar radiation reaching its surface.

However, it is only intrinsic variations in solar activity that can modify the solar flux sufficiently to produce general climatic effects over time scales commensurate with whole geological periods.[1] Thus the large fluctuations in temperature that have occurred since the beginning of the Tertiary period, along with the accompanying glaciations, must have been caused by such variations. Eddy (1977) has shown that the number of sunspots, a figure related to the activity of the sun itself, can be correlated with changes in the concentration of ^{14}C in the growth rings of ancient trees. The variations in ^{14}C concentration turn out to coincide exactly with climatic changes since the end of the Chalcolithic period. Maximum solar activity corresponds to maximum production of ^{14}C and coincides with warm periods, whereas solar quiescence is associated with minimum ^{14}C production and periods with severe winters (Figure 3.17).

Geophysical factors

The physical and chemical properties of the atmosphere govern the amount of incident solar radiation reaching the surface of the globe and thus become responsible for some of the climatic variations. The principal factors of this type are volcanic activity, the degree of cloud cover, any features that modify the albedo of the Earth's surface, and the concentration of CO_2 and other gases capable of absorbing radiation: all these, in varying degrees, are capable of modifying climate.

The intensity of solar radiation at the Earth's surface is given by

$$\Phi = \Phi_0 \, e^{-(\alpha_a + \alpha_d + \alpha_s)m}$$

where α_a is the coefficient of absorption of radiation by the atmosphere

α_d is the coefficient of scattering

α_s is the coefficient of reflection by the atmosphere

m is the cosec of the angle between the direction of the sun and the zenith

Φ_0 is the intensity at the upper limit of the atmosphere.

---

1. In fact, there is a varying density of interstellar gas and dust in the region of space through which the solar system moves and this is capable of affecting solar flux by absorption. Sizeable climatic changes can be produced in this way without significant variations in intrinsic solar activity. The solar system performs a complete revolution about the galactic centre once in 250 million years. At the moment it is crossing the spiral arm of Orion where there is undoubtedly some concentration of interstellar gas. This is accumulated by the sun to form a gaseous cloud around it, thus absorbing solar radiation and causing a cooling of the Earth. Since the crossing of the spiral arm takes about 50 000 years, a time which coincides with the mean duration of a period of glaciation, this could clearly provide an explanation of the recurrence of such glaciations over the principal geological periods. (See Acker, 1979, p. 32.)

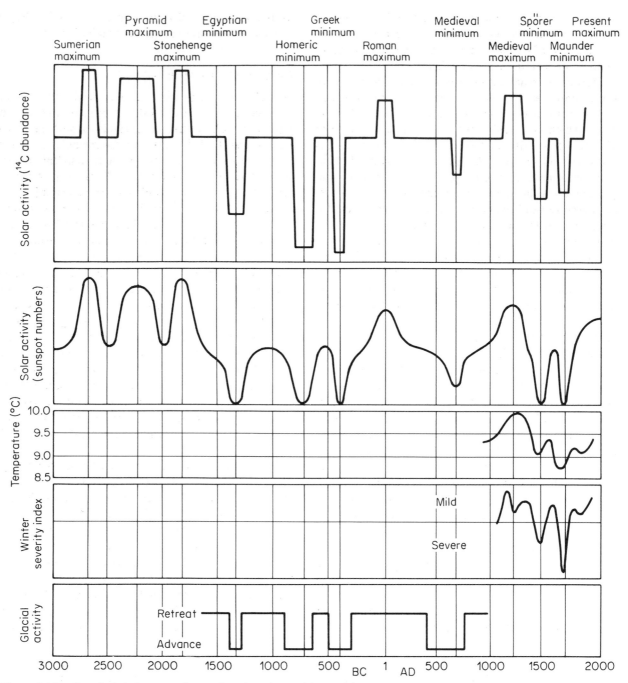

Figure 3.17 Correlations between fluctuations in solar activity and in terrestrial temperatures during the last 5000 years. Since there is a direct connection between solar activity and the atmospheric concentration of ^{14}C, isotopic analysis of carbon coupled with dating by the growth-rings of trees allows the climatic variations to be reconstructed. The minima in solar activity, marked by the scarcity or absence of sunspots, are clearly correlated with the severity of winters or the periods of glacial advance. Conversely, the periods of great solar activity coincide with those of warmer climates. (From Eddy, 1977, p. 88)

The amount of atmospheric turbulence and variations in the concentrations of gases both affect the value of the α coefficients and can thus have a great influence on climatic characteristics, but for a long time it has been volcanic activity that has been held responsible for abnormal reductions in temperature revealed by meteorological records. As long ago as 1784, Benjamin Franklin was explaining the exceptional severity of the previous winter by appealing to an excess of dust in the upper atmosphere, and it is well known that such dust can arise from the considerable quantities of ash thrown up by volcanic eruptions: ash that contains a high proportion of particles that are so small that they remain in the stratosphere for a long time without settling. In the case of first magnitude eruptions, even the lower layers of the mesosphere can be contaminated and all this, of course, leads to an overall lowering of temperatures at the Earth's surface.

During the nineteenth century, two enormous eruptions confirmed Franklin's hypothesis. The explosion of Tambora in the island of Soembawa near Java in 1815, which projected into the atmosphere more than 10^{11} tonnes of solid matter, ash and debris of various sorts, was the cause of the 'summerless year' of 1816. It snowed in June on the east coast of the USA and continuous white frost during the whole of August destroyed most of the harvest. The dust soon encircled the globe, so that Europe also experienced an abnormally cold summer. Temperatures recorded in England and Switzerland show a freezing July, as bad as the previous month in the USA: for example, the mean temperature for the month in Geneva was the lowest that has ever been recorded from 1753 to the present day. In France, the harvests were so catastrophic that massive imports of cereals were essential during the winter of 1816–1817.

The eruption of Krakatoa in 1883 injected more than 10^{10} tonnes of volcanic material into the atmosphere, of which 50×10^6 tonnes were in the form of dust which reached the stratosphere. This event produced a fall in temperature of nearly 0.5 °C over the whole of the Earth's surface (Figure 3.18).

More recently, the effects of the explosion of Agung in Bali in 1964 were carefully monitored, thanks to a network of meteorological observation balloons stationed above Port Hedland in Western Australia on the occasion of the International Geophysical Year. This eruption lowered the temperature in the troposphere but raised it simultaneously in the stratosphere because of the increased absorption of heat radiation by particles at that height.

In general, then, there is quite good correlation between small climatic fluctuations in the form of cold spells and periods of intense volcanic activity. Figure 3.18 illustrates this over the last hundred years.

The effects of human activity

Today's technology can influence the Earth's climate by its effect on the geophysical factors that govern the amount of solar radiation reaching the surface. Opencast mining, both for coal and for metallic ores; large public works involving excavation and shifting of soil; the cultivation of land with a fragile soil structure exposed to erosion by winds—all these can inject into the atmosphere considerable quantities of particulate matter which then behaves like volcanic ash in absorbing some of the solar radiation before it reaches the surface and reducing the temperature.

Particles like this do not, however, only act directly in screening the Earth's surface from radiation. They also act indirectly by becoming nuclei for the condensation of water vapour, thus increasing the amount of cloud cover. The condensation trails of aircraft in the upper atmosphere (Figure 3.19) form a radiation screen because of the creation of cirrus clouds consisting of small ice crystals located in the troposphere and lower stratosphere. The opacity of the atmosphere along the most crowded of the air corridors in the Northern Hemisphere can be significantly increased in this way.

There are, however, consequences of technology which produce the opposite effect: a general warming up resulting from CO_2 injected into the atmosphere by the combustion of fossil fuels, which is continuously increasing in spite of the so-called energy crisis.

The atmosphere is not completely transparent either to direct solar radiation or to that reflected from the Earth's surface. If it were, or if the atmosphere were completely absent, the Earth's surface could be treated as a thermodynamic black body. If that were so, it has been estimated that the surface temperature, T_e say,

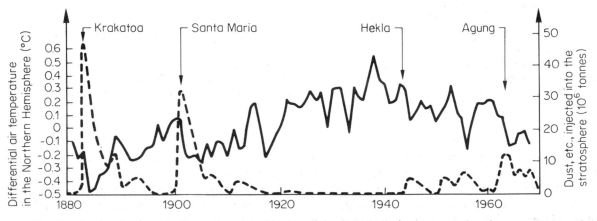

Figure 3.18   Correlation between the fluctuations in volcanic activity (broken line) and those of surface temperatures in the northern hemisphere (continuous line) during the last 100 years. Arrows indicate the major volcanic eruptions. (Adapted from various authors in Ramade, 1978a)

Figure 3.19   Condensation trails produced by aircraft flying high in the troposphere. The water vapour ejected by the jet engines condenses into tiny ice crystals which are turned into cirrus clouds by the high winds at those heights. (Photograph F. Ramade)

would be significantly lower than its actual value, T_s. On the average T_s is 13 °C or 286 K, while T_e would be only 253 K. The ratio T_s/T_e, where both are expressed in K, is called the greenhouse coefficient and has a value of 1.13. The greenhouse effect, which produces the higher temperature, is the result of the various gases in the atmosphere which absorb infra-red radiation: in particular, CO_2, water vapour and ozone. The term comes from the fact that the atmosphere is behaving like the glass of a greenhouse: during the day, solar radiation warms up the Earth's surface, which then reradiates energy outwards, principally in the infra-red. This will be trapped by the lower layers of the troposphere which are then raised to a temperature distinctly above what it would be with a transparent atmosphere.

The future evolution of world climates

Long-term predictions of climatic changes are very difficult to make because good theoretical models of the atmosphere are still not available and because there are large uncertainties in the estimated quantities of CO_2 and dust that will be produced by human activity. Moreover, future fluctuations in volcanic and solar activity cannot be foreseen and it is therefore difficult to evaluate their effects.

However, taking a still longer time scale and leaving aside the possible climatic influences of human society which are hardly in dispute nowadays, the general tendency is towards a downward drift of temperature (see above, p. 86). A study of past climatic variations indicates that the mean global temperatures experienced over the past 100 years are abnormally high, just as are those which have prevailed over the Earth since the end of the Würm glaciation. Warm climatic periods comparable with those of the past 10 000 years are not the norm: isotopic profiles show that they prevail for at most 10 per cent of the time over hundreds of thousands of years.

Core-samples of ice have also indicated a small but significant fall of temperature since the last climatic maximum and this is corroborated by the slow but continuous drop in sea levels since that period (Figure 3.20).

While there does seem to be, then, a general agreement about long-term cooling, climatologists are very divided in their opinions as to the variations to be expected over the next few decades. Estimates both of the natural tendencies in the climate and of the nature and size of human influences have varied widely. Mitchell (1970) and Bryson and Wendland (1970) consider that the fall in temperature which began at the end of the 1950s will continue until the beginning of the

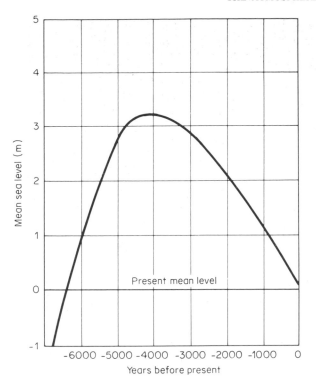

Figure 3.20   Variation in the mean sea level over the last 7000 years. (From J. Labeyrie, 1976)

pogenic factor capable of having a global influence on climate is the atmospheric concentration of CO_2: according to his analysis, the quantities of dust, etc., emitted by human activities are still too small in relation to those ejected by volcanoes to contribute to a fall in temperature. He therefore produced a theoretical model incorporating both the natural fluctuations of temperature found in the Camp Century profiles and the expected contribution from the greenhouse effect. Table 3.4 shows his estimates of the past and predicted growth of atmospheric CO_2 concentration from 1900 to 2010, together with the variation in the mean temperature of the Earth's surface that results from it. His predictions were based on the data of Manabe and Wetherald (1967), which established that a doubling of the atmospheric concentration of CO_2 would raise the temperature of the Earth by 2.4 °C. In fact, the greenhouse effect due to CO_2 is not a linear function of the atmospheric concentration but is proportional to its logarithm (Rasool and Schneider, 1971). Thus, a 10 per cent increase in CO_2 concentration is enough to produce a mean rise in global temperatures of about 0.32 °C.

Finally, Broecker predicted that the next four decades would see a general rise in temperatures greater than that observed at any time during the last 1000 years, reaching a value of 1.1 °C for the whole globe (Figure 3.21). This would imply a rise of several degrees at the higher latitudes in view of the magnification of climatic change that has been observed to occur as one moves towards the polar regions.

In the longer term, another anthropogenic factor that is at present considered negligible could cause a general rise in terrestrial temperatures with possibly catastrophic consequences: this is the environmental thermal pollution caused by conversion of fossil and nuclear fuels into heat.

World energy production in 1976 amounted to an

next century. Other experts, on the other hand, think that the tendency will be towards warmer climates and that this will be magnified by discharges of CO_2 into the atmosphere. Damon and Kunen (1976), for instance, see the general rise in the mean temperature of the Southern Hemisphere as a preliminary sign of a global warming up.

Broecker (1975), after studying the isotopic profiles from Camp Century, considers that the present influence of natural factors is acting to raise world temperatures. He also shows that the only anthro-

Table 3.4   *Past and predicted atmospheric CO_2 content and the resultant rise in temperature (based on fuel consumption data)* (from Broecker, 1975)

Year	Cumulative CO_2 from fossil fuels (10^9 tonnes)	Excess atmospheric CO_2			CO_2 concentration (p.p.m.)	Global temperature increase (°C)
		(10^9 t)	(%)	(p.p.m.)		
1900	38	19	0.9	2	295	0.02
1910	38	19	0.9	2	295	0.02
1920	97	48	2.2	6	299	0.07
1930	136	68	3.1	9	302	0.09
1940	179	68	4.1	12	305	0.11
1950	233	116	5.3	16	309	0.15
1960	312	156	7.2	21	314	0.21
1970	440	220	10.2	29	322	0.29
1980	630	310	14.0	42	335	0.42
1990	880	440	20.0	58	351	0.58
2000	1210	600	28.0	80	373	0.80
2010	1670	830	38.0	110	403	1.10

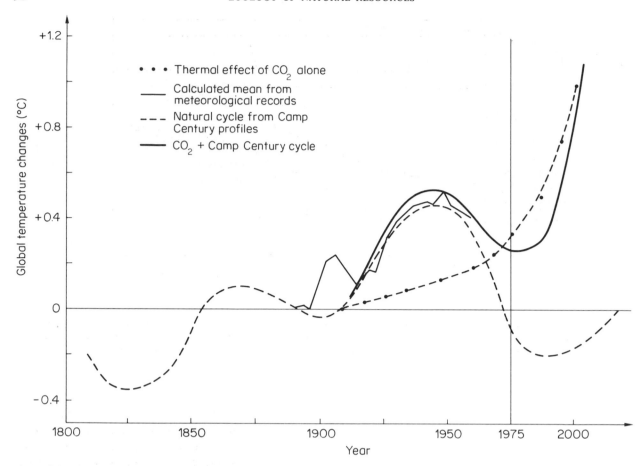

Figure 3.21   Predictions of the global temperature changes due to CO_2 from fossil fuel combustion, due to the natural climatic cycle, and due to the sum of the two. Temperatures from meteorological records over the last 100 years are given for comparison. (From Broecker, 1975, p. 461)

'installed power' (to use a term from the electrical industry) of 1.05×10^{13} W. In comparison, the total contribution from solar radiation to the ecosphere is 1.73×10^{17} W, so that the release of energy into the environment due to technological development is 1/16 500 of the solar flux. If it is assumed that energy production will grow at an average rate of 4 per cent per year (distinctly less than that observed up to 1973), then the human input of energy would reach 1 per cent of the solar input in 130 years, 2 per cent in 148 years and would be equal to it in 250 years. On the other hand, if the rate of growth itself went on increasing at the same level as it did in the 1960s, then the human contribution would be equal to that of the sun by the middle of the twenty-first century (Figure 3.22).

What climatic effects could be expected from such inputs of energy to the ecosphere? To answer that, we need to calculate the rise in global temperature that would result from a given increment of energy added to that already received from natural sources. The main source of natural energy is that received from solar radiation, which contributes 8.8×10^{20} kcal per year. Geothermal energy carried from the centre of the Earth to its surface must be added to that, but with a value of

2.1×10^{17} kcal per year it is only 1/4200 of the solar flux and so makes a negligible contribution.

What fraction of this natural energy is represented by the input from industrial civilization? In 1970, the combustion of fossil fuels released 5×10^{16} kcal into the ecosphere: that is, 1/18 000 of the solar flux. If we assume a 4 per cent annual increase in energy consumption over the next 100 years, the human contribution would amount to 1.7×10^{17} kcal per year in the year 2000 and 1.2×10^{18} kcal per year in 2050.

The Stefan-Boltzmann law[1] enables us to calculate, to a first approximation, the effect of this energy production on the equilibrium temperature of the Earth. For that purpose, we simplify matters by assuming that the anthropogenic heat distributes itself uniformly over the Earth's surface and that the radiative equilibrium of the ecosphere is quickly reestablished after any external input of energy.

Let T be the equilibrium temperature of the Earth's surface, Q the annual supply of heat from solar radia-

---

1. I should like to thank M. J. Langevin for the paper he so kindly sent me dealing with the problem of the thermal pollution of the ecosphere.

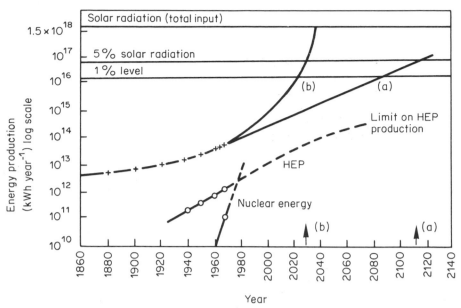

Figure 3.22   Influence of environmental thermal pollution on the overall energy balance of the ecosphere. The development of hydroelectric power (HEP) is given for comparison. Curve (a) assumes an exponential growth rate with a doubling period of 14 years. Curve (b) assumes that the growth rate itself increases as it did during the 1960s, leading to a steadily decreasing doubling period. (From Chapman, 1970, p. 636)

tion, and q the annual supply from the use of fossil fuels. If the resultant increase in the equilibrium temperature due to q is t, then the Stefan-Boltzmann law gives

$$\frac{Q+q}{Q}=\left(\frac{T+t}{T}\right)^4$$

Provided that t is much less than T and q is much less than Q, this gives

$$t=\frac{T}{4Q}q \qquad (1)$$

In other words, the percentage change in temperature is a quarter of the percentage change in energy input. From equation (1), it is possible to estimate that, other things being equal, the energy input from industrial sources would raise the mean temperature of the planet 0.014 °C by the year 2000 and 0.1 °C by 2050.

Maintaining the level of energy growth at 4 per cent per year would, in spite of the relatively low rate, make all human life impossible by the middle of the twenty-second century because of the increase in the equilibrium temperature of the Earth. At such a rate, technology would contribute 1 per cent as much energy to the ecosphere as the sun by 2082, and 5 per cent as much by 2120. It is worth noting, as pointed out by Chapman (1970), that the increase in temperature produced by a 5 per cent increase in energy input is as high as any that occurred during the climatic variations of the Pleistocene.

If the rate of energy growth itself were to increase as

it has done over recent years, the extra heat input would be equal to the contribution from the sun by the end of the next century. Because of this, Weinberg (1974) justifiably suggests that humanity has at most 50 years in which to look seriously at the limitations imposed by climatic changes due to thermal pollution.

In the real world, however, thermal pollution docs not influence the climate in isolation but acts in conjunction with the effects produced by atmospheric CO_2. The combination of increases in the greenhouse effect and in the discharge of heat of technological origin could produce quite drastic climatic changes. Meteorological records show that small changes in temperature can produce sizeable effects: the dramatic glacial retreat observed between 1880 and 1964 in tropical mountain ranges and in the temperate zones of Europe and North America was due to a mere 0.5 °C rise in the mean terrestrial temperature, for example.

According to Chapman (1970), the critical level of additional heat input is 1 per cent of that from solar radiation, taking into account the fact that thermal pollution is not uniformly distributed but is localized over certain continental regions of the Northern Hemisphere. A rise in terrestrial temperatures would produce a number of climatic consequences, all unfavourable, and associated mainly with an extension of the desert regions and with a partial or complete melting of the polar ice caps. Some authors, including Chapman (1970), have suggested that an addition of heat to the ecosphere amounting to 5 per cent of that contributed by solar radiation could raise temperatures to a level comparable with those of the Pleistocene.

Such a degree of thermal pollution combined with the release of large quantities of CO_2 into the atmosphere could raise mean terrestrial temperatures by 10 °C. This would recreate conditions reigning at the beginning of the Tertiary period with total melting of the polar ice caps and a rise of some 80 m in mean sea level. Vast areas of coastal plains would be flooded, together with the majority of the world's largest cities, including London, Paris, New York, Hamburg, Leningrad and so on, as well as practically the whole of the Netherlands and a high proportion of the Ganges and Amazon deltas.

Even without the assumption of such a large change, which implies a several-hundred-fold increase in human energy consumption, it is still essential to be aware that a rise of only 1 °C in mean global temperatures would have dramatic effects on the agriculture of many Third World countries.

In conclusion, the risk of worldwide climatic disruption over the next hundred years can only be averted by halting the growth in energy production for technological processes, and by a continuous reduction in the use of fossil fuels accompanied by a development of natural energy sources, particularly of solar energy.

Chapter 4

The Hydrosphere

Water is not only essential to life but is the predominant inorganic constituent of living matter, forming in general nearly three-quarters of the weight of a living cell. It makes up some 65 per cent of the body weight of an adult human and can form as much as 98 per cent of the mass of certain jellyfish. Organisms which contain relatively small amounts of water are generally in a dormant state or one of very slow development: seeds and certain invertebrates that live in arid environments are examples. On the other hand, high rainfall over a land mass invariably means a large biomass per unit area.

Of the three states in which water occurs in the ecosphere—gas, solid and liquid—only the last is an indispensable resource as far as human activity is concerned. The various forms of water are found in every section of the ecosphere: the atmosphere, the lithosphere and the hydrosphere (a term designating the entire system of seas and inland waters). However, in spite of the enormous quantity of the substance that exists, only a small proportion of it is actually usable by human beings (Table 4.1). The oceans alone constitute 97 per cent of the hydrosphere, and the polar ice caps, ice fields and glaciers make up another 2 per cent in the form of ice. Inland surface waters (lakes and rivers) account for barely 0.02 per cent.

Table 4.1 also demonstrates the very small quantity of water occurring as vapour in the atmosphere: if it were all condensed into a liquid and spread uniformly over the Earth's surface, it would form a thin film only 3 cm thick. Nevertheless, tiny as this amount is in comparison with the rest of the hydrosphere, the biogeochemical water cycle depends on it. In reality, of course, atmospheric water vapour is very unevenly

Table 4.1 *Masses and relative proportions of the various forms of water in the hydrosphere* (from Baumgartner and Reichel, 1979)

Form	Mass (10^{12} t)	Relative proportion (%)
Oceans	1 348 000	97.39
Polar ice caps, glaciers, etc.	27 820	2.01
Underground water, soil moisture	8 062	0.58
Lakes and rivers	225	0.02
Atmospheric water vapour	13	0.001
Total hydrosphere	1 384 120	100
Of which, fresh water is	36 020	2.6

Distribution of fresh water in the hydrosphere (%)

Polar ice, glaciers, etc.	77.23	
Underground water (<800 m)	9.86	
Underground water (between 800 m and 4 km)	12.35	
Soil moisture	0.17	
Fresh-water lakes	0.35	
Rivers and other waterways	0.003	
Hydrated minerals	0.001	
Biomass	0.04	
Total	100.00	

distributed and this produces large variations in rainfall from region to region, but even if it were uniformly distributed as regards longitude, the mean quantities condensed would be 2.5 mm at the North Pole, 10 mm at 45°N, 45 mm at the Equator, 20 mm at 45°S and 8 mm at the South Pole.

4.1 The Biogeochemical Water Cycle

The atmospheric humidity mentioned above arises from the evaporation of water from large liquid surfaces exposed to solar radiation. The subsequent movement of great masses of humid air causes the water vapour to condense into clouds and these, on cooling, return the water to the Earth's surface in the form of rain, snow, hail and other types of precipitation. Of the

total precipitation, 77 per cent falls on ocean surfaces and only 22.8 per cent on land, even though the land constitutes 29 per cent of the surface of the globe.

The water that falls on the surface of the land follows several different paths:

(a) infiltration,
(b) evaporation and/or evapotranspiration,
(c) run-off.

The first of these plays a vital role in land-based ecosystems because it ensures that the soil is kept moist by the storage of water, especially in the superficial layers rich in humus and living matter. In forests, for example, a mossy layer can be an important factor in retaining rainwater and controlling the hydrological cycle. A kilogramme of dry moss can absorb several litres of water and return it gradually to

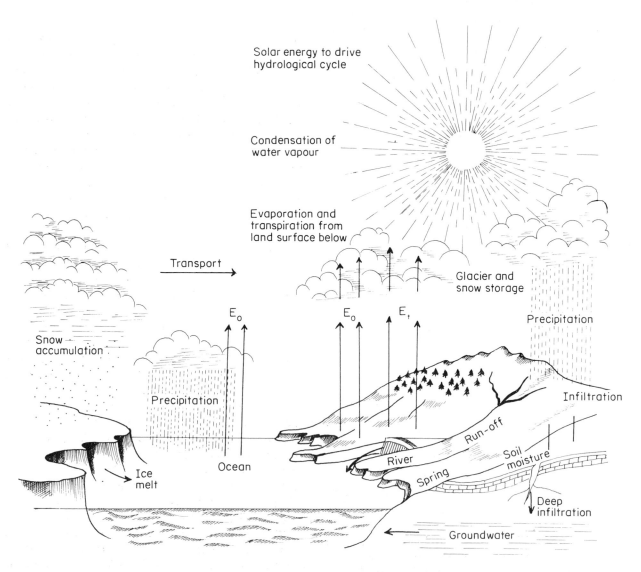

Figure 4.1   The hydrological cycle. E_0 = evaporation, E_t = transpiration. (From Pereira, 1973)

the soil. Infiltration also ensures that deposits of groundwater, underground rivers, and springs are replenished (Figure 4.1).

Evapotranspiration is the opposite of infiltration. Evaporation on its own is certainly an important process within the local hydrological system at bare land surfaces, but plants magnify the effect enormously by transpiring considerable masses of water at the level of their leaf systems. In addition, their roots, which can reach a depth of several metres in the case of trees and other woody plants, accelerate the upward movement of water in the direction from soil to atmosphere. In this way, transpiration by vegetation vaporizes considerable quantities of water: a hectare of forest at temperate latitudes, for instance, can evaporate from 20 to 50 tonnes of water per day depending on the nature of the soil and the local weather conditions.

This process of evapotranspiration, in other words the total quantity of water transpired by plants and evaporated from the soil, is thus an important factor in the water cycle of land-based ecosystems. In some regions of tropical forest (Congo and Amazon basins) it has been demonstrated that the majority of rainwater originates from evapotranspiration and only some 20 per cent comes from the transfer of water vapour from sea surfaces.

In semi-desert regions, the annual loss by evapotranspiration is greater than the gain from rainfall and the deficit then has to be made up from external sources via surface or underground circulation. The process reaches its maximum in tropical deserts (240 cm per year in the Papyrus Marshes of the Blue Nile) and in regions with a Mediterranean type of climate (160 cm per year in the Camargue). Globally, 84 per cent of atmospheric water vapour is contributed by evaporation from sea surfaces and 16 per cent by evapotranspiration from land areas.

Water vapour remains in the atmosphere for an average period of just 10 days, which gives some idea of the speed of the water cycle. Because precipitation on to the surface of the globe amounts to an average of 86 cm per year, the water vapour content of the atmosphere must be renewed as many as 30 times annually.

Distribution of rainfall

Although there is great irregularity in the strength of rainfall at any given point over many inland areas (Figure 4.2), its global distribution shows a remarkable constancy. The maximum amounts fall in Equatorial zones and in the western regions of land masses in the middle latitudes. The zones around the tropics of Cancer and Capricorn, on the other hand, are those with the largest desert areas in both hemispheres. Finally, the extreme dryness of the atmosphere combines with the low temperature to give the arctic and antarctic regions the characteristics of a cold desert.

Because land masses are mostly situated in the Northern Hemisphere, their retention of water in the form of snow and ice is greatest in the months of March and April. This causes an annual variation in sea level, which is at its lowest at the beginning of spring and rises to 2 cm above its mean height at the start of the northern autumn after the summer melting of the ice and snow accumulated during the previous winter.

The average period of time spent by surface water on land varies from 2 weeks to 3 months depending on the environment considered, but some of the rainfall is taken out of the superficial circulation system and filters down to underground waters which are renewed at a much slower rate. As a general rule, the dwelling time of groundwater increases with depth but is of the order of several months even for that near the surface. In desert regions, groundwater renewal is even slower and it can remain for as much as tens of thousands of years in certain Saharan and Australian water deposits. Finally, there are known to be fossil groundwaters, often more than 800 m deep, consisting of water taken out of the hydrological cycle during the Mesozoic and Tertiary periods.

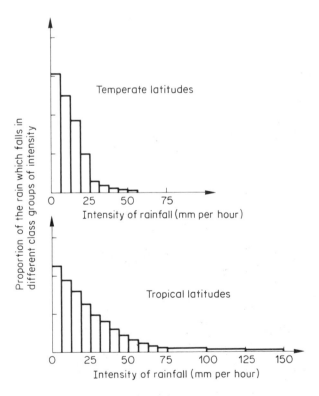

Figure 4.2 Variation in intensity of rainfall at temperate and tropical latitudes. Besides its irregularity in time, rainfall in the tropics is more varied in strength and violence and is the cause of increasing soil erosion. (Hudson, 1973)

4.2 Inland Waters

4.2.1 Supplies and consumption of fresh water

Fresh-water resources

In spite of the enormous volume of the hydrosphere, only a small proportion of it is actually available as a resource: more than 97 per cent occurs in the form of sea water whose salinity makes it useless, while fresh water makes up only 2.6 per cent and 4/5 of that is immobilized as ice and thus equally useless. In the end, the only amounts really available are those in the form of surface and underground running water.

The total volume of surface water (in rivers, lakes and marshes) is estimated at 225 000 km^3, a figure quite close to the 250 000 suggested by Hutchinson over 25 years ago. Surface fresh water thus forms only 0.02 per cent of the hydrosphere.

To that must be added the volume of groundwater situated less than 800 m in depth,[1] estimated with less precision to be 4×10^6 km^3. In practice, the usable amount of water is far more limited. Good practice in the management of water supplies does not allow the drawing off of more water from rivers, lakes and aquifers in a year than is supplied by rainfall for fear of depleting the water capital of the region. Such practice should ultimately be based on an assessment of the run-off from precipitation over the slopes that feed rivers and lakes. To add to that any flow of water from underground aquifers would obviously be a mistake since it would amount to counting the same water twice over. Indeed, if meteorological extremes such as years of drought are to be guarded against, only the lowest rates of flow should be taken into account.

L'vovich (1973) estimated that the mean annual flow in all the river basins of the world amounts to a volume of only 38 830 km^3, of which some 14 000 km^3 is a dependable amount that can serve as a basis for rational management of water resources in the biosphere (Table 4.2). There are other constraints which reduce still further the quantities that can be withdrawn. For instance, it is not possible to allow the diversion of all the water in a river to irrigate the surrounding land because the demands of navigation and cooling for power stations downstream force a minimum flow to be maintained.

Another factor to be considered is the uneven distribution of water resources. The mean annual rainfall over the Earth's land surfaces is 74.6 cm, and this would be an adequate amount in any latitude for the satisfaction of agricultural needs. Figure 4.3 shows, however, the great non-uniformity of distribution that

exists. The temperate countries of Europe and Siberia, of North America, South America and the Far East all have good water supplies, but even in these continents there is great unevenness: Lake Baykal in eastern Siberia, for example, alone contains 10 per cent of the total fresh-water stock of the whole planet. Similarly, Table 4.2 gives a falsely optimistic idea of the quantities of water per head actually usable in South America and Africa: most of the flow is that of the Amazon and Congo basins, regions where a high proportion of the soil is unsuitable for permanent cultivation and in any case does not need irrigation because of the abundance of local rainfall.

At the other extreme, the supplies of fresh water in many of the inter-tropical regions are grossly insufficient, amounting at most to a few hundred litres per person per day (compared with a world average of 10 m^3 or 10^4 litres per person per day, using the figure of 4.2×10^9 for the world population in 1979).

Excess water, too, can impede human activities for, even though floods are a natural phenomenon, they can, for instance, considerably reduce agricultural production in river deltas and other overpopulated lowlands: Bangladesh is a notorious example.

Finally, out of the total world land surface of 150 million km^2, 15 million are covered by polar ice caps and 22 million consist of permanently frozen sub-arctic soils (permafrosts), so that altogether 37 million km^2 of land in the northern and southern polar regions are unsuitable for any agricultural activity at all and are generally unfavourable for population by humans. In addition, there are 40 million km^2 of desert in the intertropical and subtropical zones of the Earth. In other words, for more than half of the total land surface in the world (i.e. 77 million km^2) fresh water is either lacking or exists in an unusable form.

Water requirements

As with all other resources, water is being used in ever-increasing quantities by our technological civilization. In the USA, the amount used in 1966, excluding that for hydroelectric power, was 1500 m^3 person^{-1} year^{-1}: in 1978, it exceeded 3000 m^3 person^{-1} year^{-1}. In fact, the bodily needs of human beings take up only a very small fraction of these total amounts: a 70 kg man requires a mere 2.5 litres a day to meet his total metabolic requirements. It is the domestic, agricultural and industrial use that accounts for by far the greatest proportion of water consumption.

Domestic consumption in France during the 1970s varied between 75 and 175 m^3 person^{-1} year^{-1}, counting both individual and community use. This amounts to an average of about 300 litres person^{-1} day^{-1} against a few tens of litres in the Third World and 600 litres in the USA. Urban environments provide the

---

1. At levels deeper than 800 m, problems of temperature make exploitation much more complicated. Moreover, the deepest water is generally too saline for normal use and is frequently a fossil deposit and therefore non-renewable.

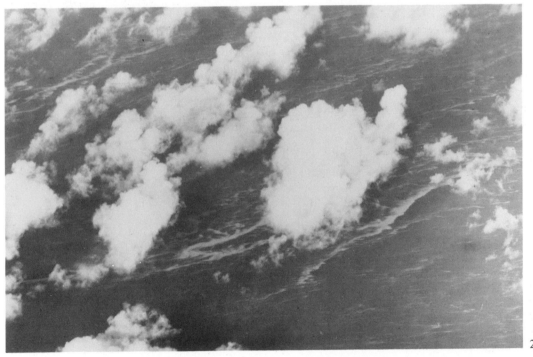

Plate VII   The hydrological cycle

1   The ice-field of north-east Greenland (Peary Land) at the time of its break-up, with many icebergs visible in this aerial view. Most of the stock of fresh water in the hydrosphere is unusable because it is locked up as ice in the Greenland and Antarctic ice fields.

2   Clouds in the process of formation above the South China Sea. Exchange of water between sea surfaces and the atmosphere plays an essential role in the regulation of climate, especially by controlling the water cycle and the composition of the atmosphere.
(Photographs F. Ramade)

Plate VIII   Productivity of inland waters

1   Lake Peyssons in the Andorran Pyrenees. This is an oligotrophic mountain lake with water that is poor in nutrient mineral salts, especially nitrates and phosphates. The relative scarcity of phytoplankton and microphytes is a sign of the low level of primary production.

2   A swamp bordering Lake Neuchâtel, Switzerland. These waters are eutrophic, rich in nutrient minerals, and show a high primary productivity. Note the abundance of algae and microphytes.
(Photographs F. Ramade)

Table 4.2 Dependable volumes for water flow by continent and the available water supplies per capita (from L'vovich, 1973, in Bethemont, 1976)

Continent	Composition of dependable flow (km³)			Total (km³)	Population (1970) (10⁶)	Available water supplies pers⁻¹ yr⁻¹ (m³)
	Running water	Controlled in lakes	Controlled by dams			
Europe	1 065	60	200	1 325	642	2 100
Asia	3 410	35	560	4 005	2 040	1 960
Africa	1 465	40	400	1 905	345	5 500
North America	1 470	150	490	2 390	312	7 640
South America	3 740	—	160	3 900	185	21 100
Australasia	465	—	30	495	18	27 500
Totals	11 615	285	1 840	14 020	3 542	3 955

highest levels: 90 litres person⁻¹ day⁻¹ in Karachi, 500 litres in Paris and more than 1200 litres in the cities of southern California.

Agriculture also consumes large amounts of water, both for irrigation and for raising livestock and growing crops. The water requirements for several important agricultural crops are shown in Table 4.3, the exact figures varying with the local level of evapotranspiration. In an area like Imperial Valley in southern California, where the annual rainfall is barely 75 mm, higher amounts than those shown will be needed because evapotranspiration is so high: for instance, lucerne grown there needs 9000 m³ ha⁻¹ year⁻¹.

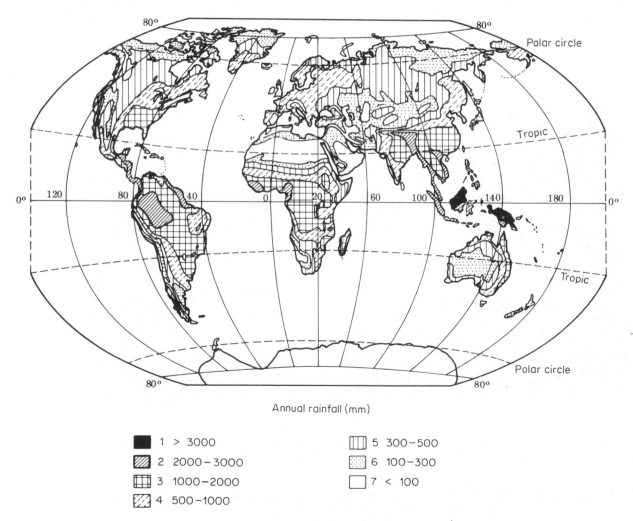

Annual rainfall (mm)

■ 1 > 3000
▨ 2 2000–3000
▦ 3 1000–2000
▧ 4 500–1000
▥ 5 300–500
⬚ 6 100–300
□ 7 < 100

Figure 4.3   Geographical distribution of rainfall over the land masses of the globe

Table 4.3   Water requirements of certain agricultural crops

Crop	Minimum rainfall needed (mm)	Mean annual supply of irrigation water needed ($m^3 ha^{-1} yr^{-1}$)
Wheat	365–760	5 200
Barley	360–700	5 000
Citrus fruits, cotton	500–600	5 500
Hay and other fodder	550–970	8 000
Lucerne	820–910	8 500
Maize	—	20 000
Rice	—	40 000

The large agricultural requirements can also be presented in a more picturesque way: the production of 1 kg of wheat takes 450 litres of water, and 1 kg of rice as much as 2000 litres. Similarly with livestock: a pig of between 34 and 57 kg consumes 7 to 25 litres of water a day. Finally, the production of 1 litre of milk needs 3800 litres of water, mainly because of the amount needed to grow the fodder consumed by cattle.

Industrial consumption is also large: figures have often been quoted like the 36 litres of water needed to produce 1 kg of cement, the 113 litres to produce 1 kg of steel and the 600 litres for 1 kg of paper. The food and drink industries are among the heaviest users: 300 litres for 1 litre of beer, for example. However, in this area the scarcity of water of sufficient purity and the recent enactment of legislation to limit pollution have led to some dramatic reductions in consumption. In the iron and steel industry, for example, new manufacturing methods using recycled water have reduced consumption from 10 m^3 kg^{-1} of steel produced to 1.5 m^3! In a similar way, oil refineries now use only one third of the amounts of water per tonne of crude oil treated compared with that in the mid-1960s.

Types of water consumption: destructive, dispersive, polluting

Three ways in which water is used can be distinguished, each with quite a different effect on the overall balance of supply and demand.

First, there are destructive uses involving the decomposition of water molecules into hydrogen and oxygen. These are limited to a very small number of industrial processes and have a negligible effect on water supplies. It is, of course, also true that primary production involves the decomposition of the water molecule but the quantities concerned, great as they are in absolute terms, are quite negligible in relation to the total volume of fresh water in the hydrosphere. In any case, the effect is exactly compensated for by the respiration of autotrophs and heterotrophs.

Secondly, there are dispersive uses which show up as net losses to the supplies of running fresh water in river basins and lakes. Irrigation is the main example of this type of usage because it causes considerable increases in losses through evapotranspiration. This is particularly evident in semi-arid regions where supplies are drawn off from groundwaters feeding downstream springs which are thus deprived of some of their flow. Cooling of industrial installations, too, particularly of electrical power stations, may reduce thermal pollution quite dramatically, but is responsible for significant losses of water to the rest of the river system. The cooling of a 1000 MW(e) nuclear reactor, for example, requires the vaporization of between 300 and 1000 litres of water per second depending on the local weather conditions.

Thirdly, there are uses of water in which it is a mere vehicle, to be later restored to downstream sections of the drainage basin in a more or less polluted state. Most current usage is of this type and in a sense it involves no actual consumption, although the reduction in quality is often so great that it is completely unsuitable for domestic and even industrial use. There is often an inconsistency here in the attitude of industrial concerns, which demand high purity in the water they use but are much less worried about the quality of their effluent that others may well have to use. Unfortunately, bad management of water resources and even great wastage are still common in industrialized countries.

At present, 90 per cent of US river water is used for the dilution of industrial effluent and the transport of waste products to the sea. The situation is much the same in Europe and in the more industrialized regions of the USSR.

In France, more than 500 km of waterways are permanently polluted by the discharge of industrial and urban effluent. The setting up of the Agences Financières de Bassin in 1964, while slowing down the growth of pollution and even stopping it in some cases, has not been able to restore an acceptable level of purity, judging by the scarcity of fish in most of the rivers. The same point is emphasized by the pollution of the Rhine through salts discharged from the potash mines of Alsace (more than 6×10^6 tonnes per year). France has not so far honoured undertakings made under international conventions, but with pressure from other countries in the EEC bordering the Rhine steps will be taken to inject the saline waste into deep groundwater deposits. Practices like this, however, are not free from risk since the accidental pollution of fresh-water systems of that type may compromise their possible use at a later date. In fact, there are generally great restrictions on the injection of any soluble wastes into deep geological strata because of the possible risks to the water stock of the region.

4.2.2 Ecological effects of hydrological development

Modifications to natural systems of water circulation and interference with several aspects of the water cycle are increasingly being practised both locally and regionally so as to make the most of the resources available. These artificial changes in fresh-water ecosystems have generally been carried out with the idea of slowing down the rate of water flow and of creating reservoirs to smooth out the irregularities in river flows during the annual cycle.

The ecological consequences of such developments may not always be as obvious as those of biological or chemical pollution of inland waters, but they can still be quite considerable.

Loss of land suitable for cultivation and for other purposes

The creation of enormous reservoirs of water involves the flooding of large alluvial areas because all the valleys upstream from any dams are submerged up to the highest shoreline determined by the designers. In France, for instance, the Serre-Ponçon and Sainte-Croix Dams (in the *département* of Alpes-de-Haute-Provence) have led to the submergence of several villages and hamlets, although the loss of arable land in these cases has been relatively limited because of the mediocre quality of the soil and the erosion that already existed in the drainage basins.

However, it is quite a different matter when the constructions become as enormous as those of the Aswan Dam on the Nile, the Kariba Dam on the Zambesi or the Volta Dam at Akosombo in Ghana. In these instances, the areas of submerged land are as large as 1700 square miles or more (Table 4.4).

Dam-building and the associated reservoirs do away with waterfalls and can completely destroy exceptionally beautiful areas. It is true that Unesco managed to save the temples at Abu Simbel and Philae from being submerged, but many fine gorges and canyons have suffered, like the Glenn Canyon in the USA and some of the deep Verdon gorges near Sainte-Croix in France. A plan which would have flooded the Grand Canyon in Colorado excited such strong emotions in the USA that its promoters had to abandon it in a hurry.

The gigantic artificial lakes behind the largest dams cause considerably greater loss of water by evaporation than would occur from the original drainage basins, particularly in semi-desert areas. In the case of the Aswan Dam, evaporation from the surface of Lake Nasser would be sufficient to reduce the mean flow rate of the Nile by 10 per cent! Another feature of arid regions is the much greater amount of water storage needed for irrigation. Thus in central India, a reservoir of 5300 m³ capacity is needed for the irrigation of every hectare, and the figure rises to 21 000 m³ in Australia and 95 000 m³ in southern California.

Further consequences of dam-building arise from the accumulation of sediments and alluvial deposits carried from the waterways upstream from the reservoir. This has two disadvantages. The first comes from the raising of the level of the reservoir bottom, causing it to be filled in more rapidly. Dams and reservoirs built in California, in the Mediterranean basin and in Pakistan are no longer reckoned to have the lifetime of several centuries they were designed for because they will fill up in the next few decades. In France, for instance, the Serre-Ponçon reservoir loses a capacity of 3×10^6 m³ each year because of the accumulation of alluvial deposits from heavy soil erosion in the drainage basin: the reafforestation of the bare mountain slopes around the reservoir that was originally planned has never been carried out (Figure 4.4). Because of that, its useful life will be only half of that initially intended.

The second disadvantage of the accumulation of

Table 4.4   Land areas occupied by lakes behind the great African dams

Name	Country	Date of completion	Area in	
			ha	sq miles
Lake Kainji	Nigeria	1968	129 000	500
Lake Volta	Ghana	1965	840 000	3 250
Jebel Aulia	Sudan	1937	59 400	230
Lake Nasser (Aswan High Dam)	Egypt	1970	464 800	1 800
Lake Kariba	Zimbabwe/Zambia	1963	444 000	1 710

Figure 4.4   Landforms caused by erosion on the banks of the reservoir behind the Serre-Ponçon dam, Alpes-de-Haute-Provence, France. Deforestation of the whole basin took place a long time ago and this has speeded up the sedimentation of the lake. The useful life of the whole construction could be only a half of that originally envisaged. (Photograph F. Ramade)

sediments behind large dams affects the downstream lands bordering the river: the fertilization by those same sediments that formerly took place no longer occurs. The Aswan Dam provides a good example. When the Nile valley used to flood, the arable soils were supplied with precious mud rich in nutrient minerals; now those minerals are lacking. In addition, the phosphates and nitrates no longer pour into the Nile delta, whose well stocked fishing grounds were once legendary. The decrease in primary production in inshore waters because of their impoverishment in phosphates and nitrates has reduced the secondary productivity: whereas 30 000 tonnes of sardines and 40 000 tonnes of inshore fish were formerly caught each year, the figures have now fallen to 5000 and 10 000 tonnes respectively since the building of the dam (George, 1972).

The conditions in a developed fresh-water ecosystem are completely changed by the construction of a dam. Upstream of the reservoir, a lotic (running-water) ecosystem is transformed into a lentic (still-water) one. Downstream the flow is reduced and is subject to large fluctuations according to how the water from the reservoir is controlled. The reservoir itself is an artificial lake which also creates new conditions: a

reduction in the oxygenation of the water and the appearance of a thermocline, producing a stratification of the lake if it is deep enough. In addition, its transformation from a river with a succession of different levels produces a considerable risk of eutrophication throughout the whole of the basin. The retardation of the flow and the accumulation of the sediments combine with persistent pollution by organic matter to produce conditions that even encourage a general dystrophication.

The present tendency in France towards a multiplication of dams in order to eliminate flooding over all national waterways is, in my opinion, a serious threat to the quality of the surface waters. The main result of such a policy will be a considerable decrease in self-purification and the production of conditions ideal for the dystrophication of the whole national hydrographic network.

Turning to the ecological disruption that can occur downstream, there is first the possible disappearance of fauna during the construction of the dam unless an adequate flow is allowed to continue while the lake is filling up. Then there are the annual summer clearances which can cause high mortality from thermal shock if the water is withdrawn from the bottom of the lake below the thermocline where the temperatures are at their lowest. Finally, there are the complete emptyings carried out from time to time intended to clear out the maximum amount of sediment. These can have catastrophic effects on downstream fauna, as in the recent evacuation in France of the Génissiat reservoir (where drainage is carried out once every 30 years), which ravaged all the fresh-water population of the Upper Rhône: the enormous quantities of fine mud (which fills the gills of both vertebrates and invertebrates) combined with the sudden deoxygenation of the water caused by reducing agents in the sediments and destroyed the river fauna as far as Lyon, nearly 140 km from the dam!

The creation of a vast artificial lake can produce great problems for communities living round it. There is the risk of eutrophication already mentioned, which is usually indicated by a proliferation of phytoplankton and microphytes, particularly in tropical or warm temperate climates. The effect is magnified by explosions in the growth of certain species of aquatic weed that are self-sown or introduced (see, for instance, Holm *et al.*, 1971). Several species of this sort have been brought to Africa from South America, among them the water hyacinth (*Eichhornia crassipes*), the water lettuce (*Pistia stratiotes*) and the water fern (*Salvinia auriculata*) (Scudder, 1972). Thus, the invasion of Lake Volta by *Pistia stratiotes* has produced not only an undesirable eutrophication of the water, but a considerable interference with navigation and a reduction in the size of catches of the fishing communities as well. Aquatic vegetation of this sort also harbours the larvae of several species of mosquito carrying filariasis and encephalitis, while the networks of irrigation canals that spread out from the lakes have encouraged the breeding of molluscs carrying bilharziasis. This last disease has become endemic among certain groups of the surrounding population since the building of the Aswan Dam, for instance.

In countries with cooler temperate climates, the artificial lakes are oligotrophic and here the reduction in productivity of inland fishing grounds after the construction of a dam arises from effects that are the opposite of those in warmer climates.

The spread of aquatic vegetation over the surfaces of the lakes increases water loss through evapotranspiration. For example, it has been estimated that such plants increase the annual water loss for the 17 western states of the USA by 2.3×10^9 m³! Herbicides can be used to shift the weeds, but they are at best a costly makeshift measure inasmuch as they are toxic to all the other communities as well.

The building of a dam creates a major barrier to the migration of fish, whether they be sea-breeders like the eel or river-breeders like the salmon. Both downstream movement to the sea and upstream migration are hindered and the latter may be impossible without fish-ladders. Even when these exist, they are often too small or ill-adapted to the strong currents. In addition, where turbines are being driven, the powerful suction draws in descending fish and, although they are protected by fine-meshed enveloping grills, the presence of strong currents causes delay in the migration and may even lead to its abandonment. An example of this has been demonstrated in some artificial fish farms in Scotland, where only 17 per cent of young salmon (smolt) produced were able to cross the barrier presented by a dam and reach the sea. As Pyefinch (1966) rightly pointed out, the necessity for salmonidae to migrate occurs during a definite period, and if the migrants are then too obstructed in their movement to breeding or spawning grounds, the instinct can disappear and cause the fish population of the waterway to die out.

4.2.3 Perturbation of the nitrogen cycle

Human activity can disturb fresh-water ecosystems even more extensively. *Dystrophication* stems from massive injections of fermentable organic matter and nutrients (nitrates and phosphates) into great volumes of water that are only slowly being renewed. One of the most worrying aspects resulting from this is the perturbation of the biogeochemical cycles of nitrogen and phosphorus through the supply of large amounts of both elements, sometimes in quantities greater than those introduced naturally.

We turn first to the nitrogen cycle which always includes a gaseous stage. Both in terrestrial and aquatic environments there is a two-way flow of the element: one is that due to nitrogen fixation and nitrification (conversion of gaseous nitrogen to nitrates and then to organic matter), and the other is that due to denitrification, which transforms nitrates and other nitrogen compounds into gaseous nitrogen. It was demonstrated only recently that a small part of the denitrification process also leads to the discharge of N_2O into the atmosphere.

The fixation of nitrogen is a biological process carried out by a restricted number of prokaryotic organisms: bacteria in the soil, and bacteria and blue-green algae in aquatic media. In the soil, the *Rhizobium*, bacteria that exist symbiotically in the root-nodules of legumes, play a major role in fixation. In aquatic environments, there are again nitrogen-fixing bacteria but various blue-green algae (Cyanophyceae) are also involved (*Nostoc*, *Anabaena*, and most of all *Trichodesmium* in marine environments). The opposite process of denitrification is essentially carried out by anaerobic bacteria.

Studies made of the nitrogen cycle in recent years have led to a substantial modification of previous estimates of rates of flow within it (Figure 4.5). The

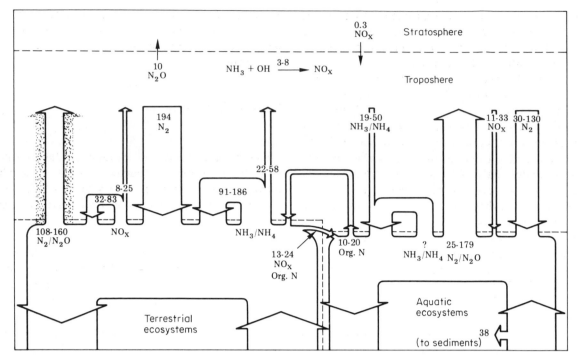

Figure 4.5   The global nitrogen cycle. Rates of flow are given in 10^6 tonnes of equivalent nitrogen per year. (From Svensson and Söderlund, 1976, p. 27)

Table 4.5   *Global circulation of nitrogen in the biosphere* (from various sources, particularly Söderlund and Svensson, 1976)

	Rate of flow in 10^6 t nitrogen yr^{-1}	
Ecosystem or environment	Partial	Total
(1) Nitrification		
Terrestrial ecosystems		
Cultivated land		
Leguminous plants	35	
Rice plantations	4	
Other crops	5	89
Grassland	45	
Forests		40
Miscellaneous		10
Total terrestrial (biological fixation)		139
Aquatic ecosystems		
Pelagic fixation		20–120
Sedimentation		10
Abiotic fixation in the atmosphere		
Volcanoes		0.2
Lightning		4–13 (?)
Grand total		173.2–282.2
(2) Denitrification		
Terrestrial ecosystems		108–160
(of which NO_2		16–69)
Aquatic ecosystems		25–179
(of which NO_2		20–80)

atmospheric stock of oxides of nitrogen (NO_x) and the rates of exchange of NO_x between the atmosphere and other sections of the biosphere have been revised downwards. Nitrogen fixation and denitrification, on the other hand, appear to be much more important than had been previously thought, both in terrestrial and aquatic environments. Finally, the fixation of nitrogen by abiotic processes (lightning discharges during storms) is less than had been assumed, while the sedimentation of organic nitrogen in marine environments is greater. Table 4.5 gives the annual rates of flow in the nitrogen cycle in the various routes.

Comparisons of the rates of nitrogen fixation due to natural biological processes and those due to human activity of various kinds show that the two are of the same order of magnitude. Particularly revealing in this respect are the figures for terrestrial environments, for whereas the various organisms were estimated to have synthesized some 139×10^6 tonnes of nitric nitrogen during 1975, technological processes produced 77×10^6 tonnes. This latter figure was made up by 53×10^6 tonnes fixed by the production of nitrogenous fertilizers and 24×10^6 tonnes by combustion of the various fossil fuels (Table 4.6). This is a considerable perturbation of the nitrogen cycle by human activity and it mainly affects fresh-water ecosystems. The nitrates used as fertilizers are certainly found in surface waters, while research carried out in North America has shown high correlation between the amount of motor fuel consumed in a given state and the concentration of nitrates in its lakes situated in non-urban, non-agricultural areas.

Isotopic methods for the determination of the $\delta^{15}N$ for nitrates of various origins have been developed and have made it possible to estimate how much of the nitrate concentration in both surface water and groundwater is due to nitrogenous fertilizers. During natural biological nitrification, the nitrates that are formed become richer in ^{15}N. The reason for this is as follows: during the decomposition of organic matter, whether in the form of excreta or decaying tissue, ammonia and ammonium compounds are produced and later transformed into nitrates by nitrifying

bacteria like *Nitrosomonas*. Some of the ammonia, which is richer in ^{14}N, evaporates and leaves the nitrates with a higher proportion of ^{15}N, according to the scheme:

$$\text{organic N} \longrightarrow \text{urea} \begin{cases} ^{14}NH_3 \quad \text{favoured} \\ ^{15}NH_4 \xrightarrow[\text{nitrification}]{} {}^{15}NO_3 \end{cases}$$

On the other hand, the nitrates in synthetic fertilizers and those formed by combustion use atmospheric nitrogen and are poorer in ^{15}N (Figure 4.6). In this way, it has been shown that in some agricultural areas the surface water contains more nitrates of technological than of natural origin.

Land clearance is another human activity that affects the nitrogen cycle. When regions of forest or other natural plant formations are turned over to agricultural land, a more rapid mineralization of the pool of organic nitrogen is produced and with it the circulation of large quantities of soluble nitrogenous minerals, principally nitrates. The combination of this effect with the use of nitrogenous fertilizers explains the very high concentrations of nitrates in the surface waters of agricultural land in relation to those found in other terrestrial ecosystems (Table 4.7).

Table 4.6 Global rates of nitrogen fixation for terrestrial systems (adapted from Söderlund and Svenson, 1976)

Process	Rate in 10^6 t nitrogen yr^{-1}
Biological fixation	139
Abiotic fixation	
Volcanoes	0.2
Atmospheric dishcarges	4–13 (?)
Technological fixation	
Nitrate fertilizers	53 } 27
Combustion	24 }

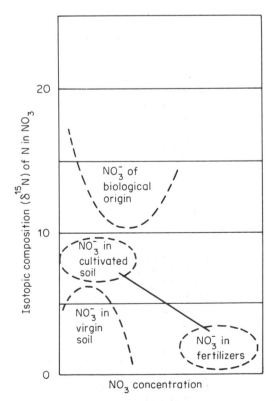

Figure 4.6 Relation between the isotopic composition $\delta^{15}N$ of nitrates and their concentration in various situations. (From Mariotti, 1977)

Table 4.7   *Global river discharges of nitrate nitrogen from terrestrial ecosystems (excluding polar and desert regions)* (from Söderlund and Svensson, 1976)

Ecosystem of origin	Area (10^6 km^2)	Rate (mg of N m^{-2} y^{-1})	Amount (10^6 t of N yr^{-1})
Agricultural	44.0	152	6.7
Non-agricultural tropical	33.9	24	0.9
Non-agricultural temperate	19.1	24	0.5
Total	97.0		8.1

4.2.4 Perturbation of the phosphorus cycle

Phosphorus passes through a sedimentary type of cycle, with the atmosphere playing only a minor role in exchanges of the element between terrestrial and oceanic ecosystems (there are still great uncertainties in the exact sizes of the various contributions to the flows). In addition, the principal stocks of phosphorus available to the biomass are contained in the lithosphere and hydrosphere (Table 4.8).

The biological importance of phosphorus cannot be overemphasized since it is an essential constituent of nucleic acids. However, while the amounts given in Table 4.8 may seem large, it is in fact quite a scarce element in the Earth's crust. In most minerals it exists merely as a trace element and it is abundant in only a few igneous rocks (apatites) and sedimentary phosphate deposits (phosphorites). The quantity of phosphorus which is capable of being extracted from these main sources under present economic conditions and with present techniques is estimated to be between 3 and 9×10^9 tonnes. More optimistic estimates have been made: 19.8×10^9 tonnes by the US Bureau of Mines and as much as 32×10^9 tonnes by Stumm (1973). However, it should be pointed out that these other estimates include some deposits with a phosphorus concentration of 8 per cent or less, which

cannot be exploited with the technology available at present, or likely to be available in the near future.

Inorganic phosphorus is carried into circulation by leaching from the surface rocks and soils containing it and by subsequent solution in inland waters. It is introduced into the fresh-water ecosystem by run-off and infiltration. Rivers then take it to the sea: the global amount thus transported has been estimated at 17.4×10^6 tonnes per year and such a never-ending supply of phosphates perpetually fertilizes estuarine regions, thus accounting for the high biological productivity observed in deltas and estuaries compared with other sea areas of the continental shelf. For the same reason, the other oceanic regions of high productivity, the upwelling zones, owe their characteristics to the ascent of water carrying high concentrations of phosphates from the depths (Figure 4.7).

Both on land and in the sea, phosphorus is incorporated into biomass by autotrophs (green plants or phytoplankton respectively). It has been estimated that between 178 and 237×10^6 tonnes of phosphorus per year are taken up in this way by primary producers on land and between 990 and 1300×10^6 tonnes in the seas (Pierrou, 1976).

The biomass of standing crops on land contains between 1.80 and 2.02×10^9 tonnes of phosphorus, while the oceanic biomass contains about 128×10^6

Table 4.8   *Global stocks of phosphorus in the main sections of the ecosphere* (from Pierrou, 1976)

System	Quantity of phosphorus (10^6 tonnes)
Biomass	
Terrestrial	1805–2020
(of this, human biomass	<1
Marine	128
Fresh-water	<1
Hydrosphere	
Fresh-water	90
Marine	120 000–128 000
Lithosphere	
Soil	160 000
Exploitable minerals	3 000–9 000
Total lithosphere	1.1×10^{13}

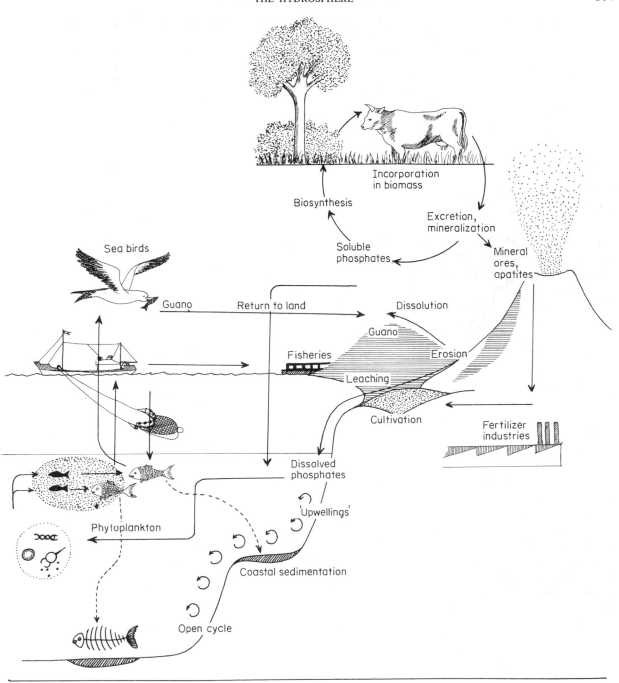

Figure 4.7   General outline of the phosphorus cycle. (From Ramade, 1978, p. 78)

tonnes. The oceanic stock is particularly small in relation to the total annual flow and this stems from the part played by organisms, the phytoplankton and zooplankton, in the production of a rapid turnover. Table 4.8 shows the quantities of phosphorus stored in the various sections of the ecosphere.

There is, in addition, some flow of phosphorus from biomass to water or soil through the death of organisms and through excreta. Decomposition of organic matter restores some 136×10^6 tonnes of

phosphorus per year to the soil, and a similar mineralization of organic phosphorus in the seas is estimated to involve some 10^9 tonnes per year.

The phosphorus cycle is an open one and is not completed. There is a general tendency for mineral phosphorus to be carried away continuously from the land to the deep sea. Sedimentation takes some 13×10^6 tonnes per year out of the sea water and causes it to accumulate in the ocean depths. Since the amount of phosphorus carried from land to sea by rivers is about 17.4×10^6 tonnes per year, it can be seen that this sedimentation removes most of the element reaching the seas as a result of leaching from soils.

The transfer of phosphorus in the other direction, from sea to land, is quite derisory in comparison and is brought about in two main ways. The most important arises from vast colonies of sea birds that produce accumulations of guano rich in phosphates along certain favoured coastal zones: some 10^5 tonnes per year is transferred in this way. The other mechanism, involving quite negligible quantities on a global scale, arises from the migration of anadromous fish from the sea, where they live, to inland waters, where they spawn and then die. In fact, this can assume considerable importance locally: in certain lakes of eastern Siberia, the supply of phosphorus from the dead bodies of salmon (*Oncorhynchus nerka*) contributes from 20 to 40 per cent of the total flow of this element into the ecosystem concerned (Krokhin, 1975, in Pierrou, 1976).

Human activity is affecting the phosphorus cycle in several ways. Sea-fishing, for instance, transfers some 50 000 tonnes of the element per year from sea to land, a figure based on a total quantity of human food supply from the oceans in 1975 of 50×10^6 tonnes. On a global scale, this is a relatively insignificant amount of phosphorus.

On the other hand, activities which increase the transfer from land to sea involve much greater quantities. World consumption of phosphates had risen to 12.5×10^6 tonnes of equivalent phosphorus per year by 1972, and a reasonable estimate of the annual growth since then would be about 4 per cent, giving a global consumption by the end of the decade of 16×10^6 tonnes per year. About 85 per cent of this is dispersed over agricultural land in the form of fertilizers and the rest is used in the chemical industry largely as an additive detergent material of which 80 per cent is a mixture of ortho- and poly-phosphates. These phosphates, from both the fertilizers and the detergents, find their way into fresh-water systems.

Another form of activity also perturbs the phosphorus cycle by accelerating the rate of transfer from land to sea: deforestation and the cultivation of vast areas of land using unsuitable agricultural methods increase the erosion of soil around river basins. As a result, human activity causes an extra 9.1×10^6 tonnes of phosphorus each year to be carried out to sea in river sediments out of a total outflow of 13.7×10^6 tonnes.

To sum up, then: the effects of modern civilization on the nitrogen and phosphorus cycles are principally manifested by increases in the quantities of nitrates and phosphates present in fresh water, either in solution or as particles. Although this book does not aim to analyse the ecological consequences of such pollution, I must emphasize the high level of dystrophication[1] which could result from it, both in certain lentic ecosystems and, in the longer term, in lotic ones as well. This would lead to a marked reduction in the quality of fresh water even in areas where primary pollution by toxic wastes and/or organic materials is not very significant. And where water resources are already subject to both excessive demand and a reduction in quality from industrial pollution, the perturbation of the nitrogen and phosphorus cycles described above can only make the situation much worse.

4.2.5 Optimization of water resources

Continual growth in the domestic, industrial and agricultural use of water is creating acute problems of supply even in countries with high rainfall and adequate hydrological resources. The problems are still more serious in semi-arid regions and steppe where large quantities of water are needed for irrigation.

The first type of remedy for insufficient water supplies consists in the diversion of the flow from areas with excess to those where it is scarce, a remedy whose ecological impact is often considerable but which is in fact the one that has been favoured until now. In France, for instance, the construction of the Provence canal has, for the time being, enabled the serious problems of supplying Bouches-du-Rhône and Var with water to be overcome, but only at the price of an almost complete drying up of the lower Durance during the period of low water in the summer.

Some enormous projects on a continental scale were conceived at the end of the 1960s in the USA and the USSR. Some 10 years or so ago, for instance, the North American Water and Power Alliance (NAWAPA) was contemplating taking water from the Yukon, MacKenzie and Columbia rivers and from others of less importance in Alaska and the coastal regions of Canada and the north-western United States, and conveying it to dryer areas like southern California, the Great Lakes, and even as far as New York! Protests from the environmental conservation movement, particularly in Alberta, coupled with considerable difficulties over investment, combined to defeat the monstrous proposal.

At about the same time, a huge hydrological development programme for central Siberia was being proposed in the USSR (Simmons, 1974). This involved the building of enormous dams on the Ob, the Yenisey and their tributaries in order to create an inland Siberian Sea, and the construction of canals to divert water for the irrigation of the central Asian steppes: an

---

1. A detailed analysis of dystrophication is found in Ramade, 1978a, pp. 352ff.

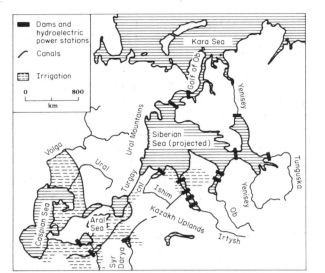

Figure 4.8 Project for the construction of large dams on the Ob, the Yenisey and their tributaries so as to create a vast inland sea for the irrigation of Central Siberia. (From Chorley, 1969, p. 538)

area stretching from the east of the Caspian Sea to the Pamirs, particularly the steppes of Kazakhstan (Figure 4.8). The inevitable and incalculable climatic effects of such environmental manipulation; the loss of several tens of thousands of square kilometres of forest through flooding; the transformation of still greater areas of the lowest-lying land into swamp; and finally, the discovery of oil and natural gas in the very area to be covered by the future Siberian Sea—all these combined to cause the abandonment of the whole titanic project.

It appears that deficits in the water supplies of some parts of the world can hardly hope to find an ultimate remedy in the transportation of excess resources from elsewhere. What is worse for the numerous regions with water problems is that the pumping of supplies from groundwater deposits should really be reduced immediately since the continual decrease in pressure in them shows that outflow is exceeding replenishment by infiltration.

Under such conditions, the only solution to the problem of deficient water supplies is the alternative one of putting a stop to wastage, particularly that caused by uncontrolled pollution, coupled with a better utilization of local resources. The reduction in pollution would have considerable advantages in the domestic and industrial spheres and it is quite obvious that it should take place at source: prevention should take precedence over decontamination. However, there should also be a large investment in purification plant for urban communities, a field in which there are serious shortcomings in both quantity and quality.

In France as in other developed countries purifica-

tion plant is still largely non-existent, and even when it does exist it is often poorly designed, of insufficient capacity, and lacking in personnel competent to make it function at full efficiency. In addition, there is the formidable problem of the varying load on the installation in tourist areas, where the population may increase tenfold in the high season. Again, serious mistakes in design are sometimes revealed, as on the Languedoc coast of southern France: not only does the purification plant in some of the new resorts not operate at normal efficiency but the effluent, supposedly free from organic matter but in fact bacteriologically harmful, is discharged into ponds which are occasionally close to breeding grounds of shellfish.

A final point is that domestic and industrial effluent should be separated before treatment and not mixed together as is, alas, too often the case at present.

The outlook for a more efficient use of water by industry is very promising and there should be an immediate and systematic imposition of an obligation upon all manufacturing concerns to purify effluent significantly. Various processes already in operation, like those of the petro-chemical industry, show that there are many methods that can be adopted to rid the water of phenols and other highly contaminating organic substances (for example, the purification plant at the Fos-sur-Mer refinery). Ultimately, however, the solution lies in the development of manufacturing methods using recycled water. The technology is already available to set up closed-circuit processes in both the iron and steel industry and the chemical and food industry.

Agriculture alone accounts for some 76 per cent of world water consumption, so that the development of new and more economic techniques of irrigation would seem to be as much a priority as the reduction of pollution. The old technique of irrigation using gravity, with the water in open ditches, consumes enormous quantities of the precious liquid and encourages erosion of the soil through leaching. This is now being replaced by spraying methods, and the most sophisticated form of these using a central bore-pipe is already widespread in the USA and other developed countries. The technique uses a main vertical pipe at the centre by which groundwater is pumped to the surface level. This pipe then feeds a horizontal tube which can be as long as 800 m and which is mounted on wheels fitted with pneumatic tyres in such a way that it can turn around the central pipe even on sloping terrain (Figure 4.9). The spray holes are positioned at regular intervals along the horizontal tube and are fitted with sprinklers. The time for a complete revolution of the mobile tube can be varied and is typically 1 or 2 weeks: it is generally operated at night to cut down losses by evaporation. This technique uses far less water than that using gravity-fed channels and it also reduces soil erosion because the flow from the whole

Figure 4.9   System of irrigation using a central bore-pipe feeding a horizontal tube having a regularly spaced row of water sprays. The horizontal tube moves on wheels round the central pipe once every week or two. Photograph taken in the State of Washington near Ritzville, USA. (Photograph F. Ramade)

set of sprinklers is adjusted to produce hardly any run-off.

More recently, an even more economical method has been developed: that of watering drop by drop through a very fine tube buried in the soil. This represents the ultimate as far as soil conservation is concerned because it does not interfere at all with its structure and the distribution of water is limited solely to absorption by the roots of the crop. Unfortunately, it is a method needing water of high purity so that its widespread use is likely to be somewhat restricted.

4.3   The Sea and its Resources

The sea, a term which embraces all the oceans and seas of the whole globe, covers 71 per cent of the planetary surface: that is, about 360 million km^2. It is the section of the biosphere in which terrestrial life began and it possesses a number of features that are noteworthy from the ecological point of view.

The first of these is its large mean depth: about 4000 metres. As much as 77 per cent of the sea surface covers depths greater than 3000 metres, compared with a mere 2 per cent of the land mass with a height of over 3000 metres.

The second is the continual circulation that occurs both horizontally and vertically, in which the masses of water that are brought into contact in this way mix very little or not at all because of their different densities. Some of the great horizontal currents such as the Gulf Stream have a major impact not only on marine ecology but even on the climates of nearby land masses. There are, in addition, vertical movements of ascending water called upwellings produced by the action of wind which blows parallel to certain coast-lines. When the rotation of the Earth causes the wind directions to be deviated towards the open sea (by the Coriolis force), the surface layers of the water are themselves dragged away from the coast and are replaced by colder water ascending from below, water rich in nutrient mineral salts which have settled there.

These upwelling zones have a high biological productivity when their waters finally reach the translucent layers near the surface.

A third feature of the sea which is important from the ecological point of view is the relative constancy of the physico-chemical conditions in it. Marine environments exhibit, for example, quite remarkably steady concentrations of the principal minerals found in them, apart from those used up by autotrophic plant life: variations of the concentrations of dissolved salts in the pelagic zone about the mean (35 per mil) do not exceed 1 per mil during the annual cycle. (All this excludes the coastal regions, particularly around estuaries, where climatic fluctuations over land cause changes in the supply of fresh water and thus in the temperature and the salinity.)

Similarly, the temperature variations in the sea are very small (Figure 4.10). The surface temperatures observed in the central North Atlantic at a latitude of 45° show a difference of at most 8 °C between the

minimum at the beginning of spring and the summer maximum. Below the zone of seasonal thermocline (the thermosphere) there is an intermediate layer whose upper limit depends on the latitude but is at a depth of 400 metres at 45 °N in the North Atlantic: no seasonal variation of temperature occurs in it. This is called the permanent thermocline, a zone characterized by a regular decrease in temperature as the depth increases until 1200 to 1500 metres. Below that are the deep layers with a temperature of 3–4 °C which is uniform from the Equator to the extreme limits of the land masses in the two hemispheres. In the deep sea under the arctic and antarctic ice caps, the water can remain liquid below 0 °C.

Sea water is generally poor in nutrient minerals, having a nitrate concentration of between 0.5 and 8.0 µg per litre and an orthophosphate concentration of between 0.1 and 0.6 µg per litre, depending on the season. The highest concentrations of such nutrients are close to the land, both in the upwelling zones and in areas supplied with biogenic elements from rivers and run-off. Because of that, marine biological resources are localized in the so-called *neritic province*, a term designating all the sea areas over the continental shelf.

The continental shelf itself is an extension of the land mass beneath the sea and consists of the region between the coastline and the continental slope, the latter marking the limit of the 'plate' to which the land mass belongs. The shelf falls in stages from zero depth to some 130–150 metres in general, and exceptionally to 200 metres, a figure that constitutes a legal limit. It covers an area of 28×10^6 km², nearly 7.8 per cent of the total sea surface, but, because of its relative richness in nutrients, it provides about 87 per cent of the total fish yield from the sea. This explains the political decision taken by a number of maritime countries during the last decade to extend their exclusive economic zone to 200 nautical miles. This amounts to putting the entire continental shelf of all land masses under national jurisdiction since the limit of the shelf itself is always less than 200 miles offshore and it is very often only 20–40 miles wide.

At the continental slope, the depth increases quite sharply from 200 metres to some 2500–3000 metres. The abrupt change in slope at the edge of the continental shelf marks the separation of the neritic province of the sea above the shelf from the *oceanic province*, which encompasses more than 90 per cent of the total sea surface. As the depth increases, various layers or zones can be distinguished, each corresponding to a different set of ecological conditions (Figure 4.11). In the oceanic province we have:

(a) the *epipelagic or euphotic zone*, with a lower limit defined by the minimum daylight illumination needed for photosynthetic activity, on the average the first 50 or 60 metres;

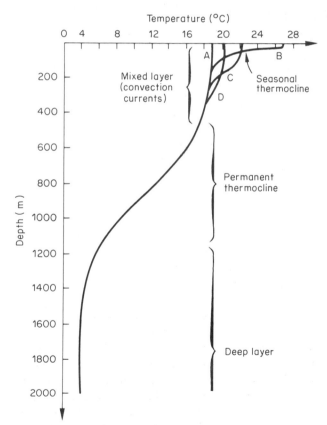

Figure 4.10 Variation with depth of the temperature in the North Atlantic at middle latitudes. Seasonal changes in the upper layers are indicated: A (April), B (August), C (December) and D (February), and here convection currents cause considerable movement. Below that is a permanent thermocline from 400 m to 1200 m in depth situated over the deep layers at uniform low temperatures. Below 400 m the temperatures remain constant over an annual cycle. (From Iselin, in Clarke, 1959)

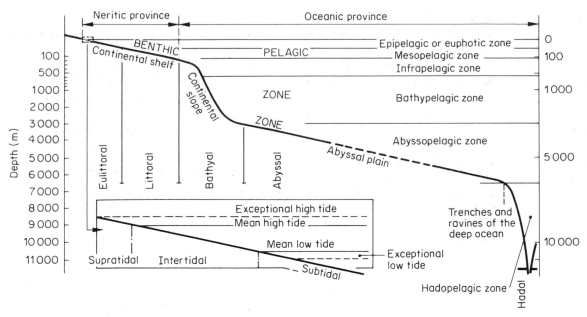

Figure 4.11   The principal horizontal and vertical zones of the marine environment. (From Pérès, 1976, p. 10)

(b) the *mesopelagic zone*, where mixing and temperature variations still occur, with a lower limit at about 200 metres;

(c) the *infrapelagic zone*, between 200 and 500–600 metres, the lowest level reached by plankton in their daily vertical movements;

(d) the *bathypelagic zone*, whose lower limit between 2000 and 2500 metres corresponds to the 4 °C isotherm at middle latitudes;

(e) beyond those are the *abyssopelagic zone* extending to a depth of 6000 metres, and a *hadopelagic zone* going as far as the bottom of the deepest ocean trenches.

Organisms which populate the open sea, whatever zone they occupy, are described as *pelagic*, in contrast to *benthic* species that live on the bottom (benthic species are referred to collectively as *benthos*).

The vertical layers of the benthic zone are based more on ecological criteria than on depth, but the various levels as described by Pérès (Figure 4.11) do nevertheless correspond quite well to the vertical limits of the appropriate pelagic zone.

In contrast to terrestrial environments, the sea is continuous, with all its various regions interconnected and undergoing incessant exchanges of masses of water. For that reason it is more difficult to define the exact limits of its various ecosystems, although the temperature, salinity and depth often prove to be great obstacles to the free circulation of organisms in the sea and thus provide boundaries of a sort.

Apart from its great biological wealth, the sea is also a potential reservoir of enormous mineral resources. Nearly all the elements in the periodic table are to be

found in it, generally only as trace elements but sometimes in sizeable concentrations (see Table 4.14 on p. 133). A significant proportion of world oil supplies is currently being extracted from offshore deposits situated on the continental shelf, and the subsoil there also contains substantial sedimentary deposits rich in metals. The sea-bed of the abyssal plain also contains considerable mineral resources, particularly the well-known ferro-manganese nodules that also include significant quantities of a whole series of noble metals.

We consider the biological and mineral resources of the sea in turn.

4.3.1   Biological resources of the sea

Rational management of the sea's biological resources needs a precise knowledge of global oceanic productivity: of its fluctuations with time, of its variation as a function of depth, and of the geographical location of the fishing grounds concerned.

Primary marine production

The bulk of the primary marine production is due to phytoplankton, even though there is a considerable contribution from benthic macrophytes in the littoral zone of some regions (for example, *Macrocystis* on the Pacific coast of North America). Phytoplankon include a great variety of unicellular algae belonging to very diverse taxonomic groups as well as some autotrophic phytoflagellates. Although diatoms, dinoflagellates and coccolithophores form the dominant groups, there are other organisms belonging to families that are less well-known to the non-specialist which nevertheless play an important ecological role: these are the nanoplankton (less than 30 μm in size) and ultraplankton (less than 5 μm) which account for the major part of primary pelagic production.

Primary marine production is mainly determined by two ecological factors: luminous intensity and the concentration of nutrient salts.

The intensity of light depends on latitude and depth. In the most translucent waters, such as the blue tropical seas, the light penetrates slightly more than 150 metres below the surface but as a general rule it does not reach depths beyond 100 metres. In coastal waters the turbidity caused by influxes of water from the land reduces the maximum depth of penetration to 30 or 40 metres and sometimes even less, except for the areas of coral which are noted for their exceptional transparency.

Other conditions being equal, light intensity in the sea decreases as depth increases according to the relation:

$$\Phi = \Phi_0 e^{-kp} \qquad (1)$$

where Φ is the intensity at depth p, Φ_0 the intensity at the surface, and k the coefficient of absorption which depends on the wavelength of the light. Ultra-violet wavelengths are absorbed within a few millimetres and infra-red within a few decimetres. The region of the spectrum where maximum transmission occurs varies according to the purity of the water: it is in the blue region for pure water but moves into the green for water that is rich in certain organic matter.

The variation in light intensity with depth produces a corresponding variation in the degree of photosynthesis. Unlike terrestrial plants, phytoplankton are inhibited in their photosynthetic activity by high levels of illumination containing a lot of radiant energy at the red end of the spectrum. Since this is precisely what the light is like in the surface layers of the sea, photosynthesis occurs less readily there and in fact increases with depth until a maximum is reached when the light intensity has fallen to one third of its surface value (Figure 4.12). Phytoplankton tend to concentrate around this depth, which has a mean value of 25 metres in tropical seas but becomes less when the absorption increases for any reason such as greater turbidity.

At some lower depth, photosynthesis will decrease to the point where, over a 24-hour cycle, it is only just balanced by respiration: this is known as the *compensation depth*. The layer of sea water above this is called the *euphotic zone* and it is here, in the first 50–100 metres, that all the primary marine production is concentrated. The sea below that, the *dysphotic zone*, is divided into (a) the *oligophotic zone*, extending to the

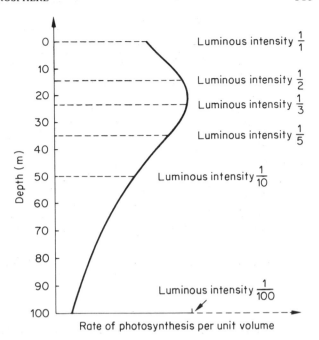

Figure 4.12 Variation of photosynthetic activity with depth. (From Pérès and Devèze, 1963)

limit of light penetration and in which respiration exceeds photosynthesis and phytoplankton cannot survive very long; and (b) the *aphotic zone*, permanently devoid of light and comprising more than 95 per cent of the total oceanic volume.

Turning to the variation of primary production with geographical location, we find that this is largely determined by the concentration of nutrient elements, particularly of phosphorus. The optimum ratio for carbon, nitrogen and phosphorus concentrations is $100:15:1$. A decline in the concentrations of nitrogen and phosphorus from these values slows down the growth of phytoplanktonic algae and reduces the primary productivity. Carbonate concentration is relatively constant from one region to another, and so is not a limiting factor for photosynthetic activity. However, considerable variations both with geographical location and time of year occur in the concentration of nitrogen and even more in that of phosphorus, so that the latter is the major limiting factor in primary marine production.

The particular group of phytoplanktonic organisms that predominates in a given region depends both on the average individual concentrations of nitrogen and phosphorus and on their ratio. Thus, diatoms form the predominant phytoplankton colonies in northern and southern eutrophic coastal waters and in upwelling zones, while dinoflagellates are the dominant group in oligotrophic water and coccolithophores in tropical waters very low in nutrients.

The net primary productivity of phytoplankton (P_n) can be calculated from the autotrophic biomass present

(B_p) and the rates of photosynthesis (R_p) and respiration (R_r) according to the relation:

$$P_n = B_p(R_p - R_r) \qquad (1)$$

Here, if P_n is expressed in grammes of carbon m^{-2} day^{-1}, B_p will be in grammes of carbon m^{-2}, and R_p and R_r will be the daily rate.

Studies of the density of chlorophyll and of the photosynthetic activity of phytoplankton in sea water have demonstrated both the small biomass and low productivity of marine ecosystems, particularly of those in the central regions of large oceans. With rare exceptions like coral reefs, both the biomass and the primary productivity are distinctly lower than those of inland ecosystems (Table 4.10, p. 115, and Figure 4.13) and they also show large variations from one region to another and, in middle and high latitudes, from one season to another.

Over the whole globe, four types of marine environment can be distinguished by the level of their primary productivity.

First of all, there is the whole neritic province consisting of the continental shelves. These have a high productivity of over 100 g of carbon m^{-2} year^{-1} because of the large quantities of phosphates and other nutrients originating from the land and dispersed via estuaries into coastal waters, and also because the shallowness makes it easier for nutrients in the benthic mud to rise to the surface.

Secondly, there are certain fairly limited regions where water from intermediate depths rises to the euphotic zone and brings with it nitrates and phosphates produced by the mineralization of organic matter. These are the upwelling zones produced by such phenomena as seasonal currents and the divergences of permanent ocean currents, and among the most productive of such zones are the Peruvian coast, the south-west coast of Africa, the Equatorial Divergence and so on. The productivity of these regions produced by the ascendance of deeper waters rich in nutrients can exceed 200 g of carbon m^{-2} year^{-1}

Thirdly, there are the temperate and sub-polar oceanic regions with a primary productivity that averages between 50 and 100 g of carbon m^{-2} year^{-1}, but which has a sizeable peak in the spring and which is limited in winter by low light intensity and in summer by a scarcity of nutrients.

Finally, there is the entire oceanic province of the tropics between latitudes 30° N and 30° S, covering enormous areas of the Atlantic, Pacific and Indian Oceans. Here, there is a very pronounced and permanent thermocline which considerably limits the supply of nutrients to the euphotic zone. As a result, blue tropical seas are the marine equivalent of terrestrial deserts as far as their biological productivity is concerned. Even though there is a high photosynthetic rate of 0.1 per day spread uniformly over the year because of the good climate, and even though the

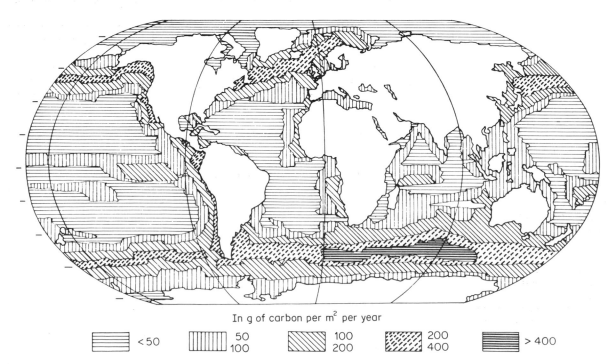

In g of carbon per m^2 per year

<50   50 100   100 200   200 400   > 400

Figure 4.13 Distribution of primary productivity in the oceans expressed in grammes of carbon per m^2 per year. The most productive regions are the continental shelves and the colder areas of the Atlantic, Pacific and Indian Oceans. (From Bunt, in Lieth and Whittaker, 1975, p. 174)

Table 4.9   Primary productivity of the ocean (Adapted from Whittaker and Likens in Lieth and Whittaker, 1975)

Region	Area $(10^6 km^2)$	Primary productivity (g of dry org. matter $m^{-2} y^{-1}$)	(g of carbon $m^{-2} yr^{-2}$)	Total production $(10^9 t$ dry org. matter $yr^{-1})$	Fixation of energy $(10^{18} cal yr^{-1})$
Open ocean	332.0	125	42	41.5	109.2
Continental shelf	26.0	360	122	9.6	44.0
Estuaries and coral reefs	2.0	1 800	611*	3.7	16.6
Upwelling zones	0.4	500	169	0.2	1.0
Totals	360.4	155	52.6	55.0	170.8

* A good proportion does not come from phytoplankton (*Cyanophyceae* and benthic diatoms).

euphotic zone can penetrate to depths of up to 200 metres because of the high transparency of the water, there is such a low planktonic biomass (less than 2 g m^{-2}) that the productivity only reaches between 10 and 50 g of carbon m^{-2} year^{-1}.

Figure 4.13 and Tables 4.9 and 4.10 demonstrate that a far from negligible proportion of marine production is concentrated in upwelling zones and the neritic province, which together constitute less than 10 per cent of the total sea surface of the world.

Comparison of the autotrophic biomass and productivity of marine environments with those of terrestrial ecosystems (Table 4.10) shows how small primary marine production is in relation to that of the land. Various current estimates agree in putting the annual fixation of carbon in the seas at 50 or so g m^{-2}, giving a photosynthetic conversion efficiency of about 0.03 per cent, whereas on land the average productivity is of the order of 300 g m^{-2} year^{-1} with a photosynthetic efficiency of 0.13 per cent.

The entire land mass of the globe, with an area of 149×10^6 km^2 and thus less than half of that of the seas, produces 100×10^9 tonnes of dry organic matter per year or 40×10^9 tonnes of equivalent carbon. The

Table 4.10   Comparison of the standing crop biomass and primary productivities of various terrestrial and marine ecosystems. (Data (1) from Bolin, Degens, Duvigneaud and Kempe, 1979; (2) from Tett, 1977; (3) adapted from Sournia, 1977)

Ecosystem	Standing crop biomass (g carbon m^{-2})	Chlorophyll (g m^{-2})	Mean primary productivity (mg carbon $m^{-2} day^{-1}$)	(g carbon $m^{-2} yr^{-1}$)
Terrestrial (1)				
Tropical rain forest	18 900	3.0	2 800	1 000
Temperate deciduous forest	9 000	2.0	1 850	675
Temperate grassland	728	1.3	962	351
Desert and semi-desert	353	0.5	176	65
Extreme desert	40	0.02–0.2	106	39
Marine (2)				
Phytoplankton of temperate coastal waters:				
winter minimum	0.1	2×10^{-3}	10	—
spring maximum	10.0	0.3	1 500	100
summer mean	2.0	4×10^{-2}	200	—
Phytoplankton of tropical waters:				
equatorial Pacific mean	1.0	10^{-2}	10–200	35
Sargasso Sea minimum	0.05	5×10^{-4}	—	—
Temperate intertidal seaweed	400	4 (?)	4 100 (max)	1 500 (max)
Subtropical eelgrass bed	600	3 (?)	—	—
Coral reef algae:				
mean	250	1	6 800 (1?)	2 600 (1)
maximum	—	—	7 200 (3)	2 700 (3)

seas, from an area of 362×10^6 km^2, produce only 55×10^9 tonnes of dry organic matter per year: in other words, a total primary production of 19×10^9 tonnes of equivalent carbon.

Secondary marine production

The transfer of energy from the autotrophic producers of the marine food chain already described to the herbivorous zooplankton at the next trophic level (that of the primary consumers) is of fundamental importance since it controls the whole secondary production of the seas and therefore of even later stages in the trophic pyramid.

The food chains of the pelagic and benthic zones of the sea follow the general pattern outlined in Table 4.11. From this, it can be seen that trophic levels I, II and III include nearly all the plankton except some of the megaloplankton like the medusae (tiny jellyfish), while levels III, IV and V (and sometimes VI!) are occupied by nekton and nektobenthic species (that is, those able to navigate freely).

The dominant group among the herbivorous zooplankton are filter-feeding crustaceans, the copepods. However, the euphausids (krill) of the sub-polar seas also form a significant proportion of the organisms at this level, as do the Appendiculariae, the salps, and other planktonic metazoa. During the time when ultraplankton and the smallest nanoplankton are proliferating, it can happen that these tiny algae cannot be consumed directly by crustaceans, and the second trophic level (herbivore) is then occupied by certain ciliate protozoa, which in turn are ingested by the crustaceans (thus becoming carnivores).

All herbivorous zooplankton behave as filter-feeders, taking in their food non-selectively by filtration of quantities of water that are quite large, considering the size of the organisms. The herbivorous copepods, the Appendiculariae, and the euphausids indiscriminately ingest unicellular planktonic algae and particles of organic matter suspended in the sea water. Only a fraction of the volume absorbed is properly digested, the rest being rejected with faecal pellets which descend to deeper levels and serve as food for other planktonic species or benthic organisms.

We have seen that the intermediate position in the food chain occupied by copepods can depend upon what is on offer from primary production. Certain species possess prehensile mouths that enable them to be predators on other copepods, so that they then permanently occupy a position midway between strict herbivores and first-order carnivores. Other species of copepod, however, which are somewhat larger, are strictly predators. The third trophic level in the marine food chain is mostly occupied, nevertheless, by macroplankton (chaetognaths, polychaete worms, larvae and adults of various crustaceans and pelagic gasteropods) as well as by microphagous nekton (various decapods and plankton-feeding fish).

Trophic levels IV and V are occupied by nekton or by nektobenthic organisms: cetaceans, predator fish and cephalopods.

The secondary productivity of the higher trophic levels depends on the efficiency of energy conversion at all stages of the marine food chain. The interactions between phytoplankton and zooplankton, therefore, have a major influence on the fish-harvesting potential of the various sea areas.

In general, the correlation between the abundance of phytoplankton, that of zooplankton and the productivity of commercial fisheries is excellent: coincidence between population peaks ensures optimum efficiency for the conversion of phytoplanktonic biomass to animal biomass. However, in some cases the density of phytoplankton is found to be out of phase in space and/or in time with that of zooplankton, with a spring peak in the first being succeeded by a summer maximum in the second (Figure 4.14). This may be the result of excessive 'grazing' of phytoplankton by copepods and other filter-feeders during the warm season. It is also often due, particularly in colder seas, to the greater speed of growth of phytoplankton and the much slower development of zooplankton.

The occurrence of either effect, and sometimes of both (producing a very high turnover of the unicellular algae in phytoplankton), accounts for a phenomenon peculiar to marine environments. This can be described as a kind of inversion of the biomass pyramid during the warmer months: the biomass of the standing crop of phytoplankton reaches at most 2 g of carbon per

Table 4.11   Organization of marine food chains

Trophic level	Function	Group of organisms
I	Primary producers	Phytoplankton (macrophytes)
II	Herbivores	Zooplankton (filter-feeders)
III	First order carnivores	Zooplankton, microphagous nekton*
IV	Secondary carnivores	Macrophagous nekton (fish, cephalopods)
V	Tertiary carnivores	Superpredators (fish, odontoceti)

* Some micronektonic organisms and a few species of microphagous fish can be placed in the second trophic level.

m^2, while that of the copepod *Calanus* can be five times greater in the same place at the same time. However, whereas diatoms divide once a day and can be ingested by *Calanus* as soon as they are produced, the *Calanus* themselves need several months (cold seas) or several weeks (warm seas) to grow from egg to adult.

Secondary marine productivity thus depends primarily on the ecological efficiency of energy production by the herbivorous zooplankton (that is, their production divided by that of the phytoplankton they eat). This in turn determines the growth efficiency of the zooplankton defined by

$$R_c = \frac{\text{total amount of tissue} + \text{eggs produced by zooplankton}}{\text{total amount of phytoplankton eaten by zooplankton}} \times 100$$

This can be as high as 30 per cent under favourable conditions, but is generally found to be between 5 and 20 per cent with a mean of about 15 per cent. Thus, in the case of the copepod *Calanus*, which is the dominant filter-feeder among the zooplankton of temperate seas, secondary productivity is of the order of 15 g carbon m^{-2} year^{-1} in water where the annual primary production is 100 g carbon m^{-2}.

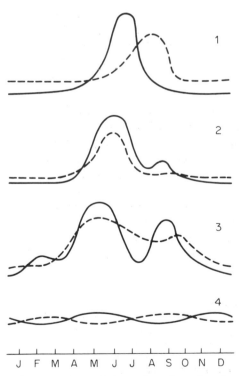

Figure 4.14 Annual variations in the abundance of phytoplankton (continuous lines) and zooplankton (broken lines): 1, glacial basin (Arctic); 2, open sea (Bering Sea); 3, open ocean in temperate zone; 4, open ocean in the tropics outside upwelling zones. (From Pérès, 1976, p. 41)

At the higher trophic levels, secondary productivity can be evaluated by taking a simplified food chain like that of the herring, a plankton-eating fish of great economic importance (Figure 4.15). If, to a first approximation, the efficiency of energy conversion for the zooplankton eaten by herring is assumed to be comparable with that of *Calanus* (that is, 15 per cent), then the 15 g carbon m^{-2} year^{-1} produced by *Calanus* will give only about 2 g carbon m^{-2} year^{-1} of fish.

In fact, the figure is not far out. The metabolism of herring consumes 9 g m^{-2} year^{-1} and four further units are unassimilated and rejected with the excreta (Figure 4.15). In other words, the transfer of energy from trophic level II (zooplankton) to III (herring) is accompanied by a loss of 60 per cent in respiratory metabolism and 27 per cent in unassimilated food and excreta. The mean growth efficiency of herring will therefore be only 13 per cent, a value that is still high, nevertheless, compared with estimates for macrophagous predators and superpredators of the higher trophic levels.

In the end, therefore, the potential yield of the food chain with the herring at level III will be

$$\frac{15 \times 13}{(100)^2} \approx 2\%$$

Thus, in temperate waters, where the average primary productivity is 100 g carbon m^{-2} year^{-1}, the annual secondary production of herring will hardly reach 2 g carbon m^{-2}.

If the energy transfers are considered more generally over all the food webs of the pelagic zone, even lower average efficiencies are encountered. Pérès (1976) estimates that the ecological efficiency at the level of predator fish that are of interest to the fishing industry (trophic levels IV and V) is about 0.25 per cent (Figure 4.16).

In the extreme case predators like tuna-fish, which are situated at the end of a chain at level V, and sometimes VI, of the type:

phytoplankton → zooplankton →

　　　　I　　　　　　　II

Clupeidae → Scombridae → Thunnidae

　　　III　　　　　　IV　　　　　　V

the ecological efficiency will be very low, with an energy yield of the order of 10^{-4} or 10^{-5}.

Global secondary productivity of the seas

Great progress has been made during the last 10 years or so in our knowledge of the overall secondary productivity of the seas. The various calculations of it that have been carried out are based on estimates of the global primary marine productivity combined with

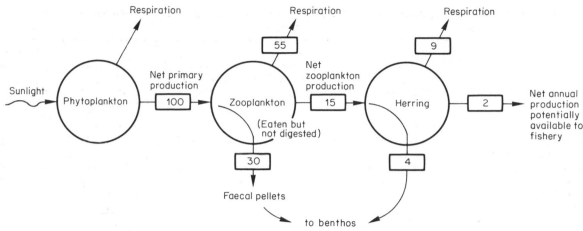

Figure 4.15   Food chain of the herring to illustrate the idea of ecological efficiency. (From Tett, 1977, p. 23)

those of the average ecological efficiency, i.e. the efficiency of energy transfer between the trophic levels involved.

Bogorov (1975) estimates that the total ecological efficiency of the food chain leading to nekton harvested by humans is 3.6×10^{-4}. This means that the global primary productivity of 550×10^9 tonnes wet weight per year would yield about 200×10^6 tonnes per year of fish and other large nekton.

Two other estimates of secondary marine productivity are given in Table 4.12: those of Ryther

(1969) and Tett (1977). The quite remarkable agreement between the various figures for the biological potentiality of the seas suggest that any rational development of these resources ought necessarily to be based on such ecological data. Thus the secondary productivity of the world oceans in fish and other marine species of economic value is estimated to lie between 180×10^6 and 240×10^6 tonnes wet weight per year. The difference between the extreme values represents quite a low margin of error and, according to Pérès, arises on the one hand from slight over-

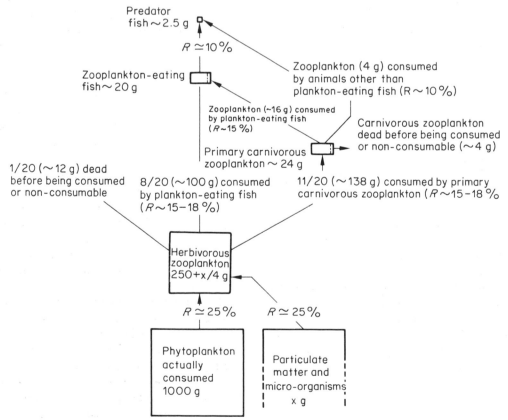

Figure 4.16   Simplified theoretical pattern of energy transfer in the food web of the pelagic zone (excluding phytoplankton-eating fish and cetaceans). R = efficiency of transfer of stored energy. (From Pérès, 1976, p. 84)

Table 4.12   Estimates of the secondary productivity of the seas and of the potential yield (limiting capacity) from world fisheries

Region	Total primary production (10^9 t carbon)		Mean number of trophic levels		Mean ecological efficiency (%)		Potential fish yield (10^6 t yr^{-1})		Maximum exploitable by fisheries (10^6 t yr^{-1})	
	a	b	a	b	a	b	a	b	a	b
Upwelling zones	0.1	0.12	1–2	2–3	20	15	120	70	—	50
Seas of continental shelf (neritic zone)	3.6	3.6	3	3–4	15	10	120	110	—	55
Temperate and sub-polar oceans	}16.3	11.3	}5	4–5	}10	10	}1.6	36	—	15
Tropical oceans		6.3		5		8		3		1
Totals	20.0	21.3	—	4	—	10	242	219	—	121

a According to Ryther (1969).
b According to Tett (1977).

estimates until recently of the ecological efficiency at low trophic levels in the neritic zone and, on the other, from underestimates of the ecological efficiency of open oceans.

The maximum catch that should be made by fisheries is necessarily lower than potential secondary production if the stock of the principal species of economic value is to be maintained or even increased. On a global scale, the difference between actual yield and total secondary production is made all the greater by the fact that only a fraction of the large nektonic species is currently consumed by humans. Moreover, some fish populations are too dispersed to make harvesting profitable, while most of the infrapelagic and all the bathypelagic species cannot be caught using present techniques. Taking all these factors into account gives a final estimate of about 100×10^6 tonnes per year for the maximum catch which can be taken without risk of overexploitation.

Table 4.13 gives estimates made by the FAO of the potential maximum fish resources of both traditional and unconventional types. The figure fixed by the FAO for conventional fisheries is for a maximum catch of about 100×10^6 tonnes per year, and this seems rather optimistic since it represents nearly the whole of the secondary productivity of the species concerned. Non-conventional resources are represented by cephalopods, which are not exploited very much at the moment, and by fish of the mesopelagic zone and deeper, for which no satisfactory methods of catching exist at present. These two groups between them give a further secondary productivity of about 100×10^6 tonnes per year (Table 4.13).

The fisheries

In 1977 the world production of all fisheries amounted to 73.5 million tonnes, of which some 63 million (85 per cent) came from the sea and the rest from fresh-water and estuarine sources or from fish farms. Most of the marine production consisted of fish, with crustaceans, cephalopods and other molluscs constituting only 8 per cent of the total. The continental shelf is of overriding importance in this production, providing 96 per cent of the total (64 per cent pelagic, 36 per cent benthic) with only 4 per cent coming from the open ocean.

World production of sea fisheries grew at a steady rate of about 7 percent per year from 1950 to 1970 and then slowed sharply at the beginning of the 1970s: the 74.5 million tonnes of 1978 exceeded the 1970 production by only 6 per cent, whereas production had doubled between 1958 and 1969 (Figure 4.17).

Out of the 73.5 million tonnes produced by all fisheries in 1977, 53 million went for human consumption and the other 20.5 million for industrial purposes, although it should be noted that most of the latter amount was used in the preparation of fishmeal for livestock feed so that only a small fraction of total production was not in the end used for nutrition in one form or another. Since the world production of meat is approximately 120 million tonnes, it is clear that in the 1970s the fishing industry supplied a significant proportion of human requirements in animal protein. In fact, 20 per cent was supplied by fish and 40 per cent by meat.

It is also worth noting that the proportion of the total catch destined for direct human consumption is continually increasing. In 1970, it amounted to 63.2 per cent of the total, 26.3 per cent being consumed in the form of fresh fish, 13.9 per cent was deep frozen, 11.5 per cent was canned and the other 11.5 per cent went for salting, curing, etc. The 36.8 per cent not consumed directly was supplied to industry, principally for the production of fishmeal. By 1977, the proportions had become 72 per cent of the total catch directly con-

Table 4.13   Potential sea-food resources, current landings and predicted growth of marine fisheries. All figures in 10^3 t year^{-1}. (Adapted from FAO, 1979)

Species of group of nekton	World secondary productivity	Total world catch 1977	Potential growth from	
			Improved stock management	Increased effort of fisheries
Conventional resources				
Salmon	650	480	170	0
Halibut and cod	6 700	3 786	2 500	400
Clupeidae (herring, etc.)	15 600	13 800	1 801	0
Crustaceans (prawns, etc.)	1 670 (?)	1 542	25	100
Thunnidae	2 260	1 592	100	550
Other pelagic fish	40 200	28 655	3 000	8 500
Other benthic fish	37 100	17 097	4 000	16 000
Sub-totals	104 180 (?)	66 952	11 596	25 550
Non-conventional resources				
Cephalopods	50 000 (?)	1 165	—	50 000 (?)
Fish from mesopelagic zone and lower	50 000 (?)	—	—	50 000 (?)
Sub-totals	100 000 (?)	1 165	—	100 000 (?)
Total large nekton	204 180	68 117		
Phytoplankton-eaters (trophic level II)				
Filter-feeding molluscs	?	3 011	—	?
Euphausia superba (krill)	50 000 (?)	123	—	50 000
Sub-totals	?	3 134	—	?
Grand total, this table	>254 000	71 251	—	>175 000

sumed by humans, and only 21 per cent used in non-food industries.

From the ecological standpoint, it is much more efficient to consume fish directly than to feed livestock with it. Indeed, even the best breeding techniques and the best animal species from the point of view of energy conversion (pigs and chickens in feedlots or batteries) produce at most an ecological efficiency of 25 per cent. This means that the 20 million tonnes of fish converted to fishmeal would be expected to produce little more than 4 million tonnes of meat.

Overfishing

The virtual stagnation in the size of the total catch since the beginning of the 1970s indicates that the principal species of fish and other sea-food of economic value are now being overexploited in most of the world's great fishing grounds.

The populations of these fish are in fact distributed

among independent and integrated groups, each known as a *stock*. A group is characterized by its geographical distribution, its breeding area, the age structure of its population, its morphological and physiological peculiarities and, finally, by the degree of harvesting to which it is subjected. Thus, to take the example of herring in the North Sea: the total population is distributed among three stocks, the Downs, the Dogger and the Buchan, each performing its own migrations with the possible exchange of a few individuals.

Overfishing can be defined as the landing of an excessive number of a given stock of any species of fish during a given time, and can be identified by a number of criteria:

(a) a diminution in the volume of the catch in spite of increasing effort on the part of the fishery,

(b) a reduction in the size of the individual fish that are caught, resulting from a scarcity of older fish whose life expectancy is lowered by increased fishing,

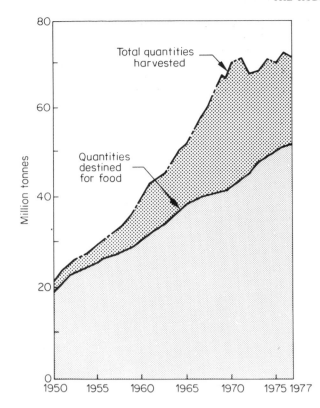

Figure 4.17 Variation in world fish production between 1950 and 1977. Notice the levelling-off in the total catch from the beginning of the 1970s. (From FAO 1979)

(c) an increase in the rate of individual growth because the young are favoured by a reduction of competition within the species, thus making more food available to them and accelerating their development.

Overfishing concerns a relatively small number of the marine species in the world. Out of a total of some 30 000, only about 1 000 are commercially exploited and more than 90 per cent of the landings are confined to some 20 or so species belonging to five taxonomic groups: the Clupeidae (anchovy, herring, sardine), the Gadidae (cod, haddock, whiting, hake), the Scombridae (mackerel), the Thunnidae (tuna and bonito) and the Pleuronectidae (plaice, halibut, sole, etc.).

Figure 4.18 shows the sizes of the catch for each of the major groups of species in relation to their maximum potential resources, and illustrates unequivocally the overfishing of the dominant groups. It is all the worse in that the possible increases in landings from better organization of fishing (the slightly more densely shaded parts of the diagram) are calculated on the assumption that there would be a reduction of effort on the part of the fisheries in order to allow the replenishment of stocks.

Historically, the first cases of overfishing occurred in the North Sea and north-east Atlantic, near the most densely populated areas of Western Europe. At the end of the last century, stocks of plaice were showing obvious signs of it, as were those of other species that had suffered heavy catches. Bernard, a French oceanographer, writing on the subject in 1902, declared that 'it would be better . . . if the trawler had never existed at all'. During the first quarter of the twentieth century, the main North Sea species of economic value were seen to be declining in abundance: halibut in 1905, cod in 1920 and so on.

Since that period, symptoms of overfishing have continued to appear in nearly all the species harvested in the North Atlantic and more recently have been occurring at a greater and greater rate (Figure 4.19). Thus the cod stocks of the continental shelf of Greenland and on the Grand Banks of Newfoundland showed a marked decline between 1950 and 1960; herring were affected in their turn by overfishing both in those areas and in the Norwegian Sea during the 1960s; and in 1976 clear signs began to appear in the stocks of whiting and capelin in the sub-arctic regions of the Atlantic.

The economic collapse of the fishing industry over the last few decades has been dramatic. In chronological order, there was first the effect of particularly drastic reductions in the population of Pacific sardine, and after that the same with Peruvian anchovy; finally, there was the decline of herring and cod over the whole of the North Atlantic. These have meant a catastrophic reduction in associated fisheries and have therefore been a strong inducement for maritime countries to extend the limit of their exclusive economic zones to 200 miles.

The Pacific sardine, Sardinops coerulea, provides a particularly good illustration of the drastic consequences of overfishing, both economically (for the fishing communities whose livelihood is at stake) and ecologically (for the populations of the fish species concerned). It lived in large groups, consisting of several thousands of millions of individual fish, on the north-east-Pacific coast of North America from California to the southern coast of Alaska. Fishing first started at the beginning of the century off British Columbia and catches grew at a remarkable rate, culminating in the landing of 850 000 tonnes between San Diego and Vancouver in 1937. This was then supplying a canning industry that brought in some $10 million a year. This prosperity lasted until 1942, when the first signs of overfishing began to appear, and after that there was a sharp fall in the size of catches which forced the gradual closure of canning factories, first in British Columbia and later on the Pacific coast of the USA. Commercial fishing for this species finally ceased in 1953, a year when only 80 tonnes were landed from the trawlers of San Francisco (Figure 4.20).

As Pérès (1976) has pointed out, it is almost certain

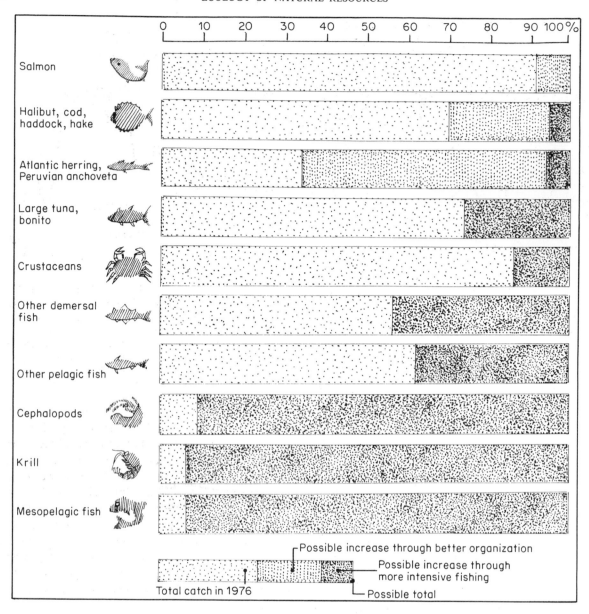

Figure 4.18   Annual landings of fish as a percentage of the maximum catch potentially available. Some stocks have been so overexploited that the catches are much lower than they would be if there had been better management and regulation. This is the case for the first five categories illustrated: for the others, exploitation would increase if new fishing methods were developed. (From FAO, 1979)

that the sudden collapse in the stocks of *Sardinops coerulea* was brought about by a combination of two events: excessive landings by the fisheries coupled with an abrupt modification of ecological factors in the breeding grounds due to natural causes. For normal development, the Pacific sardine needs water with a temperature at or above 13 °C. Between 1943 and 1957, temperatures in the Californian Current suddenly dropped below this value and stayed instead at around 11 °C, the optimum value for a competing species of anchovy, *Engraulis mordax*. It seems very probable that this new species, favoured by the cooler waters and by the reduction of interspecific competi-

tion from *Sardinops coerulea* due to overfishing, eventually occupied the ecological niche of the latter sardine and prevented any significant renewal of its population.

The exhaustion of the stocks of Peruvian anchovy (*Engraulis ringens*) provides an example of the consequences of gross overfishing that is perhaps even more dramatic than the previous one. This species, known to the local fishermen as anchoveta, became considerably scarcer after 1970, again through a combination of overfishing and an environmental change associated with the ocean currents.

There exists along the Peruvian coast what is

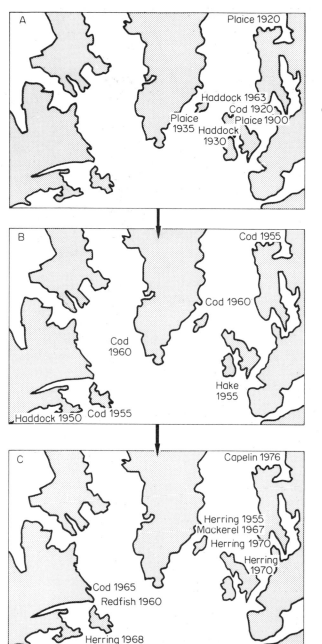

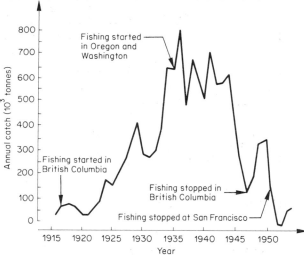

Figure 4.20   Variations in the annual landings of Pacific sardine (*Sardinops coerulea*). The collapse in population at the end of the 1940s is a prime example of the catastrophic effects of overfishing. There is still no evidence of a significant renewal of stocks more than 25 years after the cessation of industrial exploitation. (From California Department of Fish and Game, 1957)

Figure 4.19   Overexploitation of the North Atlantic fishing grounds. In A, when the stocks had begun to be exploited, the areas of intensive fishing contained a great abundance of most species of economic value. By the dates in B some of these were being sought in areas far from their markets, culminating in full exploitation and even overfishing by the dates given in C. (From FAO, 1979)

undoubtedly the most powerful upwelling in the world. Winds that blow incessantly towards the Equator drag the waters of the Humboldt Current out to the open sea and thus allow colder and deeper water to rise to the euphotic zone, carrying with it a high concentration of nutrients. As a result, the region has an exceptional primary productivity, sometimes exceeding 500 g carbon m^{-2} year^{-1}, and this in turn produces a high secondary productivity. The high density of fish populating the coastal waters has encouraged the development of enormous colonies of sea-birds, the source of the famous guano upon which the prosperity of Peru was once based: some 30 million fish-eating birds were counted in 1955, the year when commerical landings of anchoveta began. The colonies are dominated by the Bougainville cormorant (*Phalacrocorax bougainvillei*) which almost religiously abstains from eating anything but the anchoveta. Other prominent species are those of some gannets (*Sula variegata* and *S. peruviensis*) and a pelican.

Engraulis ringens flourishes in the cold waters of the Humboldt Current because of the extraordinarily high productivity and the remarkable efficiency of its energy take-up from primary producers: its diet, consisting of zooplankton when young and phytoplankton in adulthood, is one that makes optimum use of primary production.

Catches of anchoveta expanded enormously between 1955 and 1970. Starting from a negligible tonnage at the beginning of the 1950s, they rose to 10 million tonnes by 1968. The FAO was by that time worried about the risk of overfishing and wanted to limit the total catch to 9.5 million tonnes. Discussions took place but were unable to prevent the landings from increasing to 12.5 million tonnes in 1971. The feared collapse was not long in coming: in 1972, the catch fell to 4 million tonnes, and in 1973 to 1.5 million tonnes, a figure that has not been exceeded since (Figure 4.21). The economic consequences were

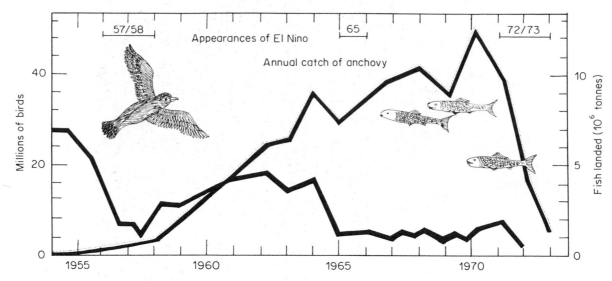

Figure 4.21   Correlation between the appearance of warm coastal currents (El Niño) in Peru, the size of annual landings of anchovy (*Engraulis ringens*) and the fish-eating bird population. (From Wyrtki, 1979, p. 1213)

disastrous and included the laying up of an oversized fleet of trawlers with an absurd total capacity of 55 million tonnes per year.

The shrinkage in the stocks of *Engraulis ringens* and the non-appearance of any significant renewal were once more caused by two factors acting together: the detrimental effects of overfishing were magnified by a naturally occurring change in the sea currents that occurs from time to time around the upwelling zone. A coastal current of warm surface water, known as El Niño (the Christ Child) because its appearances occur around Christmas time, flows in a southerly direction along the Peruvian coast, invading the colder water and weakening the upwelling. This, of course, produces a reduction in primary productivity and a scarcity of fish. Peruvian fishermen had for a long time correlated the massive mortality rates observed in the colonies of sea-birds with the appearance of El Niño at the end of December.

The 1956 appearance of El Niño, at a period when the anchovy fisheries were not yet developed, had already reduced the population of the guano bird colonies to 4 million (Figure 4.21). Subsequently, there was a recovery until the next appearance in 1965, when another massive reduction in the bird population occurred. However, because the landings of anchoveta had not become excessive, their stocks were able to recover.

It was quite different by 1971, when the combination of El Niño and the effects of overfishing caused a final collapse in the anchoveta stocks and a decline in the bird population to less than 3 million, a level never before observed. Since then, neither the anchovy nor the birds have shown any signs of renewed growth in their numbers and a resurgence in either population seems a more and more doubtful prospect.

Overfishing of North Sea herring (*Clupea harengus*) provides yet another example of the detrimental effects of overfishing on stocks of great economic importance. With world catches amounting to 2.5 million tonnes per year (3.1 per cent of the total marine production), herring is clearly a species of some significance to world fisheries. Between the end of the Second World War and the middle of the 1960s, landings of herring from the North Sea remained at a relatively stable 850 000 tonnes per year. This was in spite of a continual increase in fishing effort through greater numbers of trawlers and improvements in performance due to new fishing techniques. However, after that stable period, there was a temporary increase to a maximum catch in 1968 that exceeded 1 million tonnes. Subsequently, landings declined sharply until the catastrophic year of 1974 when barely 340 000 tonnes were caught.

North Sea herring are distributed among three separate stocks whose spawning grounds are located around the periphery of the region (Figure 4.22). The Downs stock, which has been the subject of most research, breeds at the end of the autumn period when each female deposits anything between 10 000 and 80 000 eggs in coastal waters near the bottom of the sea, generally at a depth of about 50 metres. The young fish live first on their inbuilt reserves, then on phytoplankton and later still on zooplankton. At about 6 months, the young herring are carried by currents to estuarial zones, the 'nurseries', where they remain for about a year. With the approach of sexual maturity after 2 or 3 years, they leave the nurseries and join adult populations in the open sea between Great Britain and the German and Dutch coasts, and between Denmark and the southern coast of Norway.

Since 1974, the stocks have suffered a remarkable

Plate IX   Coastal waters

1   An estuarial zone with sedimentary mud exposed at low tide (Cook Inlet, near Anchorage, Alaska). Such environments, with both a large diversity and a very high biological productivity, are nurseries for numerous species of fish of major importance to maritime fisheries.

2   Salt lagoons in the Camargue nature reserve. In addition to their fascinating aquatic life, such habitats also harbour unusual bird life belonging to the Anatidae and Charadriidae families.
(Photographs F. Ramade)

Plate X   Coastal waters (continued)

1   The Pacific coast of northern California in the Redwood National Park. This area is shrouded in an almost perpetual mist during the summer months because of the cold Californian Current. The current produces a strong upwelling which gives the continental shelf a high biological productivity and this in turn is exploited by the flourishing commercial fisheries of California, Oregon and the state of Washington.
(Photograph F. Ramade)

2   Nektobenthic fish on a bed of *Posidonia* in the Mediterranean (Port-Cros national park). In spite of its relatively small area, the continental shelf is by far the most important region of the seas from the point of view of fish and other sea-food. Its primary productivity is very high in both the pelagic and benthic zones. This is particularly so with macrophyte beds in intertidal and sublittoral zones, where the productivity is comparable to that of terrestrial ecosystems having a very high level of photosynthetic activity.
(Photograph J. C. Moreteau)

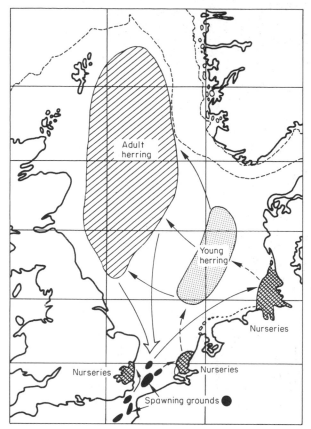

Figure 4.22   The movements of young and adult herring belonging to Downs stock in the North Sea. Broken lines indicate the 200 m depth contour. (From Tett, 1977, p. 28)

possible to assess the ecological impact of overexploiting herring in this region, particularly when account is taken of the other species of economic value that also inhabit it. In fact, in spite of the establishment of a quota for herring catches from the beginning of the 1970s, countings made of larvae suggest that adult stocks are not being replaced and that the spawning population is barely a tenth of its size at the end of the Second World War.

The last example we shall analyse to show the absurd results of chaotic management of an important natural resource is that of the overfishing of cod (*Gadus morrhua*). This is a benthic fish inhabiting vast areas of the continental shelf and continental slope of the North Atlantic between 45° N and 66° N, and even extending beyond those limits to the Barents Sea as far as Spitzbergen and Novaya Zemlya. Over the whole of this region, cod is widely overexploited, just as is a similar species, haddock (*Melanogrammus aeglefinus*) (see Figure 4.23 which shows data for both).

Although there was a slight falling off in cod landings after 1970, the annual catches between 1955 and 1975 remained relatively constant (Figure 4.23). Nevertheless, profound changes in the pattern of fishing were taking place during this period: a decline in the stocks of some of the traditional fishing grounds (Iceland, Newfoundland) was masked by the simultaneous opening-up of new areas and an increase in overfishing. Thus, cod landings in the region covered by the International Commission for Atlantic Fisheries (Newfoundland and Labrador), which amounted to 1.9 million tonnes in 1968, fell to 1 million tonnes in 1972 and at present are still below that figure.

decline, particularly those of Downs and Dogger. Characteristic signs of overfishing have appeared: 3-year-old fish entering their first productive year, which hitherto had formed only a small proportion of the catch, were found to be constituting up to 70 per cent of adult stock. Moreover, the age of recruitment, at which the young individual at maturity joined the adult population, went from 4 years to 3. Finally, an increase in the speed of individual growth was observed in parallel with the reduction in the average size of the catch.

Landings of herring in the North Sea amounted to some 1.2 million tonnes per year at the end of the 1960s. Assuming a mean ecological efficiency of 10 per cent and using the fact that herring on the average occupy the third trophic level, the figure for the total catch corresponds to a primary productivity of 1.2×10^{13} g of equivalent carbon per year. Now the species inhabits a sea area covering some 400 000 km^2 having a mean primary productivity of 100 g carbon per m^2 per year and thus a total of 4×10^{13} g carbon per year. Thus the food chain of the herring that are caught alone consumes 1.2/4 or 30 per cent of the primary productivity of the North Sea.

Figures like these, more than anything else, make it

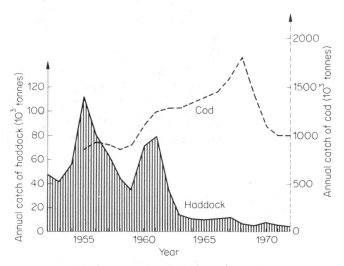

Figure 4.23   Decline in stocks of cod (broken line) and haddock around the Canadian coast from the Grand Banks of Newfoundland to Greenland and the south of Baffin Bay due to overfishing during the 1960s. (From Pinhorn, 1976)

Although world cod landings were maintained at the very high level of nearly 3 million tonnes a year throughout the 1970s, nearly all stocks showed clear signs of overfishing even if partly concealed by the increased efficiency of fishing techniques. However, while the absolute value of the catch did not fall, there was a decrease in the quantity landed per unit fishing effort (Pinhorn, 1976). Just as in the north-west Atlantic, the Icelandic fisheries suffered a distinct recession in the 1970s, catches falling from 471 000 in 1970 to 350 000 in 1975. At that time, it was estimated that the cod population aged 7 years or more (the time of onset of sexual maturity in that region) would fall by 1982 to a level that was barely 15 per cent of that in 1975. Here again, there is a reduction in the size of the reproducing groups, a sure sign of overfishing. It is to be hoped that the establishment of a quota for all cod stocks and the extension of exclusive economic zones to 200 miles will prevent cod from suffering the fate of the Peruvian anchovy.

There are many other commercially important species subject to overfishing at the moment. Haddock, for instance, whose world landings were nearly 500 000 tonnes at the beginning of the 1960s, have become considerably more scarce over the last 15 years or so. The Newfoundland stock (Figure 4.23), overfished since 1955 when the catch exceeded 100 000 tonnes, finally collapsed in 1964 and landings at present are less than 1000 tonnes per year. Hake, capelin, mackerel and several of the Pleuronectidae (e.g. halibut, plaice) are just a few examples of the many populations subject to excessive pressure from fisheries.

It is the same for *marine mammals*. The decline in the populations of large cetaceans illustrates more vividly than that of any other marine resource, the improvidence and even the vandalism of those nations with a powerful fishing industry.

The enormous size of certain species of whalebone whales—a blue whale can exceed 30 metres in length and 120 tonnes in weight—has often caused commercial whalers to forget the difference between biomass and productivity. In fact, Mysticeti have a short life span of about 30 years in spite of their size and they reproduce late, even then having only one or at most two calves at a time. This gives them a low biotic potential. Yet paradoxically they are at the end of one of the shortest food chains that can exist and thus one with a high ecological efficiency, giving whales a very special place in the oceanic food web. Rorquals, which feed on krill, are, for example, at the end of the chain:

$$\text{phytoplankton} \rightarrow \textit{Euphausia superba} \rightarrow \begin{array}{l} \textit{Balaenoptera physalus} \\ \textit{Balaenoptera musculus} \end{array}$$

The food chain of the sei whale (*Balaenoptera borealis*), although longer, still shows a high ecological efficiency since the whale is only at the fourth trophic level:

$$\text{phytoplankton} \rightarrow \text{copepods} \rightarrow \text{amphipods} \rightarrow \textit{Balaenoptera borealis}$$

There were some 200 000 blue whales (*Balaenoptera musculus*) living in antarctic waters before the high commercial value of their meat and oil led whaling nations to organize large-scale hunting just after the First World War. Catches rose each year, culminating in 1930 with one of 29 400, a number far

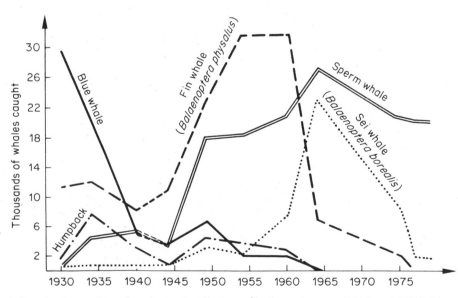

Figure 4.24   Variations in the catches of cetaceans in the Antarctic Ocean between 1930 and 1976. Protective measures of the International Whaling Commission have proved ineffective and did not prevent the collapse in stocks during the 1960s. (From Berzin and Yablokov, 1978)

in excess of recruitment, so that the stocks collapsed. The resumption of hunting after the Second World War led to such a situation that in 1964 the species was put under international protection. Unfortunately, this has been poorly respected and at present fewer than 9000 still exist.

Finding the blue whale less productive, whalers then fell back on the fin whale (*Balaenoptera physalus*) and some 30 000 of these were caught each year from 1948 until 1962, when this species in turn began to decline in numbers (Figure 4.24). The present population is less than a quarter of its level before hunting began (Martin, 1978).

The reduction in stocks of the large *Balaenopteridae* led whalers to two smaller species, the sei whale and the humpback whale (*Megaptera novaeangliae*) whose populations then also began to fall rapidly. Catches of the sei whale amounted to 20 000 in 1960 and only 6 000 in 1969, while between 1963 and 1967 its population was reduced to one third. At present, stocks of sei whale are at most only 40 per cent of their initial levels and are decreasing at an alarming rate. As for humpbacks, they have been subjected to such excessive hunting since the beginning of the 1960s that the total stocks have fallen from 110 000 before hunting began

to a current level of less than 5000, so that severe protective measures have had to be taken.

Finally, when all the species of Mysticeti with any economic value were either on the way to extinction or drastically reduced in numbers, whalers turned their attention to sperm whales (*Physeter catodon*) at the end of the 1960s. Now, these in their turn are showing signs of overfishing in spite of the relatively large size of the initial stocks and it is currently estimated that their numbers have already been halved.

Since the beginning of the 1970s, the Japanese, who must bear a heavy responsibility for the decline of the large cetaceans, have come round to hunting dolphins . . . and have discretely scrapped a fleet of factory-ships that had become superfluous—in the same way as other nations had after taking part in the carnage of the whale populations.

Recently, various plans for exploiting antarctic krill itself have been under development. Figure 4.25 shows the areas having the highest concentrations, from which the Japanese and Russians harvested some 123 000 tonnes in 1977.

Krill can be used in two different ways: directly as a human foodstuff or as a source of meal for incorporation into livestock feed. It should be emphasized in

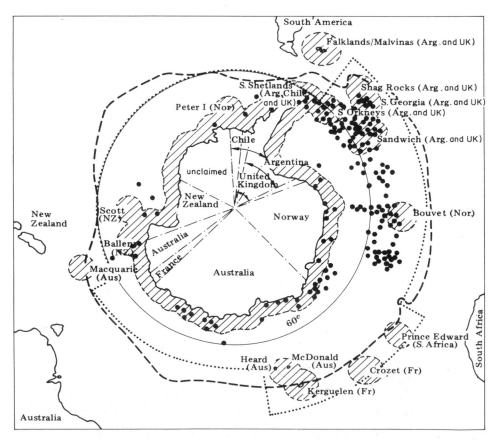

Figure 4.25   Distribution of krill in the Antarctic Ocean. Black dots indicate the main areas of high concentration. Shaded areas show the limits of the continental shelf. (From Mitchell and Sandbrook, 1980, p. 18)

either case that such exploitation is contrary to the best ecological principles.

Frozen krill is sold in Japan at $700–1000 per tonne, a price that is high compared with that of chicken and comparable with that of beef from cattle raised in feedlots. Moreover, there is a marketing problem since consumers are being placed in the same situation as they would be if cattle were eliminated and they were asked to indulge in the pleasure of eating concentrated protein extracts from grass! As far as the use of krill in cattle food is concerned, the efficiency of energy conversion is poor compared with that of the food chain involving whales. Again, the consumption of energy in harvesting krill is relatively high because of its wide dispersion in the sea and the distance between its habitat and the consumer markets, and the costs here can only be expected to grow in the future with the increase in fuel prices. The cost is still further increased by the complex treatment needed to make it suitable for incorporation into cattle food. Finally, the ecological efficiency of cattle is low at about 10 per cent. Since 100 kg of krill produce 5 kg of whale, it is highly probable that in the end the best way of exploiting it would be through a wise management of the stocks of large antarctic cetaceans.

The rational management of marine resources

Overfishing raises the general problem of the rational management of biological marine resources. The question to be answered is: how can we achieve the maximum catches compatible with the long-term protection of stocks of adult fish? Only through a precise knowledge of the ecology of fish populations can the main controlling factors be quantitatively assessed and a calculation be made of the optimum catch for each species: that is, the landings that will maximize the stock without jeopardizing the long-term future of the fisheries. Paradoxically, a properly adjusted fishing effort can *increase* the secondary productivity of a commercially valuable species.

The regulation of a population is governed by factors associated with intraspecific and interspecific competition and with predation. Tett (1977) deals with the complex problem of such regulation by using two simplifying concepts: those of the 'prudent' predator and of the 'efficient' prey. A prudent predator lives in equilibrium with its prey: in other words, it takes no more than can be replaced through reproduction by the adult and growth of the young. The concept of an efficient prey is illustrated by looking at the variation in the number of fertile descendants (recruitment) as a function of the number of parents. The larger the number of descendants per adult, the larger the catch that can be made. However, for a given stock or species, there is an optimum adult population for which

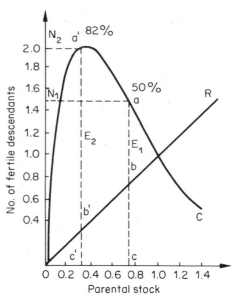

Figure 4.26   Ecological basis for the rational management of a commercially valuable fish stock. The straight line R represents a stable population with expansion above it and contraction below. The curve C shows the actual number of fertile young produced per adult for various total populations. The stock is maintained as long as the catch is less than that represented by the vertical distances ab, a′b′, etc., between C and R, the rest (bc, b′c′, etc.) being the stock left for reproduction. In the case illustrated, the greatest surplus available for the catch occurs for a parental stock equal to Oc′: the catch E_2 (= a′b′) which can be made while still maintaining the population is then 82 per cent of the number of fertile descendants. (From Smith, in Ramade, 1978a)

the potential catch has a maximum (Figure 4.26, curve C). An efficient prey will be one that reacts strongly to a small decrease in its adult population: in other words, one for which a moderate rate of predation will produce a sizeable growth in the size of catch.

In the end, therefore, we come to the following question: under what conditions of exploitation (that is, what size of catch) will the biomass of a stock be a maximum? The answer is provided by Figures 4.26 and 4.27. These illustrate the variations in the main ecological population parameters as a function of fishing effort for a given species or stock. In Figure 4.26, the straight line R from the origin of slope 45° separates two regions, the upper in which the population is expanding and the lower in which it is contracting, irrespective of the absolute value or density of the adult population. The curve C shows the total number of fertile young produced per adult (i.e. the recruitment) for various sizes of population. The intersection of R and C determines the size of the adult population that corresponds to the limiting capacity of the environment. What is remarkable is that for a smaller population than that, a greater recruitment per fertile adult is obtained. Indeed, the optimum recruitment

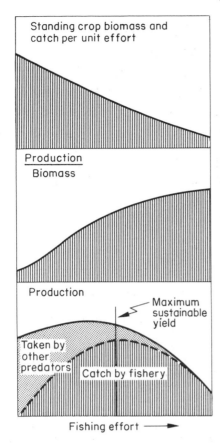

Figure 4.27   The effect of fishing effort on the standing crop biomass, on the ratio of production to biomass, and on the absolute value of the production. (From Tett, 1977, p. 33)

achieved in Figure 4.26 at a' gives a surplus production amounting to 82 per cent of the fertile descendants.

Figure 4.27 shows the effect of fishing effort on the standing crop biomass, on the ratio of production to standing crop and on the absolute value of the production. The first curve shows that the biomass and catch per unit effort decrease with increasing effort, while the second one shows that the ratio of production to biomass increases. The third curve, showing total production, indicates that there is an optimum effort for which the maximum yield sustainable by the stock is obtained. This value corresponds exactly to the effort which would adjust the adult population to a level giving the maximum number of surplus fertile descendants (a'b' in Figure 4.26).

It may be asked why a moderate degree of predation should increase the net secondary productivity: why it is that a suitably adjusted catch should allow the whole stock to transform the energy absorbed from lower trophic levels with maximum efficiency. The reason is very simple: predation, in the form of fishing effort, increases the proportion of young fish relative to adult reproducers and the young use the energy from food more efficiently than adults, who merely consume it to

meet their metabolic requirements. Secondary productivity will therefore be greater in a population that includes a higher proportion of juveniles.

Several mathematical models have been constructed which describe the evolution of a fish population in terms of the numbers and weight of individuals in the various age groups (year classes). These are four factors which govern the equilibrium of a population of a given species and, *ipso facto*, the balance of the energy flow entering and leaving it. These are:

(a) The growth of the individual with time, particularly the increase in its weight. This is of prime interest to fisheries since, for a given total weight, a catch consisting of a large number of small individuals is very different from one with a small number of large individuals (Figure 4.28). An equation due to Von Bertalanffy gives the variation with time of the weight of an individual:

$$P_t = P_\infty [1 - e^{-k(t-t_0)}]^3 \qquad (1)$$

where P_t and P_∞ are respectively the individual weight at time t and a theoretical time t_∞, t_0 is the initial time and k a coefficient characteristic of the species concerned.

(b) The recruitment. This is the size of the new contingent of young fish incorporated into the stock.

(c) The mortality F which occurs after recruitment and depends on the fishing effort.

(d) The natural mortality, m.

If the total mortality $F + m$ is put equal to M say, and if M is assumed constant, then the decrease in the population dN in a time interval dt is given by

$$dN = -MN\,dt \qquad (2)$$

where N is the population at time t.

The total number of fish caught C between the age of the first capture and the maximum age t_{max} attained by the fish in the stock will be given by

$$C = \int_{t_c}^{t_{max}} FN\,dt \qquad (3)$$

The weight of the catch, P_c, will be given by the expression:

$$P_c = \int_{t_c}^{t_{max}} FNP_t\,dt \qquad (4)$$

If the natural mortality and the fishing effort remain constant, it can be shown that the whole annual production of all the year classes in a given stock is equal to the production of a single year class during its whole existence.

The variation in the weight of a complete year class in the course of time follows a normal curve with a maximum at a critical size or age, t_k (curve B, Figure 4.29). The optimum catch is obtained if it is made at

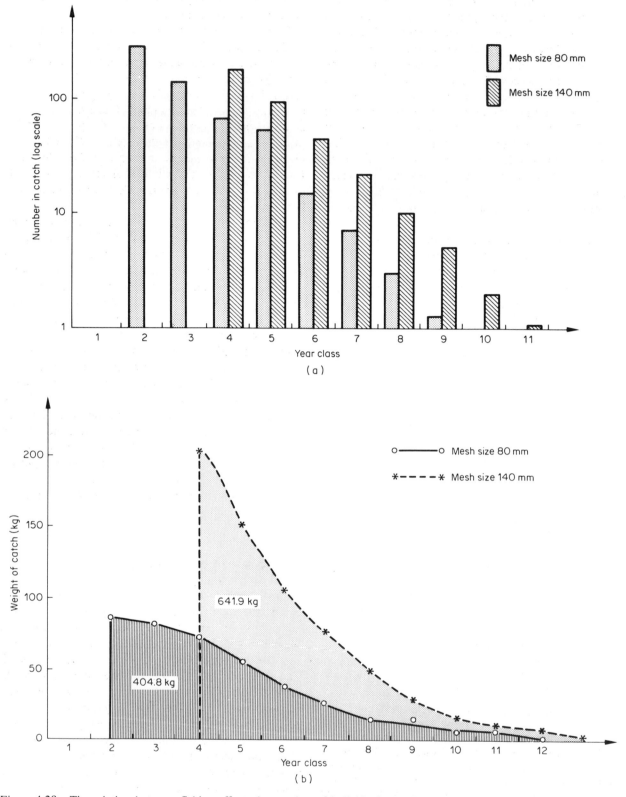

Figure 4.28   The relation between fishing effort, the number of individuals caught, and the weight of catch. Diagram (a) shows the variation in the *number* caught in the various year classes of the stock for two different fishing efforts. The greater effort (80 mm mesh) gives a catch with the larger number of individuals over all classes. Diagram (b) shows the *weight* caught in each year class for two different mesh sizes. Here it can be seen that a lower effort produces a larger catch, not only in total weight but in average individual weight as well

time t_k and in numbers that are in excess of those needed to achieve a stable population. As a basis for a method of fishing, this is clearly only a theoretical possibility since the year classes in one stock are all mixed together and the surplus individuals of just one class cannot be selectively removed. For that reason, specialists in the ecology of fish communities have long ago developed the idea of a 'best minimum size' at which fishing must begin to produce a catch of maximum volume (see, for example, Ricker, 1945). Two other characteristic sizes (or ages) have also been defined: that of recruitment, t_r, and that at which capture first occurs, t_c (Figure 4.29). The latter depends on one parameter only: the mesh size chosen for the net or trawl.

Considerations like these have served as a basis for the development of several mathematical models enabling a determination of the maximum yield sustainable by a stock to be made: that is, a calculation of the maximum annual volume of catch compatible with the endurance and stability of the population. The theory,

to use a crude analogy, seeks to maximize the interest yielded by a given capital.

One of the classic models, that of Beverton and Holt (1956), consists of a collection of mathematical expressions whose representative curves (known as eumetric curves) enable a calculation to be made of the yield per recruit for a given degree of fishing effort and a given minimum size of fish captured. Eumetric curves will give the maximum production of a fishery as a function of the volume of landings (the mortality F), and for any given degree of fishing effort will give the value of t_c. Finally, the curves also allow a calculation to be made of the maximum sustainable yield beyond which overfishing occurs.

For large nektonic species like tuna, there are other more appropriate models based on different ecological considerations. These models use data from statistics of fishing effort and total annual catches, and make it possible to estimate the maximum equilibrium catch for a population: that is, the equilibrium catch which adjusts the number of adult reproducers to a value giving the maximum rate of growth.

Our knowledge of the ecological principles relating to the populations of commercially valuable species of fish is now sufficiently precise to make a rational management of marine resources quite feasible. A combined use of statistical data on the main stocks of commercial interest and of mathematical models for the calculation of optimum fishing effort would, if applied systematically over the world fisheries, put an end to overfishing and would in some cases allow catches to be increased after a latent period of a few years.

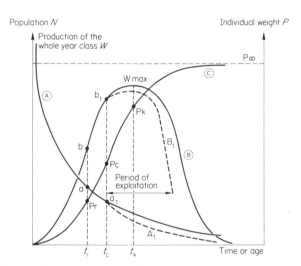

Figure 4.29 Curves showing the principal parameters involved in the theory of fishing.
Curve A: Variation with time of the total number of individuals N belonging to the same year class.
Curve B: Variation with time of the total weight W of a year class.
Curve C: Variation with time of the individual weight P.
t_r = age of recruitment in the stock being fished; t_c = age at which catches are first made; t_k = critical age. P_r, P_c and P_k are respectively the mean individual weights, and hence the sizes, corresponding to t_r, t_c and t_k. From t_c onwards, because of mortality, curves of population A_1 and of production B_1 fall below those describing the situation in the absence of fishing. The points a, b and a_1, b_1 correspond to the population and production of the year class at t_r and t_c. Theoretically, all the individuals in a given year class should be caught at t_k. Since that is in practice impossible, there is a period of exploitation which starts at t_c, the minimum age at which fish can be caught with the chosen mesh size.
(Adapted from Gulland, 1969)

The potential of existing fisheries

At the present time, the temperate waters of the Northern Hemisphere are being fished to their maximum capacity. The Atlantic, given its area and special ecological characteristics, is the most productive of the oceans but also the most heavily exploited. The Pacific supplies about half the total world catch, but the stocks of most of the major species—*Engraulis ringens*, halibut, hake (*Merluccius productus*), salmon, large tuna and bonito—are already either fully harvested or are overfished. The Indian Ocean perhaps offers the possibility of some increase in landings, although here, too, most of the commercially valuable stocks are overexploited, particularly prawns. Only the Antarctic still has any possible room for further increases in production, particularly of krill (with all the reservations that can be made about it) and of various benthopelagic and mesopelagic species of fish or cephalopods. Nevertheless, it should be remembered that the hunting of whales, which were once flourishing residents of the Antarctic Ocean, has seriously depleted most of the stocks.

One method of increasing catches would be the harvesting of families or groups that have been neglected so far. Cephalopods are probably relatively underexploited on a world scale and a significant growth in landings of mesopelagic and bathypelagic species might be obtainable. Geistdorfer (1975) has demonstrated the sizeable populations of species occupying the continental slope: for example, Gadidae in the bathyal zone. However, the techniques needed for catching such species have still to be developed since they are bound to be more complex than the traditional ones used at present.

Another important contribution to increased yield would be the avoidance of waste and loss after capture. Prawn fisheries, for example, catch from 4 to 20 times more fish than crustaceans, most of which, amounting to some 5 million tonnes per year at least, are thrown overboard! On a world scale, fisheries throw back into the sea some 15 to 20 million tonnes per year of small pelagic species immediately after being caught. Fish of such a small size deteriorate rapidly and it is not profitable to gut them on factory-ships. A final cause of wastage is the poor organization of the distributive network, particularly in the means adopted for storage. This produces an annual loss of 10 million tonnes of fish, mainly in the Third World.

An important factor in maintaining the productivity of maritime fisheries is the protection of spawning grounds and coastal nurseries. These biotopes are essential to the life cycle of many commercially valuable species, but they are often situated in estuaries where they are exposed to changes in their environment from the development of upriver industrial regions and the resultant pollution of coastal waters. Technocratic 'developers' too often forget the essential role of estuaries in the productivity of maritime fisheries when they encourage unregulated industrialization near the sea in so-called 'developed' countries.

In France, industrial development in the Gulf of Fos, in the estuaries of the Gironde and the Loire, not to mention that of the Seine, have caused a significant and sometimes dramatic decline in coastal fishing.

Finally, another way of using aquatic ecosystems more extensively for food production is through the development of aquaculture or fish farming in lagoons and/or estuaries of the coastal regions. At the moment, only a small fraction of the areas that could be used for this activity are in fact devoted to it. Appropriately managed, aquaculture does not alter the environment in any way and it enables habitats that are not suitable for any other purpose to be developed economically. World production from various forms of fish-farming, which amounted to 7 million tonnes per year at the end

of the 1970s, could reach 30 million tonnes per year if it were given the benefit of the scientific, technical and economic aid necessary to enable it to flourish.

4.3.2 Mineral resources of the sea

These can be divided into three categories: those which are found dissolved in sea water; those occurring in the benthic sediments or in recent sedimentary deposits of the continental shelf; and lastly, those which occur in the sedimentary layers on the ocean bed beyond the continental shelf.

Sea water

Sea water contains nearly every chemical element in the various substances dissolved in it although mostly only as traces. Table 4.14 lists the most abundant of these in order of decreasing concentration. Only a few are present in sufficient amounts to be economically recoverable: sodium chloride in particular has long been extracted from saltpans. Sea water is also the source of most, if not all, of the world supplies of iodine, bromine and magnesium.

However, even in the case of elements that are relatively quite abundant, the extraction cannot be carried out directly but only through an intermediary such as a biological concentrator or some other device. For the halogens, great use is made of certain macrophyte seaweeds like the Laminariae which concentrate these elements within themselves at levels much higher than those in sea water. Magnesium is extracted by a complex chemical process in which roasted oyster shells are mixed with sea water to form a precipitate of magnesium hydroxide, a method which now provides half the world supplies of this metal.

The other elements are too dilute for any recovery to be economically feasible. It is quite common to hear it said that 4.34 kg of gold could be extracted from 1 km^3 of the sea, but one wonders whether those who say this have considered the energy consumed in such an operation. In any case, a process making such recovery possible has yet to be discovered. . . .

The ocean bed

The deep ocean bed is covered in places with polymetallic nodules or with metal-bearing mud. The discovery of the nodules rich in metals excited great interest a few years ago when it was found that certain abyssal regions of the Pacific contain about 120 000 tonnes of manganese per km^2 as well as various other metals of prime importance like cobalt, nickel and titanium (Table 4.15). However, it ought to be pointed out once more that techniques for extraction need

Table 4.14   List of the principal elements contained in sea water in order of decreasing abundance (from Cloud *et al.*, 1969)

Element	Concentration (kg per 10^6 litres)	Chemical form	Value ($ per 10^6 litres)
Chlorine	19 920	NaCl	244.1
Sodium	11 040	Na_2CO_3	100.0
Magnesium	1 416	Mg^{++}	1091.0
Sulphur	930	S^{--}	26.7
Calcium	420	$CaCl_2$	39.6
Potassium	398	K_2O (equiv.)	24.0
Bromine	68	Br_2	50.2
Carbon	30	Graphite	negligible
Strontium	8.4	$SrCO_3$	0.53
Boron	4.8	H_3BO_3	0.79
Silicon	3.1	—	negligible
Fluorine	1.3	CaF_2	0.09
Argon	0.6	—	—
Nitrogen	0.5	NH_4NO_3	0.26
Lithium	0.18	Li_2CO_3	9.5
Rubidium	0.12	Rb	33.0
Phosphorus	0.06	$CaHPO_4$	0.02
Iodine	0.06	I_2	0.26
Barium	0.04	$BaSO_4$	0.002
Indium	0.02	In	1.1

much more development before it can become economically feasible to exploit these sources.

The continental shelf

This is at present the only sea area whose mineral resources are really accessible. Moreover, since it is merely the underwater extension of the continental crust its geology is well known. Thus, oil and natural gas can now be extracted from depths of up to about 300 metres and such offshore deposits are currently providing a quarter of the world's supplies. Similarly, there are other examples of exploitation of the minerals in the continental margins such as the underwater veins of tin in South-East Asia and the gold-bearing 'placers' in Alaska.

To sum up, then, it is clear that the exploitation of marine resources both in energy and minerals can make a sizeable contribution to world supplies. However, here too there is no question of having unlimited resources, and the technical difficulties and resultant high cost of extraction reduce the quantities available still further. Lastly, the impact on the marine environment is not necessarily negligible: the catastrophes that occurred at the offshore oil wells of Santa Barbara and more recently at Ixtoc 1 in the Gulf of Mexico illustrate the magnitude of the risks involved.

Table 4.15   Average composition of nodules from various regions of the Pacific (from Cronan, 1972, in Kesler, 1976)

Metal	Region of the Pacific		
	North-East	Central	South
Manganese (%)	22.33	15.71	16.61
Iron (%)	9.44	9.06	13.92
Nickel (p.p.m.)	10 800	9 560	4 330
Cobalt (p.p.m.)	1 920	2 130	5 960
Copper (p.p.m.)	6 270	7 110	1 850
Lead (p.p.m.)	280	490	730
Molybdenum (p.p.m.)	470	410	350
Titanium (p.p.m.)	4 250	5 610	10 007

Chapter 5

Agricultural Ecosystems and Food Production

As the end of this century approaches, humanity is facing for the first time three interdependent world-scale problems that have to be resolved simultaneously: a population explosion, underdevelopment, and an insufficient production of food. Among the various factors that ultimately limit the world's population, it is undoubtedly the shortage of food which is most obvious to the lay person and which first produces recognizably disastrous effects, some of which have already been witnessed.

It was a comparison between the potential growth of the human population and the possible increase in food supplies, the first exponential, the second at best linear and even then only temporarily so, which led Malthus to write his famous *Essay on the Principle of Population*. As long ago as 1798, this author was insisting on the fact that 'the possibilities for growth in a human population are infinitely greater than those of the soil for the production of the necessary food'. It is true that in some self-styled 'progressive' circles the term 'malthusian' was used pejoratively and even somewhat scornfully, but ecological analysis shows that in fact his greatest mistake was to be born ahead of his time. In spite of the obscurantism associated with certain religious persuasions and of the dogmatism permeating certain philosophical movements, some awareness began to develop of the dangers to which a population explosion would expose the whole of humanity because of the necessarily limited quantities of natural resources available.

Such awareness became particularly evident during the 1970s. The world population, estimated at 2000 million in 1930 and 4000 million in 1975, would become 8000 million by 2015 at the current rate of growth. This rate—nearly 1.8 per cent per year in 1979—corresponds to a doubling time of 40 years. This means that in order to maintain food production at its present volume *per capita*, already clearly insufficient on a world scale, there would have to be an increase in total production over the next 40 years equal in amount to its growth from the beginning of Neolithic times to the present. To put it another way: an exponential increase in the amount of agricultural production is a necessary counterpart to an exponen-

tial growth in human population. In reality, the need for the increase in food supplies is more urgent than that, given the present precarious situation for three-quarters of humanity and the malnutrition endemic among nearly 1000 million people.

In short, then, agricultural output has to double over a brief period of at most 30 years if the world food situation is to improve, a demand that means an increase during that time equal to its total growth from its beginning some 10 000 years ago. This is without doubt the most formidable challenge that will confront humanity on the threshold of the twenty-first century.

5.1 Human Food Requirements

The nature and magnitude of human food needs, whether of individuals or populations, are determined by various external (environmental) and internal (physiological) factors, so that absolute values for such requirements cannot be established except as a statistical abstraction. It is quite obvious that minimum needs vary greatly with age, sex, bodily state, occupation, latitude and so on. Even within a given age class, two individuals of the same weight in identical conditions would have different metabolic rates and thus different dietary needs. For a whole population, the amounts of various foods required are all distributed normally: that is, they follow a Gaussian or normal curve.

However, given that there is this individual variability, it is possible to establish the minimum dietary content needed per day for each basic nutritional category (energy, proteins, vitamins, etc.). This minimum applies to 95 per cent of the individuals in any given age class and includes, as well as an overall energy content, quantities of protein, fat, carbohydrates and various organic and mineral substances in smaller amounts. The national and international authorities that specialize in the field of nutrition, like the joint FAO–WHO committee, have from time to time established minimum daily intakes of each of these food categories.

The total quantity of *energy* necessary depends a lot

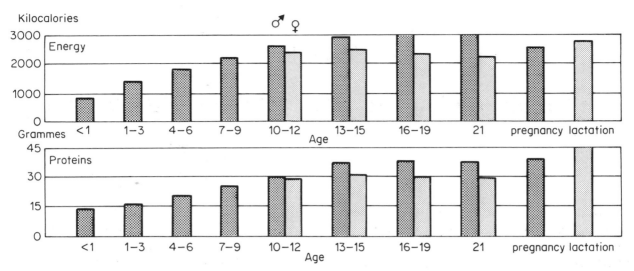

Figure 5.1   Recommended daily needs in energy and proteins for humans from birth to adulthood. (From Scrimshaw and Young, 1976, p. 60)

on age, sex, climate, physical activity and so on, and lies between 800 kcal per day for a very young child and 3000 kcal per day for an adult man (Figure 5.1). However, those in professions demanding great physical effort naturally need much more: Scandinavian lumberjacks, for instance, use up over 5000 kcal per day, while sportsmen and sportswomen performing exceptional feats of endurance (cyclists in the Tour de France) exceed 6000 kcal per day for the duration of the event.

Daily intake must also include a minimum quantity of *proteins*, but here there is some disagreement between different specialist committees. In particular, the norms prescribed by the FAO seem especially low in comparison with those recommended by public health authorities in the USA and Canada (Table 5.1). According to the North American standards, the minimum daily allowance for a baby under 1 year lies between 2.0 and 2.2 g kg^{-1} of body weight, for an adolescent it is about 1.05 g kg^{-1} day^{-1}, and for an adult of either sex it amounts to 0.8 g kg^{-1} day^{-1}. There is thus a progressive decrease in the amount of protein needed per unit body weight as age increases, except for the higher demands of women during pregnancy (1.34 g kg^{-1} day^{-1}) and lactation (1.18 g kg^{-1} day^{-1}).

However, a daily diet containing proteins in amounts equal to or even greater than these norms may turn out to be insufficient if it does not include in addition appropriate proportions of essential amino-acids which the human body cannot synthesize. There

Table 5.1   *Recommended minimum quantities for the daily intake of proteins according to various specialist authorities* (from Ehrlich *et al.*, 1977)

Category	Age (years)	Weight (kg)	Protein (g day^{-1})		
			FAO–WHO	NAS(US)*	Canada†
Babies	0–1	3–9	1.9 kg^{-1}	2.0–2.2 kg^{-1}	1.9–2.2 kg^{-1}
Children	1–9	10–30	14–25	20–39	18–37
Adolescents					
Male	10–19	31–64	36–38	40–54	38–54
Female		31–54	26–30	40–48	38–43
Adults					
Male	20–80	65–70	37	56	56
Female		55–58	29	46	41
Pregnant			38	76	61
Lactating			46	66	65

* From US National Academy of Sciences/National Research Council.
† From the Canadian Bureau of Nutritional Sciences.

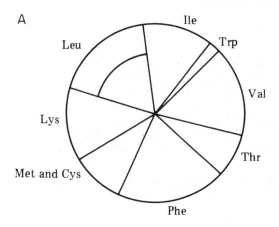

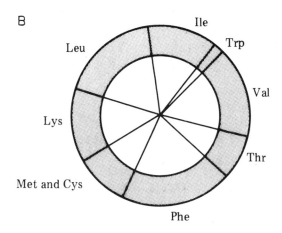

Figure 5.2   Nine essential amino-acids that cannot be synthesized in the human body but must be present simultaneously and in the correct relative proportions in the food supply for protein synthesis to occur. If one or more of the amino-acids are partially deficient (e.g. Leu as in A), all the others will be utilized in the same reduced proportion as well (grey areas in B). Leu = leucine, Val = valine, Phe = phenylalanine, Thr = threonine, Ile = isoleucine, Lys = lysine, Cys = cystine, Trp = tryptophan, Met = methionine.
(From Scrimshaw and Young, 1976, p. 56)

are nine of these: lysine, tryptophan, methionine, threonine, etc., and the proteins of animal origin (fish, meat, dairy products, eggs) themselves contain satisfactory proportions of them. A diet that is predominantly vegetarian, on the other hand, even though adequate in overall amounts of amino acids, will nevertheless be deficient because they are not in the correct proportions (Figure 5.2).

Finally, the human diet must also contain a certain quantity of *lipids*—particularly sterols and unsaturated fats—and a minimum daily supply of *vitamins* and *mineral salts*.

Food shortages and malnutrition

At present, nearly three-quarters of humanity are inter-

mittently or permanently faced with a shortage of food. If the FAO–WHO standards for minimum energy intake are used (2200 kcal per day for an adult woman, 2800 kcal per day for an adult man, with an average of 2500 kcal per day) then such amounts are available to barely one half of the world's population at best. If the norm of 2800 kcal per day used by European authorities is taken as the standard, then the shortage is even more evident and affects virtually every continent with the exceptions of North America, Europe and Siberia, and Australasia (Table 5.2, Figure 5.3). Similarly, no more than half of humanity have available the minimum daily quota of animal protein.

There is, however, a whole series of stages between a modest food shortage and a state of temporary or endemic famine. Nearly 1000 million people at present suffer from malnutrition, a condition recognized by the appearance of physical disorders linked to a deficiency of calories, or animal protein, or both. In less severe cases, malnutrition produces a growing susceptibility to infectious or parasitic diseases and a certain amount of dwarfing. In the worst cases, two formidable diseases develop: kwashiorkor, resulting from protein deficiency even with an adequate calorie intake, and marasmus, due to combined protein–calorie deficiency.

Children are clearly the most vulnerable to the effects of malnutrition. When there is severe undernourishment of the pregnant mother and of the child after birth, not only does it increase the risk of infant mortality but it also causes anomalous neuron development that may permanently impair the mental faculties. The brain is formed in the maturing foetus and during the first 2 years of infancy, a period when interconnections are being established between the originally independent neurons. Malnutrition at this stage causes a retardation of such development which cannot be remedied later on, with consequences for the individual and society that can readily be imagined (see, for example, Cravioto et al., 1966).

Generally speaking, malnutrition has such dramatic effects on very young children that the index of infant mortality is a more faithful guide than any other statistic to the overall food situation in a country (Table 5.2).

Although the essential causes of malnutrition are deficiencies in energy-giving food and proteins, lack of vitamins and of inorganic trace elements can also result in serious diseases. While it is certainly true that scurvy and pellagra have virtually disappeared and that beri-beri has become quite rare, other signs of vitamin deficiency continue to occur. Cases of rickets are still observed in Muslim populations among women leading highly secluded lives. Blindness from vitamin A deficiency is particularly widespread in India, southeast Asia and South America. Goitre of the thyroid due to lack of iodine is still to be found in numerous

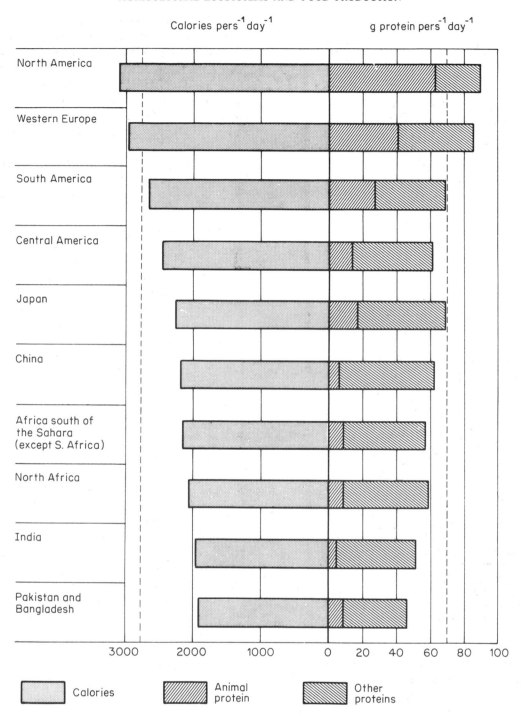

Figure 5.3   Average calorie and protein intake for various regions of the world. The dotted lines indicate the minimum amounts considered essential in the developed countries of the West. Notice the widespread deficiency in animal protein in the many tropical and subtropical areas

mountainous regions of the Himalayas, the Andes and Africa: according to the WHO, 5 per cent of these populations suffer from cretinism, an irreversible effect caused by shortage of iodine in the maternal diet before and during pregnancy. Similarly, in various tropical regions of Asia, Africa and America, the WHO estimates that 10 per cent of men and 30 per cent of women suffer from anaemia through lack of iron.

Finally, at the extreme limits of the range of food shortage are nearly 500 million people who live on the edge of acute famine and of whom some 12 000 die every day....

Table 5.2   Amount of energy and protein in the daily food intake for various regions of the world
(UN International Task Force on Child Nutrition, in Mayer, 1976)

Region	Energy in kcal pers⁻¹ day⁻¹	Protein in g pers⁻¹ day⁻¹	Index of infant mortality per thousand†
Developed countries			
Western Europe	3130	93.7	11
North America	3320	105.2	11
USSR and Eastern Europe	3260	99.3	24
Average, developed countries*	3150	96.3	15.9
Third World			
Africa	2190	58.4	120
Far East	2080	50.7	52
Latin America	2530	65.0	65
Middle East	2500	69.3	99
Average, Third World	2210	56.0	93

* Including those not in the above groupings.
† Source: Population Reference Bureau data sheet, Washington, 1983.

5.2   Changes in World Food Production During the 1970s

The world food situation has significantly worsened during the 1970s. According to the FAO's World Survey of Food (1977) covering 71 developing countries, in only two of them was the growth in agricultural production equal to or greater than the increase in population. In fact, there was some progress in Third World agriculture during the 1960s and 1970s when it achieved a rate of growth of food production comparable with that of the industrialized world. However, whereas the food available *per capita* had tended to increase in the latter case, it remained steady or even declined in the Third World.

The continual increase in the grain deficit illustrates the way in which the world food situation is deteriorating (Figure 5.4). In 1935–1936, most regions were either net exporters of grain or were self-sufficient in it, the exceptions being Western Europe, and China (where intermittent shortages undoubtedly occurred at that time). By 1975–1976, the whole world was in deficit except for North America, Australasia, Argentina and France (not shown in Figure 5.4).

A new feature that arose in the mid-1970s was the grain deficit of the USSR and Eastern Europe, who were formerly exporters. The deficit in the USSR amounted to several million tonnes each year from 1975, and reached 25 million tonnes in 1979–1980. In fact, the cereal shortage in these countries was largely the result of a change in eating habits to a diet containing more and more meat. Since the ecological efficiency of secondary production has an average

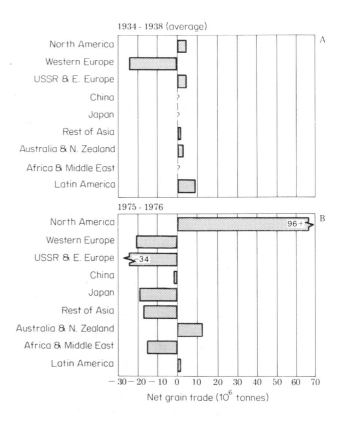

Figure 5.4   Increase in world cereal deficits over the last few decades. Diagram A shows the trade balance in cereals for various regions before the Second World War and B the same in 1975–1976. Whereas most continents (Western Europe excepted) were self-sufficient 50 years ago, only North America, Australasia, temperate South America and France (not shown) were net exporters in 1975. (From Wortman, 1976, p. 37)

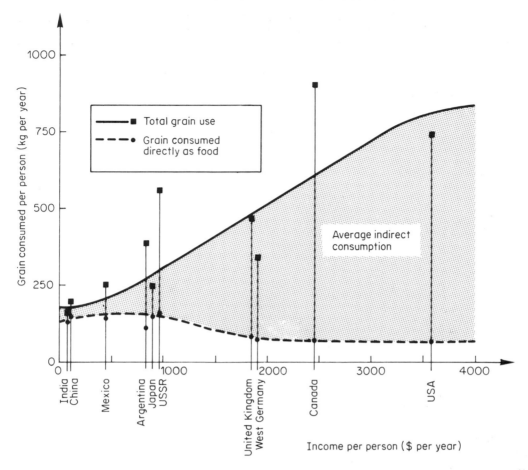

Figure 5.5   The relation between direct and indirect grain consumption and the mean income *per capita* (1960 values) for various countries. Notice that the higher consumption of animal products in developed countries greatly increases the indirect consumption of grain because of the low ecological efficiency (efficiency of energy transfer along the food chain). (From Brown, 1973)

value of at most 20 per cent for cattle raised in feedlots, the consumption of more meat in an affluent society tends to cause considerable increases in the indirect consumption of cereals (Figure 5.5), and this was certainly the case in the USSR.

The world grain deficit has thus grown continuously for 20 years or more (Figure 5.6). Already around 42 million tonnes in 1975–1976, it will lie anywhere between 75 and 100 million tonnes per year over the period 1985–1990, depending on the assumed growth in demand, and this in turn depends upon both the increase in population and the change in income *per capita* in Third World and developed countries.

Table 5.3 shows forecasts made by the United Nations of the increases in overall food requirements and agricultural production from 1970 to 1985, extrapolated from data for 1960–1975. The main feature emerging from this table is that in developed countries with relatively stable populations the increase in agricultural production will outstrip that of the population, whereas the reverse is true in underdeveloped countries.

Whatever future demographic policies may be followed by developing nations—even if we assume the adoption of voluntary and efficient programmes of birth control leading to early and massive reductions in the rate of population growth—world agricultural production must be substantially improved over the next few decades. It is not only a question of putting a definite end to famine, but of increasing the diet of most inhabitants of the Third World, whose daily food quota is still precarious even when there is no severe shortage.

What, then, are the ecological factors determining whether and by what amount increases in world food production can be achieved? Synecology shows that the productivity of agricultural ecosystems depends essentially on three of the principal ecological variables: energy, matter (in the form of inorganic nutrients available) and the species that make up the total population (that is, the nature of the plant species and the varieties chosen for cultivation plus the breeds of animal chosen for secondary production). Finally, the primary and secondary production of an

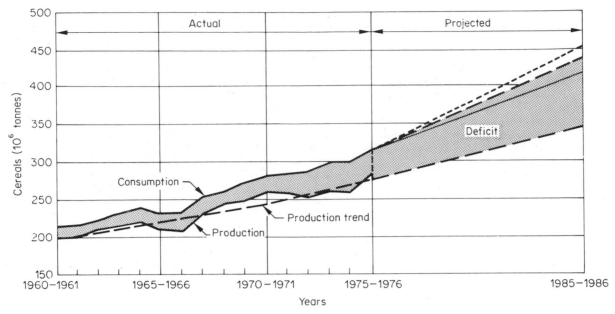

Figure 5.6   Increase in world grain deficit from 1960 to 1985. From less than 20 million tonnes per year in 1960, it will reach at least 70 million tonnes in 1985 and might even be as great as 100 million tonnes if there is a large increase in *per capita* incomes in the Third World (dotted line). (From Wortman, 1976, p. 36)

Table 5.3   *Forecasts of world food requirements and agricultural production in 1985 relative to those of 1970* (source: United Nations)

Region	Volume rates of increase (% per year)			Total volume by 1985 (1970 = 100)	
	Food requirements	Agricultural production	Population	Food requirements	Agricultural production
Western countries	1.4	2.4	0.9	124	143
USSR + E. Europe	1.7	3.5	0.9	130	168
Average, developed countries	1.5	2.8	0.9	126	151
Third World	3.6	2.6	2.7	170	146
Africa	3.8	2.5	2.9	176	145
South-East Asia	3.4	2.4	2.6	166	143
South America	3.6	2.9	3.1	170	152
Middle East	4.0	3.1	2.9	180	157
Asian communist countries	3.1	2.6	1.6	158	146
Average, developing countries	3.4	2.6	2.4	166	146
World average	2.4	2.7	2.0	144	150

agricultural ecosystem depend on space: that is, on the total area of cultivated land and grassland that is available.

5.2.1   Cultivated and potentially arable land areas

The primary factor determining world agricultural production is the soil and, with it, the amount of land with soil suitable for cultivation. What areas of such land are available in the various regions of the world? The estimates quoted by most authors are those con-

tained in a report of the US Department of Agriculture published in 1974, which calculated the maximum potentially arable land area of the world as 3.2×10^9 hectares and the area already cultivated as 1.4×10^9 hectares (Table 5.4 and Figure 5.7).

Table 5.4 shows that there are large regional differences in the availability of new land areas suitable for cultivation, many of which in any case are already in use either for raising livestock or for forestry. Whereas in Europe and Asia more than 80 per cent of the potentially arable land is already exploited, the

Table 5.4   Areas of land suitable for cultivation in various regions of the world

Region	Population* in 1979 (millions)	Land areas (10^8 ha)†			Area already cultivated per person (ha)	Area potentially arable per person (ha)	Cultivated area as % of potentially arable
		Total	Potentially arable	Already cultivated			
Africa	457	30.2	7.33	1.58	0.34	1.6	21.6
Asia	2498	27.3	6.28	5.18	0.20	0.25	82.5
Australia and New Zealand	17.6	8.2	1.54	0.16	0.91	8.75	10.4
Europe	483	4.8	1.74	1.54	0.32	0.36	88.5
N. and Central America	333	21.1	4.66	2.39	0.72	1.34	51.3
South America	264	17.5	6.80	0.77	0.29	2.56	11.3
USSR	264	22.3	3.56	2.27	0.85	1.34	63.8
Total	4317	131.4	31.91	13.89	0.32	0.74	43.5

* From *World Population Data Sheet 1979*, Population Reference Bureau, Washington DC 20036.
† From *The World Food Problem*, Economic Research Service, US Department of Agriculture, Foreign Agricultural Economic Report 298, Washington DC, 1974.

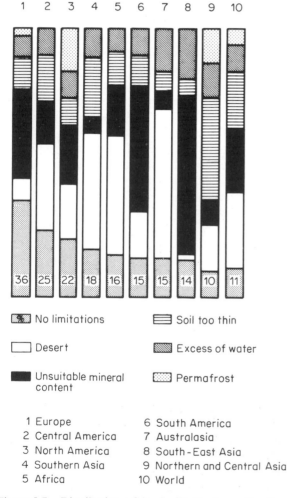

Figure 5.7   Distribution of land suitable for cultivation in the various regions of the world. The figure at the bottom of each column gives the area of soils that can be freely cultivated as a percentage of the total area. Note that such soils cover just 11 per cent of the total land surface of the world (From IUCN, 1980b)

corresponding figure is only 11.3 per cent for South America and 21.6 per cent for Africa.

According to the data provided by the US Department of Agriculture, the area per person already cultivated has an average value of 0.32 hectare over the whole world, whereas the area that is potentially usable would be 0.74 hectare per person: since only 43.5 per cent of the land suitable for agriculture is in fact cultivated, the arable land area could apparently be multiplied by a factor of 2.3. Personally, I consider such an estimate to be entirely utopian, since it takes no account of a whole series of ecological and/or economic factors which make lasting agricultural development of much marginal land quite impossible, particularly since it often has a fragile soil structure and is situated in areas of irregular rainfall.

The US Department of Agriculture's estimates have not made allowance for competition between the potential agricultural development of marginal land and its present use. In tropical or temperate regions, large expanses are often already given over to livestock, raised either by nomadic tribes like the Masaï of East Africa or by ranchers like the cattle-breeders of Western USA or Australia. In either case, the primary production is already used to its maximum and it is not easy to see how the secondary productivity of the land could be increased. The remaining soil considered capable of development as arable land by the US Department of Agriculture is already covered with tropical or mountain forests. Their clearance would mean the end of the economic exploitation of forestry at a time when the production of wood and its price are both rising as a result of increased demand.

It should also be noted (cf. Chapter 1) that the amount of arable land available not only decreases as the population increases, but does so at a greater rate than mere inverse proportionality. At present the total area covered by all the towns in the world exceeds the overall area of France, but in addition they are generally built on the best agricultural land. For example, the new towns which have been allowed by a technocratic administration to develop around Paris represent a lavish waste of some of the world's richest soil. There is also the construction of roads, airfields and military sites, the flooding of valleys for hydroelectric schemes and so on—all of which are at the expense of land that is already cultivated or is suitable for cultivation.

The cost of clearing underdeveloped land that is suitable for cultivation is another factor. The investment involved would be prohibitive: current costs of clearance range from $150 to 5000 per hectare with an average of $1100. If we suppose that we wanted to increase agricultural production simply by developing virgin land at a rate comparable with that of the growth in world population, and make the rather unrealistic assumption that 1 hectare of new land could feed 4 people, then a world investment of $20 $\times$ 10^9 would be involved.

All the undeveloped land considered by economists as suitable for cultivation proves to have shortcomings in one or more aspects when all relevant data are taken into account: soil, climate and so on. It is thus very uncertain whether it is capable of exploitation at all. In fact, it turns out that nearly all the cultivable land has already been developed, sometimes for a very long time. The area currently estimated as under cultivation (1.41 $\times$ 10^9 ha) represents nearly the whole of the potential agricultural land area, estimated by specialists of the FAO in 1972 to be about 1.45 $\times$ 10^9 ha (Table 5.5). Another FAO report, in 1973, emphasized that 'in Southern Asia, in several regions of the Far East, in North Africa, and in certain areas of tropical Africa and Latin America . . . there is

at present no possibility of increasing the amount of arable land . . . in the driest regions it will even be necessary to reconvert to natural pasture areas whose yield is marginal or sub-marginal'.

The Nairobi conference organized by the UN Programme for Environment and Development in September 1977 even pointed out that some large areas of steppe and savanna—land which the US Department of Agriculture would like to develop—ought to be rapidly taken out of all use, even for pasture, if a short-term conversion to desert was to be avoided.

These difficulties undoubtedly explain why the agricultural development of marginal land is proceeding at a rate of only 0.15 per cent per year, or twelve times less than the rate of growth of the world population over the period 1950–1970. In fact, cultivation of this sort of land, two-thirds of which is steppe or savanna, poses a formidable threat to the stability of the soil. Evidence of this has already been provided in the great plains of western USA which were devastated by horrifying dust-bowls during the 1930s (see p. 145), following the development of cereal monoculture on fragile soil. More recently, Soviet agronomists have been faced with considerable problems over the vulnerability of the soil following the clearance of virgin land in Kazakhstan at the end of the 1950s.

In the same way, the agricultural development of the 6 $\times$ 10^9 hectares of forest considered as potentially arable by the US Department of Agriculture seems just as utopian. Not only is much of this area already being exploited for its wood, but its soil is generally quite unsuitable for cultivation, being ferrallitic and hydromorphic. A recent study of the Amazon basin, for example, showed that only 3 per cent of the land at present covered by tropical rain forest was suitable for clearance and subsequent cultivation. In spite of that, the Brazilian Government is speeding up deforestation along the Amazon, greatly risking the conversion of

Table 5.5 Land suitable for cultivation worldwide (from FAO *Production Yearbook 26*, 1972)

Continent	Land area (10^6 ha)	Area suitable for cultivation (including temporary grassland and itinerant farming)	
		(10^6 ha)	(% of total)
Europe	493	145	29.4
Central and N. America	2 242	271	12.0
South America	1 783	84	4.7
Asia	2 753	463	16.8
USSR	2 240	227	10.1
Oceania	851	47	5.5
Africa	3 031	214	7.1
World total	13 393	1451	10.8

Plate XI   Space as a resource

1   View of Victoria on the island of Hong Kong.

2 and 3   Old buildings in a heavily populated district of Hong Kong and high-rise blocks under construction in a residential area. Hong Kong provides a better and more extraordinary example than anywhere else in the world of the type of urban environment produced by overpopulation of a very limited amount of space. Nearly 6 million people are crowded into about 900 square kilometres of a mountainous archipelago creating a population density of over 200 000 per square kilometre in some districts.

4   Agricultural land around Paris undergoing uncontrolled urbanization. The absence of rigorous environmental planning during the last few decades has allowed towns in France to spread out in a disorganized way. As a result, a confused jumble of agricultural and urban areas has been created whose main consequence is an accelerating loss of some of the most fertile land in the world.
(Photographs F. Ramade)

Plate XII Soil erosion

1 Soil erosion through slipping of the land surface following the clearance of mountain forests in the Nyambeni Hills, Kenya.

2 Water erosion in the Great Plains of the USA. Here and there, examples of contour cultivation can be seen following the curves of equal land height. This is designed to stop the destruction of the land by erosion due to run-off.

3 Wind erosion in Colorado. The bands of lighter colour which can be seen in this aerial view have a general direction perpendicular to that of the prevailing wind. They correspond to patches where the surface layers have been eroded.
(Photographs F. Ramade)

the region into a new Sahel within a few decades (Friedman, 1977).

Various physico-chemical changes cause soils to deteriorate when they are cultivated or when their natural plant cover is destroyed through human activity, and we now turn to an examination of some of these.

Degradation of soils, 1: salinization

The irrigation of marginal land having inadequate and uncertain rainfall in order to cultivate it has in many areas produced irreversible chemical changes in the soil. In countries with a long dry season, poor drainage and/or the existence of saline groundwater near the surface, together with high evapotranspiration, lead to high concentrations of various salts in the surface layers, making the soil unsuitable for cultivation, if not totally sterile. When saline groundwater rises by capillarity, it forms outcrops of *solonchak*, a soil having a white efflorescence due to the accumulation of sodium chloride and sodium sulphate (salinization). In some cases, where the soil is then leached by irrigation, calcium is replaced by sodium to produce an irreversible change in soil chemistry. During a dry period, the sodium migrates to the upper horizons and forms sodium carbonate. The soil becomes very alkaline with a pH value greater than 10, and this deflocculates the clayey mass and disperses the organic

matter. A black efflorescence gradually appears, indicating the destruction of the humus-clay complex and producing a *solonetz*, a soil whose fertility has completely vanished and is considered to have no chance of being restored.

In Pakistan, nearly 3 million hectares have been lost in this way because of poor irrigation techniques, and the phenomenon also occurs in India, Iraq, the southern republics of the USSR, Mali, North Africa, and in south-western USA, where more than half the irrigated land is subject to it. In Imperial Valley, southern California, the use of water from Colorado that is already loaded with salts from upstream irrigation systems, causes great problems that are made even worse by the fine texture of the soil which hinders drainage. Excessive irrigation has produced high concentrations of salts that are not easy to disperse, and this is threatening some 250 000 hectares of cultivated land (Dregne, 1977).

Degradation of soils, 2: laterization

In humid tropical regions, laterization is a formidable obstacle to the cultivation of new land by clearance of tropical rain forests. Lateritic soils are acid, with a low concentration of silica and a high concentration of iron and aluminium oxides that give them a red or yellow colour. A vertical profile of such a soil reveals an upper horizon that is ferruginous and very poor in organic

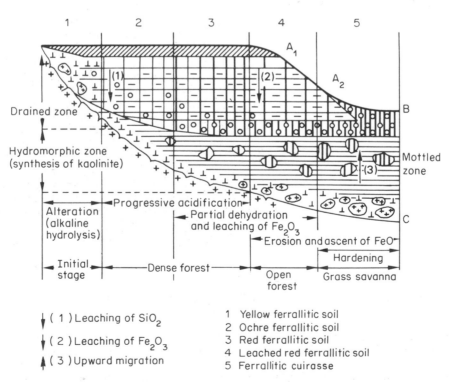

Figure 5.8   Mechanism of the formation of a lateritic cuirasse (From Duchaufour, 1965)

matter and from which silica and nutrient minerals have been leached. Below that is a horizon with the same composition, but with a porous texture that allows the silica, etc., to migrate through it. When a forest, or the secondary vegetation that has replaced it, is cleared in order to cultivate the land, the exposure of the soil to the sun and the rain produces a lateritic cuirasse either by migration of iron oxides to the upper horizons or by erosion of the surface layers (Figure 5.8).

Because of changes of that sort, clearance of tropical forest ecosystems has very often transformed densely covered forest areas into denuded and sterile land. In West Africa, for instance, the lateritic cuirasse at present occupies an area of more than 1 million km² stretching from Cameroun to Guinea, with just a thin veneer of soil here and there between the outcrops of haematite or bauxite which represent the final stage of the conversion to laterite. Eastern India, south-east Asia, Madagascar and South America also have lateritic soils over the greater part of their land surfaces.

In view of all this, great care ought to be exercised in the development of large expanses of land around the Amazon where the soils are for the most part ferrallitic (Hammond, 1977). At Iata in the north-western part of the Mato Grosso, the clearance of a large afforested area at the beginning of the 1960s transformed a magnificent rain forest into a sterile lateritic cuirasse in less than 5 years (McNeil, 1964). In spite of that, most Third World governments continue to encourage the clearance of virgin forests for cultivation and in doing so expose the soils both to laterization and, in the case of mountain forests on steeply sloping ground, to erosion.

Degradation of soils, 3: erosion

Deforestation and unsuitable cultivation of soil that is fragile (with too poor a structure or having too steep a

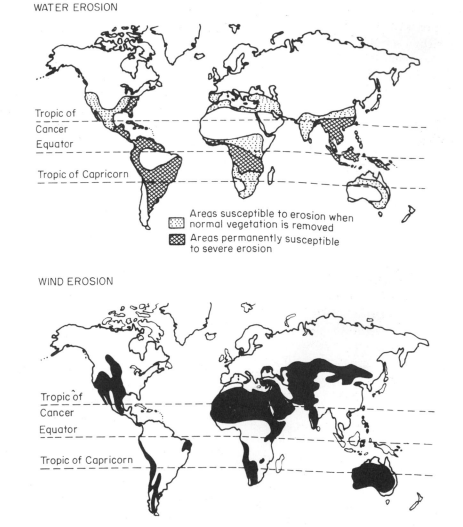

WATER EROSION

Tropic of
Cancer
Equator

Tropic of Capricorn

Areas susceptible to erosion when normal vegetation is removed

Areas permanently susceptible to severe erosion

WIND EROSION

Tropic of
Cancer
Equator

Tropic of Capricorn

Figure 5.9   World distribution of soils susceptible to rainfall and wind erosion. (From Tivy, 1975, p. 32)

slope) are bad agronomic practices which can lead to catastrophic erosion of the surface layers. By the end of the Neolithic period, immense tracts of land both in the Middle East and in China had already been ruined by the cultivation of ground that should never have been deprived of its natural plant cover. At present, 34 per cent of Chinese territory consists of eroded land whose possible rehabilitation is highly questionable. Undoubtedly, the risk of erosion is the main obstacle to the cultivation of marginal land considered as 'potentially arable' by some economists. Its two main agents are water and wind (Figure 5.9).

Water erosion This type of erosion is most prevalent in regions with high relief, although it can occur even in land with quite moderate slopes if it is planted with row crops. The protective power of plant cover has long been appreciated and was already known to the ancient Greeks several centuries before the beginning of the present era.

Soil erosion does not occur in forests and is only very slight on grassland, even if it slopes quite steeply. The continuous carpet of litter and moss in afforested areas acts like a sponge: 1 kilogramme of dry moss can absorb 5 litres of water so that 1 hectare of Mediterranean forest, for instance, retains some 400 m³ of water after a violent storm. Part of this is lost through evapotranspiration and the rest seeps slowly downwards and gradually replenishes the groundwaters below. There is no run-off.

Continuous plant cover of any sort slows down raindrops so that they lose a lot of their kinetic energy and reach the soil at very small speeds. Moreover, any run-off that does occur is held in check and its destructive effects nullified. Deforestation or overgrazing, on the other hand, increase erosion by allowing much more violent impact of the rain on the bare surface and a greater run-off.

The collapse of ancient civilizations in what is now Iraq was undoubtedly the result of soil degradation due to overgrazing and deforestation in the upper drainage basins of the Tigris and Euphrates. These rivers originate in the Armenian plateau, a region which has long supported dense pastoral populations and their large flocks of sheep and goats. The land degenerated because of uncontrolled and excessive cultivation and rearing of livestock, and there was in addition increasing deforestation of the slopes of the drainage basins brought about by the demand for wood for heating and construction and by the need to create new areas where the overabundant flocks could roam. All these factors combined to generate massive soil erosion and thus a continual surfeit of sediment collecting in the Tigris and Euphrates. During the Sumerian civilization, the inhabitants of the river valleys were able to control the great quantities of sand and silt, but succeeding

empires found it more and more difficult to maintain the gigantic irrigation network intact, fed as it was by a growing mass of river sediment, even with armies of workers and slaves labouring without respite to dredge the channels. Alluvial deposits, due mainly to erosion in the drainage basins of the Tigris and Euphrates, have been so great over the past few millennia that the river mouths have advanced some 250 kilometres into the Persian Gulf since the Sumerian period (Dasmann, 1976).

These tracts of land remained verdant and fertile as long as the channels and ditches were kept in good repair. The final collapse, however, came in the twelfth century when the hordes of Genghis Khan wiped out the Mesopotamian civilizations and destroyed the irrigation network, thus opening the door to desertification of the land.

The development of modern industrial agriculture based on a very restricted number of cultivated crops, or even on monoculture (groundnuts or coffee in the tropics; wheat or maize in temperate regions), has become a sizeable contributor to soil erosion. Pimentel *et al.* (1976) showed that the cultivation of maize in the USA has been accompanied by an annual loss of soil amounting to between 6.6 and 200 tonnes per hectare, the exact figure depending on the locality and slope of the land (Table 5.6). The data allow a comparison to be made between the protection against erosion afforded by forest cover or grassland and the catastrophic effect of growing a row crop like maize. It should also be noted that rotation of crops gives the soil better protection.

Wind erosion A substantial amount of wind erosion occurs in steppe-like regions where the soil has a sandy texture or consists of fine periglacial alluvium (loess, for example). Such erosion has ruined vast tracts of land in Asia, in the Mediterranean basin, in the savannas of the Sahel, and in North America, particularly in the Mid-West of the USA.

The clearing of the 'prairies'—grasslands with very fertile but fragile chernozem-type soils—has produced enormous devastation in Texas, Oklahoma, eastern Colorado and North and South Dakota (see Plate XII). A cereal monoculture subjected to prolonged periods of drought, particularly during the 1930s, increased the rate of wind erosion and created the famous dust-bowl. Gigantic tornadoes, whirlwinds of dust originating over areas that sometimes covered almost 500 000 km², were carrying hundreds of millions of tonnes of soil into the atmosphere and leaving only the bare substratum of rock. This devastation accentuated the crisis already being faced in rural America during the interwar years and had socio-economic consequences that have been graphically described in John Steinbeck's *The Grapes of Wrath*.

Table 5.6   Annual loss of soil from various regions of the USA for different crops and vegetation (from Pimentel *et al.*, 1976)

Crop or type of vegetation	Region	Slope (%)	Loss (tonnes per hectare)
Maize (monoculture)	Missouri (Columbia)	3.68	48.6
	Wisconsin (La Crosse)	16	219.0
	Mississippi (northern)	—	53.8
	Iowa (Clarinda)	9	69.9
Maize (disc harrow)	Indiana (Russell)	—	51.6
	Ohio (Canfield)	—	30.1
Maize (conventional)	S. Dakota (eastern)	5.8	6.6
Maize (monoculture)	Missouri (Kingdom City)	3	51.8
Maize (contour)	Iowa (south-western)	2–13	52.8
	Iowa (western)	—	59.2
	Missouri (north-western)	—	59.2
Cotton	Georgia (Watkinsville)	2–10	47.1
Wheat	Missouri (Columbia)	3.68	24.9
	Nebraska (Alliance)	4	15.5
Wheat (with pea rotation)	Washington (Pullman)	—	13.8
Wheat	Washington (Pullman)	—	17.1–24.4
Bermuda grass	Texas (Temple)	4	0.07
Native grass	Kansas (Hays)	5	0.07
Forest	North Carolina	10	0.005
	New Hampshire	20	0.02

Most of the North American badlands are in fact the result of an uncontrolled cultivation of soils which ought never to have been cleared at all and which, if ecological factors had been properly considered, would have been left to a more natural exploitation through extensive or semi-extensive rearing of livestock.

Degradation of soils, 4: other effects connected with cultivation

There are other practices which can in varying degrees jeopardize the long-term stability of agroecosystems. One way in which biotopes are destroyed is particularly evident in France and is connected with agricultural operations having only one aim: to promote the development of an intensive industrial agriculture whatever the ecological or economic consequences may be.

One of the main factors responsible for the destruction of biotopes—and of the landscape—in France has been *remembrement* (the regrouping of cultivated plots of land) in the areas of *bocage* that stretch across the country from the mid-Atlantic coast to the Paris basin and beyond. This operation was originally justified by the fact that single farms had become fragmented into tiny plots, sometimes distributed over several communes, a situation which meant that farmers spent more time in travelling from one piece of land to another than they did in cultivation. However, the regrouping that has increased the sizes of the plots belonging to the same farmer has at the same time become essentially anti-ecological in its nature and scale of operations.

Similar regrouping in Great Britain and Germany had been carried out in a way that showed good judgement, so that its maximum impact on the landscape and biological equilibrium was limited. In France, on the other hand, it has assumed a brutal and even offensive quality because of its ecological effects. In certain areas, enormous stretches of land with over 100 hectares to a single owner have replaced the former mosaic of small fields and meadows, each surrounded by living hedges or enclosed in embankments planted with various types of tree—environments that harboured diverse populations, including numerous insect-eaters and other life-forms of benefit to agriculture (see Plate XIII).

What agronomist or ecologist worthy of the name could assert in good faith that the elimination of hedges, the levelling of embankments and the filling-in of ditches, all to help the movement of large agricultural machines, would have no effect on the microclimate and would not be an ecological disaster of the first magnitude? Not only that, but this fanatical levelling of the land is accompanied by a 'rectification' of waterways using mechanical diggers, an operation justified by some so-called ecologists using arguments that are at the very least misleading; it is an operation with disastrous hydrobiological consequences.

Apart from the effect on the microclimate, one of the

most obvious results of eliminating embankments planted with trees serving as windbreaks is a considerable increase in run-off. Some of the disastrous floods in the south-west of France and in Brittany during the 1970s must be directly attributed to the regrouping of land. Apart from flooding, however, the introduction of enormous areas planted with maize or other row crops that leave the soil denuded of vegetation encourages erosion by surface water or through leaching even where there is only low relief.

Another factor capable of producing long-term changes in soil structure is the use of various chemical products in agriculture, particularly of mineral fertilizers which have replaced the now abandoned organic manure. Commoner (1972) has pointed out that the application of excessive potash or nitrate chemical fertilizers can modify the granularity of the soil by concentrating certain of the elements in the clay-humus complex: in particular, certain clays can become deflocculated.

Agricultural machinery also modifies the soil structure. Heavy tractors, gang-ploughs and so on, driven over clay-rich soil, can compress it to form a 'plough-pan' or 'plough-sole', a relatively impermeable layer that inhibits proper drainage and sometimes has a disastrous effect on the fertility of the land.

The fight against soil erosion

Unrestricted cultivation of fragile soil and the use of unsuitable techniques are the main causes of erosion, both in tropical and temperate regions, and are a constant threat to the stability of agroecosystems. In the USA, for instance, it has been estimated (Dasmann, 1976) that at the beginning of the 1960s, out of a total of 193 million hectares of land considered to be cultivable, 139 million were already under cultivation and 125 million were showing more or less pronounced signs of erosion. In all, some 20 million hectares of land and at least one third of the topsoil have been completely destroyed in the USA over the last 200 years by erosion arising from cultivation.

Even in France, as we have already seen, erosion is not confined to areas of high relief (that is, the 4.5 million hectares south of a line from Andorra to Modane, east of Grenoble). The development of intensive cereal farming is causing erosion by water as soon as the slope of the land reaches a gradient of 4 per cent, and this is happening in Aquitaine, in Brittany, and even in the Ile de France. Although the effects are not as serious as in less temperate countries, they are in fact aggravated by the large areas of open farmland from which all hedges and embankments have been removed.

The agronomic value of a soil and its resistance to all forms of erosion depend to a large degree on its physico-chemical structure. Soil texture can be represented on a *texture triangle* (Figure 5.10) in which each apex corresponds to one of the three classes of granularity: clay (particle size below 2 μm), silt (between 2 μm and 20 μm) and sand (above 20 μm)—although the size at which the division between silt and sand occurs varies among different systems of classification and is often placed at 50 to 60 μm. Soils deficient in clay and/or organic matter do not retain water very well and have a structure that is sandy and liable to crumble so that they are highly susceptible to erosion. Excess clay, on the other hand, makes the soil impermeable to water and inhibits its aeration.

The structure of a soil determines most of its properties and its suitability for cultivation: its permeability to moisture, its capacity to retain water, its degree of aeration, its ability to make available to plants the nutrients stored in the clay-humus complex, its ability to withstand mechanical working of the topsoil and, finally, its ability to support a permanent plant cover to reduce the amount of erosion. In general, soils suitable for cultivation without any precautions being taken are those with reasonable proportions of sand and silt bound together by a well-flocculated clay-humus complex. With this type of texture, such a soil can absorb any excess rainfall and thus cut down surface run-off. Moreover, the strong cohesion between particles provided by the clay-humus fraction protects it against erosion by either wind or water.

The protection of cultivated land against erosion involves a combination of civil engineering and biological techniques. On land of only moderate slope, ploughing has to be carried out with the furrows running along lines of equal height (contour ploughing), while on more steeply sloping ground steps or even terraces are essential. Terracing in particular is a technique that has been widespread since the earliest times in the Mediterranean region, in China, and in the Inca civilizations of America (Figure 5.11). Very often, the plough is discarded and replaced by the disc harrow, which does not turn the soil and leaves the plant debris on the surface to reduce the kinetic energy of the raindrops.

Biological methods for increasing resistance to erosion either involve reinforcement of soils by the addition of litter, manure or other organic matter, or else they involve the crops themselves through the adoption of different techniques of cultivation. Crop rotation with successive plantings of row crops and cover-crops considerably reduces erosion. On steep slopes, the worst effects of run-off can be avoided by strip-cropping: the alternation of row crops, cereals, legumes and fodder plants, and even artificial grassland, greatly increases soil stability.

Finally, the planting of hedges and tree belts

Plate XIII   Rural landscapes in central France

1   Bocage in the Monts du Lyonnais.

2   Orchards around Saint-Genis-Laval, Rhône.
(Photographs F. Ramade)

3   Regrouped land in Nièvre, now much more uniform in character. Notice the heaps of twigs and roots from former hedges now destroyed.
(Photograph F. Terrasson)

4   Water erosion of the soil at a point where a former hedge was uprooted during a regrouping operation.
(Photograph F. Terrasson)

1

Plate XIV   Soil conservation

Terraced ricefields near Tasikmalaya in Central Java.
(Photograph F. Ramade)

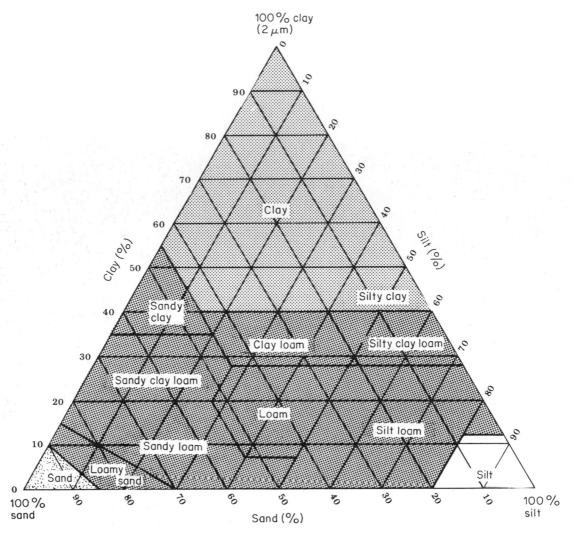

Figure 5.10   Triangular diagram for the classification of soils by texture in terms of the relative proportions of sand, silt (particle size less than 20 μm) and clay (particle size less than 2 μm). (From US Department of Agriculture Soil Survey Manual)

provides good protection against wind erosion: in the American Mid-West and in the USSR veritable bands of forest have been created to act as windbreaks.

Loss of agricultural land from other causes

Biotopes are being destroyed in large numbers by urbanization and industrialization with the most productive agroecosystems suffering the greatest damage from incursion by concrete and tarmac (Brown, 1970). The reckless waste of high fertility land through the expansion of cities and from the building of factories, communication networks, airfields, etc., is reducing the potential food production of every country. Like the growth of population, it is decreasing the area of cultivable land available *per capita* so that this latter quantity diminishes more rapidly than the population increases (cf. Chapter 1, p. 14). In short, the agricultural area *per capita* is drastically reduced by

any increment in the population because of the associated urbanization and industrialization.

From the earliest times, human beings have habitually built their towns in the most fertile river basins and coastal plains, and indeed the first cultivated territories were those with the greatest potential productivity. As long as urban populations remained small, this practice was not of great significance, but it has become catastrophic since the beginning of the century under the combined pressure from general population growth and the drift from the land generated by industrial development.

In 1925, 28 per cent of the world's population lived in urban areas: the figure now is more than 40 per cent. In France, the urban population has grown from under 50 per cent in 1939 to 80 per cent now. In 1960, there were in the world 109 cities with populations of more than 1 million: now there are 220! By 1985, the largest city in the world will be Tokyo with nearly 25 million

Figure 5.11   Terrace cultivation in the Spanish province of Murcia. (Photograph F. Ramade)

inhabitants (15 million in 1970), followed by New York with 18.8 million and Mexico City with around 18 million.

It is not difficult to imagine the great loss of fertile land brought about by the growth of a city like Tokyo, in a country where only one fifth of the land area is cultivable and where the best soil is concentrated in narrow coastal plains. Similarly, the urban area of Cairo, with more than 5 million inhabitants, is continually expanding at the expense of the most fertile soil in the Nile delta—and that in a country which already has nearly 1000 inhabitants per km² (the density in the cultivable land of the Nile valley).

Even in France the alleged policy of regional development conceived by technocrats and economists ignorant of ecological matters has produced a shocking waste of natural resources. The new towns in the Paris region have already been mentioned. The expansion of the urban areas of Lyon, Marseille and many other large cities has occurred at the expense of market gardening activities around their periphery on land of exceptional productivity. The Huvaune valley near Marseille and the area around Saint-Genis-Laval south of Lyon, at one time famous for their early crops of vegetables and their orchards, have both been rendered virtually barren by uncontrolled urban and industrial development. Similar cases are occurring over the whole country, with local authorities selling off high-quality agricultural land—and the prosperous farms working it—at bargain prices, under the pretext of 'creating employment'. It is debatable whether this is economically sound: for one thing, it does not take into account the loss of agricultural jobs resulting directly or indirectly from industrialization.

In the USA it is estimated that urban sprawl has sterilized 132 000 km² of cultivable land, to which must be added a further 129 000 km² taken up by the road network, with 3000 km² for parking alone! Opencast mining is also taking over 200 000 hectares of land per year, 60 000 by the excavations themselves and the rest by soil erosion and the discharge of sterile waste on neighbouring land. In all, 80 per cent of the urbanization in the USA between 1949 and 1969 took place at the expense of land in categories I to III (see p. 151): that is, of soil with maximum fertility (see Pimentel *et al.*, 1976).

In France the consumption of good cultivable land during the last 30 years has been even more senseless. In the absence of any precise studies of the scale of such losses, it does not seem unreasonable to take a figure of 100 000 hectares per year as the area lost to concrete and tarmac. The road network outside urban zones already occupied 3500 km² in 1975. Car parking alone is estimated to need 2500 hectares per million vehicles, which means that it needs a total of 500 km² at the present time.

However, the potential expansion of the road system appears even more worrying in prospect. 'Experts' in this field suggest that the minimum road area per vehicle needed for safe travelling is 600 m² and this leads to an estimate of more than 10 000 km² as a response to the 'needs' of the current car population. Such an increase would put an area amounting to that of two *départements* under asphalt and sterilize it: something that appears quite insignificant to some of those planning road systems!

The construction of a large number of motorways or throughways degrades large areas of cultivable land. It can never be repeated often enough that a two- or three-lane motorway together with interchanges and service areas occupies a width of some 70 to 80 metres or an average of 7 to 8 hectares per kilometre of road. In addition, further space is consumed by sizeable secondary effects which degrade the environment. The rubble and hard core needed for road construction need large numbers of sand and gravel quarries as well as dumps for excess materials and the transport of millions of cubic metres of earth. Deep gashes are cut into hills and mountains, and many small valleys of undoubted agricultural or biological value are obstructed or even filled in (Figure 5.12). Not only that, but the streams or rivers that flow along the valleys are either destroyed or turned into outlets for the enormous quantities of rainwater that drain off the asphalt surfaces. Finally, the slopes undermined by excavation become threatened by erosion.

Another factor which leads to a waste of open space and natural resources in rural areas, and sometimes to disfigurement of attractive landscapes and beauty spots, is the proliferation of secondary or holiday homes in most developed countries over recent years. If all the inhabitants of France had a second home built on a 5000 m² plot of land (surrounded by lawns, of course!) some 10 million hectares or nearly 20 per cent of the national territory would be diverted from its original use as a mainly agricultural or forest area.

Soil classification and environmental planning

If we want to encourage a new attitude to land development that takes into account the conservation of natural resources, there is a pressing need for a classification of soils based on their agricultural value. This is all the more necessary because only land

Figure 5.12 Motorway routes preferentially follow alluvial valleys and thus make high-quality arable land barren. (Photograph F. Ramade)

Table 5.7   Classification of land according to its suitability for cultivation (from Wohletz and Dolder, 1952)

Class	Soil characteristics and precautions to be taken	Suitable primary uses	Secondary uses
Group 1	*Soils suitable for cultivation*		
I	Excellent, well-drained and well-aerated soil. Level land, No restriction on agricultural usage	Agriculture	Pasture Nature reserve
II	Good land but with some disadvantages: slightly sloping ground, sandy soil or poor drainage	Agriculture Pasture	Nature reserve Tourism
III	Quite good land but with considerable limitations because of slope, soil quality or other factors. Need for long-period rotation of crops and special methods of cultivation (contour ploughing, terracing, etc.)	Agriculture Pasture Water supplies from drainage basin	Nature reserve Tourism Industrial and urban use
IV	Good land but with severe limitations because of slope or soil structure. Only partial and occasional cultivation	Pasture Woodland Industrial and urban use	Nature reserve Tourism
Group 2	*Soils not suitable for cultivation*		
V	Soils naturally suited to forest or grassland and usable without special precautions	Woodland Grazing Water supplies	Nature reserve Tourism
VI	The same but with minor limitations related to risk of erosion, thin topsoil, etc.	Woodland Grazing Water supplies Industrial and urban use	Nature reserve Tourism
VII	Adapted to pasture or woodland but with considerable limitations due to slope, soil texture, etc. Fragile soil needing extreme precautions	Water supplies from drainage basin Grazing Forests Industrial and urban use Nature reserve Tourism	
VIII	Unusable for pasture or woodland because of steep slope, poor soil, extreme climate, etc.	Tourism Nature reserve Water supplies Industrial and urban use	

classified as capable of withstanding normal agricultural usage without alteration of the soil ought to be cultivated if erosion is to be avoided.

Such a classification was in fact developed by the US Department of Agriculture some 30 years ago (Table 5.7). When land has been allocated to one or other of these classes, it is then the responsibility of public authorities to see that it is taken into account when any development schemes are being put into practice. The documentation of such schemes and all plans for land usage should be open to examination by

third parties. Only in this way shall we be able to avoid a future in which the most fertile regions, and even irreplaceable areas of forest, are abandoned to land speculation and uncontrolled invasion by concrete and asphalt.

In conclusion, it is completely misleading, if not irresponsible, to suggest that increasing the area of cultivated land could produce any substantial improvement in food production. The present amount of cultivable land available per individual is of the order of 0.35 hectare, markedly below the theoretical estimate of the

US Department of Agriculture. Ecosystems capable of being opened up to cultivation have long been exploited and any new transformation from forest or steppe would lead to their irreversible destruction in the medium term with no overall benefit to humanity.

It should also be noted that the number of food-exporting countries is becoming quite small. Except for Canada, the USA, the Argentine and New Zealand, every nation in the world is at present a net food importer or is at best self-sufficient. Contrary to a widespread belief, even France has a slight deficit in food and agricultural produce estimated at about 5 per cent at the beginning of the 1970s: cereals are certainly exported, but there are imports of soya and other high-protein crops as well as meat and various tropical products, and it only needs a season of bad weather to produce a deficit in the agricultural balance of payments as occurred in 1976 and 1977.

Food-importing countries in general will have to count much less on their traditional suppliers in future. The principal supplier, the USA, for example, will suffer a 24 per cent growth in population during the next quarter of a century, an increase which will entirely consume the food surplus that is at present exported.

The only possible way of increasing the food intake in Third World countries and of satisfying the food needs of a growing human population over the whole world is through the adoption of more intensive agricultural practices in order to obtain greater yields per unit area. We now examine the various parameters that could produce such increased productivity and the limiting factors that are capable of frustrating it.

5.2.2 The agricultural energy budget

Any increase in agricultural yield per unit area depends on three factors:

(a) *The energy supplied*, since a greater productivity automatically implies an increase in energy flow through the ecosystem.
(b) *Matter*, in the form of water supplies and nutrients (organic and inorganic fertilizers).

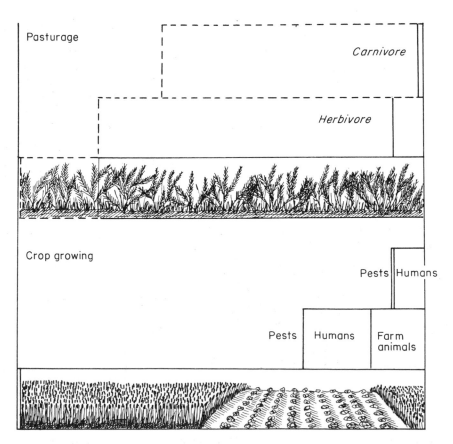

Figure 5.13   The position of humans in the food web of agricultural ecosystems. When food production is based on the extensive use of natural grassland (pasturage) the ecological pyramid is composed of a certain number of plant species, mainly grasses, a few herbivorous domesticated animal species and a single carnivore, the human being. When based on crop growing, the pyramid is even simpler, with an agriculture based on a very small number of plant species cultivated on a large scale, and with upper levels represented merely by farm animals, human beings, and pests. (From Woodwell, 1970, p. 35)

(c) *The use of cultivated varieties (cultivars) of crops and breeds of livestock that are highly productive,* equivalent in a way to an increase in the species diversity of world agricultural ecosystems.

The productivity of the biosphere is directly governed by the flow of energy through it, and agricultural systems are included in this general rule in spite of a curiously widespread belief to the contrary among the general public. This means that the amount of food available per individual depends primarily upon the proportion of the energy input to each ecosystem that can be diverted towards satisfying the food requirements of human beings.

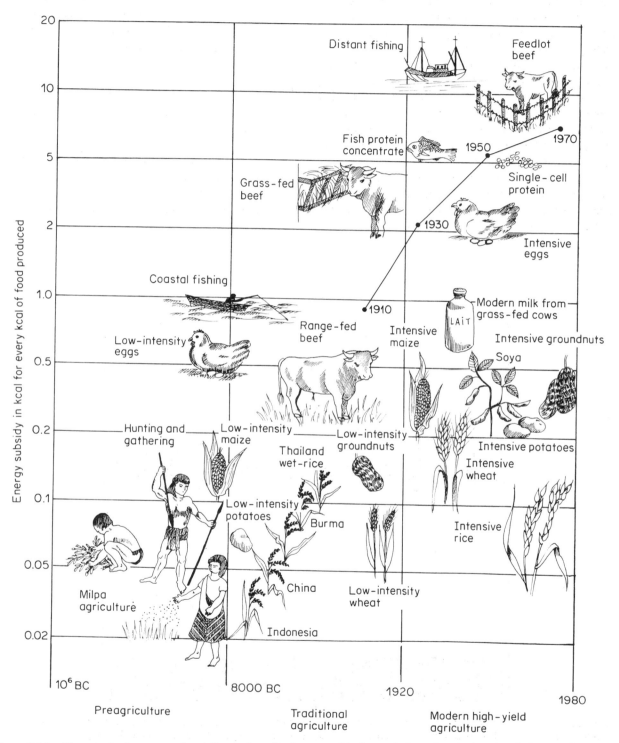

Figure 5.14   Changes in the energy input for various food crops with the evolution of agricultural techniques. The curve shows how the energy subsidy per kcal of food produced has grown in the USA due to the introduction of increasing use of fossil fuels. (From Steinhart and Steinhart, in Ehrlich *et al.*, 1977, p. 349)

In fact, choosing species of crop or breeds of livestock that are particularly well adapted to human diets is a sort of technological process designed to divert a greater proportion of the incident solar radiation to our benefit through its conversion into primary energy by autotrophic producers. Looked at in that way, agriculture can be assimilated into the overall food web with human consumption at the summit of the pyramid (Figure 5.13). Depending on circumstances, humans can occupy the second trophic level (herbivore) or the third (carnivore), and sometimes even a level between the third and fourth (for example, the consumption of battery hens in the USA partially fed on fishmeal made from Peruvian anchovy).

From the beginning of agriculture in the Neolithic age, increases in yield have been achieved by a continuous growth in energy flow through the ecosystems (Figure 5.14). The dramatic increase in productivity during the last few decades is the result of a greater and greater input of artificial energy ultimately provided by fossil fuels. A detailed analysis of the energy budget in primary production (growing of crops) and secondary production (stock-rearing) shows that the energy content of the foodstuffs produced owes nearly as much to fossil fuels as to photosynthetic conversion, and in some cases more (cf. Table 5.10, p. 155). As H. T. Odum put it in 1971: 'Human beings no longer

eat potatoes produced only by photosynthesis—the potatoes they eat today are partly made from oil.'

The energy budget of modern agriculture, 1: consumption

For ten years or more, various ecologists have been establishing the amounts of energy consumed in several types of ecosystem and have even made calculations for the complete crop production in certain countries. A classic example is that of the maize crop in the USA, for which Pimentel *et al.* (1973, 1976) have made an analysis of energy consumption between 1945 and 1970 (Table 5.8). This shows a dramatic increase in input of energy from about 2.3×10^6 kcal per hectare in 1945 to 7.1×10^6 kcal per hectare in 1970. For nitrate fertilizers alone, the energy supplied from fossil fuels increased from 145 300 kcal per hectare in 1945 to 2 345 000 kcal per hectare in 1970.

New techniques of cultivation, like the central pivot irrigation system, have made it possible to grow maize in semi-arid zones (e.g. western Nebraska) and this has led to record energy inputs per hectare. In addition, the introduction of hybrid maize, which is capable of growing and giving high yields in more northerly latitudes, has been an important factor in increasing energy consumption. For example, because of the unfavourable climate in these latitudes, the maize does

Table 5.8   Energy inputs in maize production in the USA expressed in kcal ha^{-1} (from Pimentel *et al.*, 1973)

Inputs	1945	1959	1970
Labour	30 900	18 800	12 100
Machinery	444 900	865 000	1 038 000
Fuel	1 343 000	1 790 400	1 970 000
Nitrogenous fertilizers	145 300	851 000	2 325 000
Phosphate fertilizers	26 200	60 000	116 000
Potash fertilizers	12 800	149 000	168 000
Seeds for planting	84 000	90 200	155 700
Irrigation	47 000	76 600	84 000
Insecticides	0	19 000	27 200
Herbicides	0	6 900	27 200
Drying	24 700	247 000	296 000
Electricity	79 100	346 000	766 000
Transport by rail	49 400	148 300	173 000
Total inputs	2 287 300	4 668 200	7 158 200
Maize yield (output)	8 470 000	13 453 300	20 170 000
Ratio kcal output/kcal input	3.71	2.83	2.82

Note In 1970, the energy input needed to grow 1 hectare of maize reached a value of 7.15×10^6 kcal. It follows that the efficiency of maize production (the ratio *kcal produced/kcal consumed*) decreased from 3.71 in 1945 to 2.82 in 1970.

Table 5.9   Annual energy inputs in maize production in the USA, France and Great Britain expressed in kcal ha⁻¹ (from various sources in Bel, Le Pape and Mollard, 1978)

Input	USA*	%	A	%	B	%	C	%	Great Britain‡	%
Machinery	1 420	16	955	11	1 194	8	716	15	1 591	25
Fuel	2 100	24	955	11	955	6	179	4		
Fertilizers	2 937	34	3 055	35	4 872	32	2 518	54	1 316	21
of which, N	2 429		2 687		4 299		2 150		1 070	
P	286		215		344		215		150	
K	220		153		229		153		96	
Seeds for planting	146	2	143	2	143	1	143	3	—	—
Irrigation	780	9	—	—	2 150	14	—	—	—	—
Insecticides	101	1	62	1	62	1	62	1		
Herbicides	181	2	62	1	124	1	124	3	93	2
Drying	375	4	2 579	28	3 869	25	—	—	3 296	52
Miscellaneous (transport, elec., etc.)	626	7	955	11	1 911	12	955	20	—	—
Total inputs	8 667	100	8 766	100	15 280	100	4 697	100	6 296	100
Yield in 100 kg ha⁻¹	54		60		90		50		50	
Total output	18 771		20 660		27 808		17 380		14 736§	
Efficiency of production	2.16		2.39		1.84		3.70		2.34	

(Columns under *France†*: A, B, C)

* According to D. and M. Pimentel, 'Count the kilocalories', *Cérès*, Sept.–Oct. 1977.

† According to J. Boyeldieu, 'Rendement énergétique de la production agricole', *Agriculture*, no. 386, May 1975. This author distinguishes three methods of cultivation: A, dry farming, harvested at 35 per cent water; B, crops irrigated and dried; C, cribs, direct sowing.

‡ From G. Leach, *Energy and Food Production*.

§ Leach takes only 2.9 × 10³ kcal kg⁻¹ as the energy coefficient of maize against 3.5 in the other studies.

not finish ripening and driers have to be used. These not only consume fuel directly, but they have a limited life of some 3 to 4 years due to excessive corrosion, so that their frequent renewal involves a further input of energy from their manufacture.

Research carried out both in France and Great Britain has revealed an overall energy budget comparable with that of the USA (Table 5.9). The items that consume the largest amounts of energy are nitrate fertilizers, the drying of grain, and irrigation. The irrigation of the maize crop in central France, together with drying, consumes twice as much energy from fossil fuels as the corresponding operations in the USA, according to the average figures quoted by Pimentel *et al.*

The energy budget of modern agriculture, 2: output

The energy efficiency of the principal types of crop, expressed as the ratio of digestible yield to auxiliary input, is today quite low with an average value less than 3. Calculations of such efficiencies for certain cereals and fodder plants grown on experimental plots of land have been carried out at the French Institut National de la Recherche Agronomique (INRA) and

Table 5.10   Energy efficiency (the ratio of digestible energy yield to input of auxiliary energy) for various cereals and fodder crops (from Hutter, 1976)

Crop (1 = non-irrigated; 2 = irrigated)		Energy efficiency for:	
		Total dry matter	Grain
Sorghum	1	9.1	4.7
	2	6.4	3.3
Maize	1	9.1/8.0	4.5/4.0
	2	7.8/6.9	3.8/3.5
Soya	1	8.5	4.8
	2	7.1	3.9
Sunflower	1	7.3	3.3
	2	5.8	2.5
Wheat (soft, low yield)	1	4.9	2.2
	2	3.1	1.6
Lucerne—pasture	1	3.7	
	2	2.5	
Lucerne—hay	1	1.2	
	2	1.0	
Fescue—pasture	1	1.8	
	2	1.5	
Fescue—hay	1	0.9	
	2	0.8	
Rye-grass—pasture	1	1.7	
	2	1.3	
Rye-grass—hay	1	0.7	
	2	0.7	

some of the results are shown in Table 5.10. In relation to the production of grain, the efficiencies ranged from 1.6 to 4.7 depending on the type of crop and the species, while in relation to total dry matter the figures for fodder crops were even smaller: from 3.7 for non-irrigated lucerne pasture down to a mere 0.7 for rye-grass hay.

Studies of the changes in energy efficiency with the passage of time show that it has an unmistakable tendency to decline. For maize in the USA, the output/input ratio fell from 3.71 in 1945 to 2.16 in 1977. In France, a value of only 1.85 was achieved for irrigated hybrid crops in 1975. In Iowa, eleven times more energy is needed today than in 1910 to produce 1 kcal equivalent of maize grain.

It is possible to analyse the yield from various types of crop and cropping system in areas where farming has reached different levels of industrialization (Table 5.11). In a way, this is a method of reconstituting the development of agronomic methods from subsistence farming to modern industrial agriculture. When the energy efficiencies are plotted against input energy as in Figure 5.15, it can be seen that the efficiency falls along a hyperbolic curve as the supply of auxiliary energy increases: the yield per unit input is much larger in crops receiving small energy subsidies. If the same data are plotted as actual energy yield (that is, the digestible energy itself instead of its ratio to the input) as in Figure 5.16, it is then clear that increasing the input of auxiliary energy from fossil fuels causes the primary productivity first to increase to a maximum and then to decline: additional input does not increase the yield.

Of course, plotting data for farming systems at different stages of evolution on the same diagram does not allow the contribution of auxiliary energy input to be isolated from other factors which increase yield, such as selection of plant varieties. However, other things being equal, it can be shown that the productivity P of a crop (expressed, say, in kg protein ha^{-1} $year^{-1}$) varies with the quantity of auxiliary energy E introduced (expressed, for example, in MJ ha^{-1} $year^{-1}$) according to a law of diminishing returns: Slesser (1975), for example, has shown that P and E in the units just quoted are related by $P = 1.4E^{0.718}$ for certain protein-rich plants.

To sum up, then, it is quite clear that yields from crop production do not show an increase proportional to the quantity of additional energy introduced or, more generally, to the total technological input into agriculture. It is therefore quite easy to foresee that the gains in output from farming will not be as great over the next few decades as they have been during the last 20 years.

The energy budget for secondary production

Here again there has been a clear tendency for inputs of additional energy to increase over the years, and particularly during the last decade. Research carried out on the consumption of energy by animal production shows that the efficiency of livestock is declining

Table 5.11 *Various types of crop and cropping system used for the analysis of the way energy yields and efficiencies depend on energy input.* The code numbers refer to the data points plotted in Figures 5.15 and 5.16. (From Heichel, in Slesser, 1975.)

Level of industrialization	Crop	Area	Date	Code no. (Figures 5.15 and 5.16)
I	Vegetable	New Guinea	1962	2
II	Irrigated rice	Philippines	1970	1
III	Maize for grain	Iowa	1915	3
	Maize for grain	Pennsylvania	1915	4
	Maize for silage	Iowa	1915	5
IV	Lucerne	Missouri	1970	6
	Oats	Minnesota	1970	7
	Sorghum	Kansas	1970	8
	Sugar cane (without processing)	Hawaii	1970	10
	Sugar cane (including processing)	Hawaii	1970	10^1
	Soya	Missouri	1970	9
	Maize for grain	Illinois	1969	11
	Maize for silage	Iowa	1969	12
V	Sugarbeet (without processing)	California	1970	13
	Sugarbeet (including processing)	California	1970	13^1
	Groundnuts	North Carolina	1970	14
	Irrigated rice	Louisiana	1970	15
	Winter wheat	Montana	1970	16
	Potatoes	Maine	1968	17

The low energy yield from secondary production is aggravated by the low efficiency of conversion from plant to animal biomass, in conformity with Lindemann's law. On the average, this is only about 10 per cent although experts in the field have long known that some species of domestic livestock (game birds and poultry, for instance) have a greater efficiency than others, particularly cattle.

Overall energy budget for agricultural ecosystems

Examination of the energy budget for the whole agricultural system of a country shows both of the effects described above: a continual increase in the amounts of auxiliary energy introduced and an accompanying reduction in the energy efficiency. Thus, a recent study of the French system (Table 5.12) showed a decrease of 41 per cent in the energy efficiency between 1961 and 1972. The ratio of energy yield to energy input was 6.70/4.05 or 1.65 in 1961, but by 1972 it had fallen to 1.28/1.25 or 1.02.

Overall energy budget for complete food systems

Analysis of the energy budget is now broadened in scope to take account of the complete food system in developed countries, from its sources (including the energy consumed in the manufacture of farm machinery, pesticides, etc.) to the domestic level (including deep-freezing, food preparation, etc.). When this is done, it can be seen that a significant proportion of the total consumption of primary energy in industrialized countries is absorbed by the food and agriculture sector. Table 5.13 shows that in five such countries this sector consumes between 10 and 20 per cent of the total. For France, a study carried out by CNEEMA (June 1975) calculated the amount of energy absorbed by the food system as 48.4 million tonnes of coal equivalent or 32 million tonnes of oil equivalent: about 20 per cent of the total national consumption.

There is in general an increase in the energy consumed by the food system in developed countries, both as a proportion of the total consumption and in its absolute value. Table 5.14 shows the change that has occurred in the USA between 1940 and 1970 (Steinhart and Steinhart, 1974). The energy consumption grew from 685×10^{12} kcal in 1940 to 2172×10^{12} kcal in 1970: more than a three-fold increase in 30 years. The authors make it clear, as well, that they have not included in their budget the indirect input of energy from the transport and distribution of food: the manufacture of vehicles, the maintenance of roads, the 'daylight' illumination of supermarkets, etc.

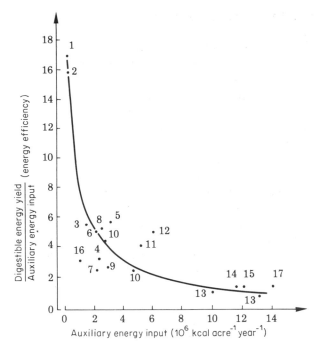

Figure 5.15   Variation of the energy efficiency (ratio of digestible energy yield to energy input) as a function of energy input for various crops and levels of agricultural industrialization. The numbers refer to Table 5.11. (From Heichel, in Slesser, 1975)

more rapidly than that of primary production. The genuine and quite ridiculous case of French calves reared in Italy on powdered milk imported from Brittany by road transport illustrates the absurdities that can result from short-sighted economic reasoning that does not take into account even the most elementary ecological concepts.

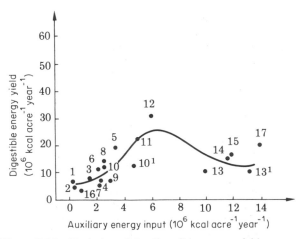

Figure 5.16   Variation of the digestible energy yield per acre per year with auxiliary energy input for various crops and levels of agricultural industrialization. Note that there is an optimum input beyond which the yield decreases. The numbers refer to Table 5.11. (From Heichel, in Slesser, 1975)

Table 5.12   Analysis of the energy budget for the whole of French agriculture. All figures are in 10^9 kcal. (From Bel, Le Pape and Mollard, 1978)

Yields	1961	1972	Inputs	1961	1972
Wheat	23 495	45 307	Fuel	15 436	63 440
Rye	119	491	Electricity	1 750	5 600
Barley	3 240	16 995	Fertilizers	14 995	36 923
Oats	—	1 159	of which, N	9 970	26 590
Maize grain	595	13 863	P	3 160	6 470
Sorghum	—	411	K	1 865	3 863
Rice	270	—	Liming	450	658
Total cereal	27 719	78 226	Agricultural machinery	5 714	13 398
Sugarbeet	8 196	12 682	of which, tractors	3 291	7 851
Potatoes	5 636	3 737	Food composites	2 122	8 453
Potato flour	241	387			
Dried vegetables	286	148	Total inputs	40 467	128 472
Fresh vegetables	1 669	1 458			
Fresh fruit	882	1 454			
Citrus fruits	—	4			
Dried fruits	21	35			
Nuts	265	160			
Total plant products	44 915	98 291			
Beef	2 365	2 604			
Veal	552	526			
Pork	4 685	5 845			
Lamb, goat	275	313			
Horsemeat	69	25			
Total meat	7 946	9 313			
Milk	12 572	15 321			
Eggs	720	924			
Poultry	566	923			
Other meat	164	361			
Offal	475	539			
Total animal products	22 443	27 381			
Total yields	67 358	125 672			

Relationship between cultivated area and energy budget

It is quite clear that the decline in the energy efficiency of farm production goes hand in hand with the pursuit of a continual growth in output. This in turn is produced by the increase in population. It can never be repeated often enough that the doubling of the world population by the year 2015 means that, in just over 30 years, food production must increase by an amount equal to that since the beginning of the Neolithic Age 10 000 years ago—and that is simply to maintain a diet already insufficient in many Third World

Table 5.13   Energy consumption in food production for certain developed countries (from Slesser, 1975)

Country	Year	Energy consumption (10^{15} J) By agricultural production	From farm to dinner plate	Total annual	Energy efficiency*	Total as % of national energy consumption
USA	1963	2310	4125	6435	1/6.4	12
UK	1972	340	300	646	1/6.5	10
Australia	1969	87	121	208	1/0.3	10
Israel	1969	33	not determined		1/1.5	20
West Germany	1960	102	386	488	not determined	12

* The energy efficiency or energy ratio is the ratio of the quantity of digestible biochemical energy in the consumer's diet to the total quantity of energy from fossil fuels needed to produce it.

Table 5.14   Energy consumption in the US food system between 1940 and 1970. All values are in 10^{12} kcal. (From Steinhart and Steinhart, 1974)

Activity	1940	1954	1970
On farm			
Fuel (direct use)	70.0	172.8	232.0
Electricity	0.7	40.0	63.8
Fertilizer	12.4	30.6	94.0
Agricultural machinery	23.4	55.6	101.3
Irrigation	18.0	29.6	35.0
Sub-totals	124.5	328.6	526.1
Industrial processes			
Food-processing including machinery	147.7	216.4	314.0
Paper packaging	8.5	20.0	38.0
Glass containers	14.0	27.0	47.0
Steel cans and aluminium	38.0	73.7	122.0
Transport (fuel)	49.6	122.3	246.9
Transport (manufacture)	28.0	47.0	74.0
Sub-totals	285.8	506.4	841.9
Commercial and domestic			
Commercial refrigeration and cooking	121.0	161.0	263.0
Refrigeration machinery	10.0	27.5	61.0
Domestic refrigeration and cooking	144.2	228.0	480.0
Sub-totals	275.2	416.5	804.0
Grand totals	685.5	1 251.5	2 172.0

countries. Since the total area of land already cultivated or potentially so is constant, this implies the injection of ever-increasing amounts of additional energy.

Figure 5.17 shows the relationship between this extra energy per individual per year needed to produce certain diets and the agricultural land that is available per individual. For a mere survival diet, the energy input needed increases exponentially as the area per individual gets smaller (note that the axes have logarithmic scales). For a European-style diet, the corresponding variation is that of an exponential of an exponential!

Some industrialized countries are in precisely that condition. Their populations have reached such high densities that they are no longer self-sufficient in food and have not been so for many years. They meet their needs by importing any food they lack, thanks to surpluses in their industrial trade balances.

Borgstrom (1969) has introduced the concept of 'phantom' land: that is, of a land area that would be needed by a country to produce the food that it imports. Thus Japan, which theoretically has available only 5.8×10^6 hectares of cultivable land, lives on a phantom area of 38×10^6 hectares (the equivalent area of imported agricultural produce and fish). The United Kingdom similarly adds 38×10^6 hectares of phantom

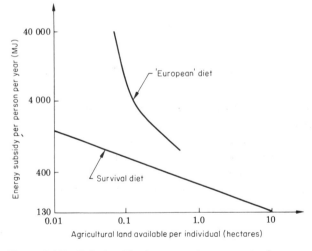

Figure 5.17   Relationship between the amount of energy needed to produce individual European and survival diets and the land area available. (From Slesser, in Lenihan and Fletcher, 1975, p. 16)

land to its own real 13.5×10^6 hectares under cultivation.

Conclusion

At first sight, the answer to the problem of world food

Plate XV   Industrialized monoculture farming

1   Aerial view of the Great Plains, Iowa, USA. Regions of intensive cereal-growing like this are heavy consumers of energy, fertilizers and pesticides, and are examples of an industrialized agriculture unconcerned about ecological equilibrium . . . or about the long-term future.

2   Olive groves in Andalusia near Malaga, southern Spain.
(Photographs F. Ramade)

Plate XVI   Tropical rain forests

1   The Forest of Yapo in the Ivory Coast. Notice the light-coloured tree trunks characteristic of such ecosystems. Notice also the absence of distinct stratification and the abundance of creepers and epiphytes.
(Photograph Y. Gillon)

2   A mountain rain forest on the slopes of the volcano Tangkuban Prahu, Java. Here, too, there is no stratification. Notice the tree-ferns in the undergrowth.
(Photograph F. Ramade)

3   A tropical rain forest 2300 metres up Mount Kenya. These mountain forests are clothed in a semi-permanent mist which encourages the growth of numerous epiphytes, particularly lichens of the genus *Usnea*.
(Photograph F. Ramade)

supplies involves a choice between two simplistic alternatives:

(a) to ensure that enough land is available in the future so that a less intensive agriculture can be developed having a low energy profile;
(b) to make inexhaustible and cheap energy resources available so that the current structure of the food systems in industrialized countries can be maintained.

The obstacles to both of these 'solutions' are too obvious to need explanation.

Because the principal raw material for world agriculture is no longer soil but oil and its derivatives, it is greatly to be feared that the scarcity of this source of energy and the high price of substitutes will make it impossible to maintain necessary levels of world food production in the long term.

In the relatively short term it is easy to show that there will be enormous economic constraints on agricultural methods involving heavy consumption of hydrocarbons. The food consumed per individual in the USA today requires 0.56 hectare of cultivated land and 0.425 tonne of oil equivalent per year. Since the world population is 4.2×10^9, the energy needed to maintain a food system comparable with that of the USA would be the equivalent of 1.8×10^9 tonnes of oil equivalent per year. In such a case, even if all the oil extracted were devoted entirely to agricultural production, the proved world oil reserves would be exhausted in 56 years. A similar calculation made using the most optimistic estimates of reserves both available and still to be discovered, that is, 300×10^9 tonnes, would still give us only 166 years. These arguments are, of course, incorrect since the world population is continuing to grow and will probably stabilize at around 15×10^9 (Frejka, 1973). On the same basis, such a population would exhaust world oil reserves in 50 years. However, it is clearly impossible to devote all the oil extracted to agricultural production and the other uses would reduce still further the useful period of this fuel.

The development of industrialized agriculture, with its high consumption of energy from fossil fuels, is therefore by no means assured for long into the future. International tensions over oil supplies to consumer countries could well foreshadow the long-term risk of a general food shortage.

The maintenance of current agricultural methods of production (basically anti-ecological) against all odds can only be achieved, even in the short-term, by continual increases in the price of food owing to the rising cost of fossil fuels and other indispensable raw materials.

Chapter 6

Forests and Woodlands

World forests and woodlands at present occupy more than 3 million hectares. Although this area is second in size only to that taken up by deserts over all the land masses, it is showing a disquieting contraction under the combined effects of overexploitation for timber and simple destruction for the creation of new arable land or pasture. Even in developed countries deforestation is generally the method of compensating for the loss of agricultural land through urban spread, industrialization, road-building and so on.

The distribution of afforested areas over the world is very uneven. Countries that have long had high densities of population such as China, India or those of Western Europe tend to possess fewer forests for a given set of other ecological conditions. In addition, the interplay of various factors connected with soil, climate and human activity has encouraged or discouraged the development of forest biomes at different latitudes and accounts for the present extent and distribution of the principal woodland areas.

It is difficult to make a reliable estimate of the world total of forest and woodland areas, since it depends on the criteria used to define them. Some authors include not only the ecologically well-defined closed forests, but also the open woodland types of plant community consisting of degraded forests and even sometimes areas of shrubland with no silvicultural value and barely fulfilling the normal ecological functions of a forest. For similar reasons, there is ambiguity in the statistics provided by many countries and even by international organizations, where official figures often include nursery areas and even orchards under the heading of forests when they are completely unrelated ecosystems.

It is easy to see, therefore, that there can well be marked divergences between various sets of data depending on whether they include only closed forest areas or all woodland irrespective of the density of tree cover. Thus, the FAO (Table 6.1) estimated world forests in the mid-1970s as covering 3.8×10^9 hectares, whereas more recent figures including only closed forests gave 3.1×10^9 hectares (Table 6.2). At the opposite extreme, Bazilevitch, Rodin *et al.* (1971) arrived at a value of 6.4×10^9 hectares, undoubtedly using very optimistic figures for tropical rain forests that did not take into account significant reductions known to have occurred over the last few decades.

6.1 Structure, Productivity and Potentiality of the Principal Forest Ecosystems

The term *forest* covers ecosystems that vary widely in their tree species, in their biological productivity and in their potentiality as sources of timber. The areas listed in Table 6.1 thus include a highly diverse collection of environments whose only common characteristic is that their dominant species are always trees. However, as the latitude increases from the Equator to the polar circles, it is possible to identify a regular progression of zones incorporating various types of forest ecosystem which we now consider in turn.

Tropical rain forests

Tropical rain forests achieve their maximum extent around the Equator and form an almost continuous band along the Equatorial regions of the Earth (Plate XVI). Their full development requires special conditions that include uniformly warm temperatures and an annual rainfall greater than 1500 mm distributed reasonably well over the seasons. They are characterized by their permanent foliage and by the adaptation of their plant life to growth with only a feeble amount of sunlight, particularly on the part of their dominant tree species in the early stages of development. They are among the oldest forests on the planet, being the only plant formations that have survived the biological upheavals during the quaternary ice ages.

The outstanding features of Equatorial forests are their high species diversity, their complex stratification and their large standing crop biomass.

As examples of their high *diversity*, the 600 species of tree found in forests of the Ivory Coast or the 2000 or more in those of Malaysia can be quoted. According to Bernhard-Reversat *et al.* (1978), the Yapo Forest of the Ivory Coast contains on the average 427 trees per

Table 6.1 Afforested areas, volumes of timber, and felling in various regions of the world
(from Bramryd, 1979)

Region	Afforested area (inc. open woodland and plantations)		Standing volume of timber (10^9 m³)	Felling in 1973 (10^6 m³)
	10^6 ha	As % of total land area		
Europe	169	34	12.8	332
USSR	880	39	78.9	383
North America	630	32.5	58.5	480
South America	742	36	101.1	278
Africa	760	25	39	309
Asia	460	16.7	40	691
of which, China	77	8	—	—
Oceania	185	21.7	8.5	27
World total	3 826	28.5	338.8	2 500

hectare belonging to about 50 different species, while forests in the Amazon basin have a mean density of 423 trees per hectare (87 species) and some in the Indo-Malaysian region possess more than 200 species per hectare (Figure 6.1).

Trees of tropical forests appear quite homogeneous in character, the various species typically having light-coloured trunks and broad leaves that are stiff and laurel-like. They are deciduous, but evergreen because of the regular rainfall and the absence of marked seasonal variations in climate.

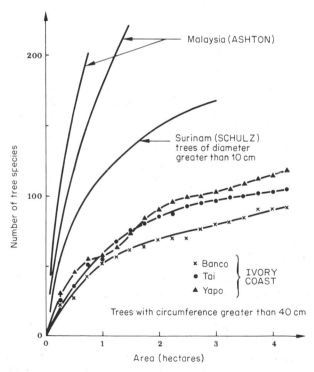

Figure 6.1 Curves giving the relationship between the number of tree species and the area for various tropical rain forests. (From Bernhard-Reversat, Huttel and Lemée, 1978)

Tropical rain forests, unlike temperate ones, have a complex and continuous *stratification* with a high density of vegetation at all levels except at the lowest herbaceous layer. Another of their distinguishing features is the great abundance of lianas climbing through the foliage and sometimes more than 100 metres in length. Finally, the constant high humidity encourages the growth of a wide variety of epiphytes (bryophytes, ferns, orchids, Araceae, bromeliads and so on) which can completely cover the surfaces of the tree trunks and branches. These plants are adapted to the various amounts of sunlight they receive at their different levels in the stratified vegetation.

In contrast to the luxuriant aerial growth in rain forests, the root development is quite limited. In the Ivory Coast, for example, it has been observed that 95 per cent of the root systems occur in the top 130 cm of soil and as a result the trees are highly susceptible to uprooting. Moreover, by far the majority of the biomass in such ecosystems is above ground.

Tropical rain forests possess the greatest concentrations of living matter in any terrestrial ecosystem, having an average *biomass* of 450 tonnes per hectare and an upper limit sometimes exceeding 1000 tonnes per hectare. (Note, however, that the average biomass in certain temperate forests can reach or even exceed these values, although it is exceptional for them to do so.) The primary productivity of the tropical forests is very high, amounting to more than 20 tonnes of dry matter per hectare per year (Table 6.2).

The paradox of these ecosystems is that, while they are the most productive of any on land, they generally flourish on the least fertile soils of the whole planet in ancient terrain based on igneous rocks. The surface is thus poor in nutrients, ferrallitic, and undersaturated in silica owing to climatic conditions that promote leaching. Because so little of the biomass is incorporated in the root systems, the majority of the

Table 6.2   Primary productivity and standing plant biomass for the principal types of forest (adapted from Ajtay *et al*., 1979)

Type	Area (10^6 ha)	Primary productivity (dry matter)			Standing plant biomass (dry matter)	
		Range (t ha^{-1} yr^{-1})	Mean	World total (10^9 t yr^{-1})	Mean (t ha^{-1})	World total (10^9 t)
Tropical						
Tropical rain forest	1000	10–35	23	23	420	420
Dry tropical forest	450	10–25	16	7.2	250	112
Mangrove	30	—	10	0.3	300	9
Temperate						
Temperate evergreen*	300	6–25	15	4.5	300	90
Deciduous + mixed	300	6–25	13	3.9	280	84
Northern						
Taiga (coniferous)	900	4–20	8	7.2	230	205
Plantations	150	6–30	17.5	2.6	200	30
Totals	3130	—	—	48.7	—	950

* Including Mediterranean forests.

nutrients occur in the vegetation above ground (Figure 6.2) but are very rapidly recycled: another peculiarity of these ecosystems. The root fibres penetrate the layer of litter and this allows the elements contained in the dead organic matter to be used directly. The mineralization of nitrogen occurs very quickly and nitrification takes place in the top 15 cm of soil. Phosphorus and potassium are similarly to be found in only very small proportions below a depth of 40 cm (Bernhard-Reversat, 1975).

All this gives rise to serious misgivings about the present methods of exploiting these forests. These entail the irreversible removal of a large proportion of the nutrient stock with a resultant diminution of soil fertility. This will have disastrous effects on future attempts at regeneration of the vegetation destroyed, not to mention the risk of laterization.

Tropical rain forests are largely confined to the zone lying between latitudes 10° N and 10° S. Further away from the Equator than this, the amount of rainfall decreases quite markedly and the dominant ecosystems become the dry tropical forests.

Dry tropical forests

These include the monsoon and thorn forests, and consist mainly of deciduous species that defoliate during the dry season mixed with sclerophyllous and evergreen trees, all of which are generally smaller than the trees of tropical rain forests. The duration of defoliation and the number of evergreen species depend on the length of the dry season, which varies according to the location. The longer the dry season, the greater the proportion of deciduous trees. The area covered by these types of forest has been considerably reduced by human activity.

Mediterranean forests or chaparral

Mediterranean forests are quite varied and complex. Their species diversity is high but lack of rainfall during the summer produces a relatively low biomass and primary productivity. Two main types of forest can be distinguished here: the *sclerophyllous oak forest* and the *evergreen coniferous forest* (human exploitation of the climax forests of oak since the dawn of history has undoubtedly encouraged the spread of pine). The standing crop biomass is significantly lower than that of tropical rain forests but can nevertheless be as high as 320 tonnes per hectare or more in the oak forests of *Quercus ilex* (Lossaint and Rapp, 1978). On the other hand, the primary productivity is limited to about 7 tonnes of dry matter per hectare per year, and even less in forests of Aleppo pine.

Temperate forests

These include two major biomes: the *broadleaved deciduous forests* and the *mixed broadleaf and conifer forests* on the northern fringes. There are also *temperate coniferous forests*, but they are of much more limited extent.

Deciduous forests formerly covered the whole of central Europe from the Atlantic to the Urals, including southern Scandinavia, as well as northern and central China and that part of the North American

Plate XVII   Temperate deciduous forests

1   Beech trees (*Fagus sylvatica*) in the forest park of Gros-Fouteau at Fontainebleau near Paris.

2   The effect of tourism in the Sainte-Baume national forest (Var) in France. A mixed woodland of pubescent oak and Scots pine has been fenced off (on left of picture), while free access to the public is still allowed alongside (on the right). Overuse of the track worn by tourists has led to the almost complete destruction of the lowest stratum of vegetation along it, while the same stratum remains highly developed in the fenced-off area.

(Photographs F. Ramade)

Plate XVIII   Northern coniferous forests

1   Taiga in the Tanajoki basin in Finnish Lapland.

2   Floating logs on Lake Inari in northern Finland. Overexploitation of northern coniferous forests has been brought about by the growing needs of the paper and cellulose industries.
(Photographs F. Ramade)

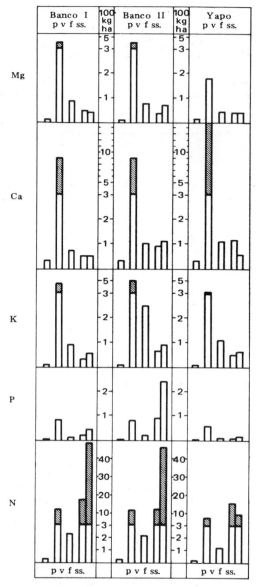

Figure 6.2   The relative amounts of various nutrients at different phases of their biogeochemical cycles for tropical rain forests in the Ivory Coast. p = input from rainfall; v =,storage in the biomass; f = flow from plant to soil; s = reserves in the soil, s₁ from 0 to 10 cm depth, s₂ from 10 to 50 cm depth. (From Bernhard-Reversat, 1975, p. 249)

Mature trees in these deciduous forests can exceed 40 metres in height, though nowadays it is exceptional to find any that are not already felled. Hardly any of these woodlands still remain in their virgin state and some were completely destroyed at the beginning of the historical period. The vestiges of them which still persist, like the broadleaved forest of north-eastern China, are so profoundly altered that it is impossible to establish what were their original plant species.

Temperate deciduous forests exhibit three distinct strata: trees, shrubs, and herbaceous plants. The species of the lower strata have a short growing period, adapted to their particular environment which is rapidly plunged into shadow at the beginning of summer as the foliage of the trees develops a canopy above it.

Northern coniferous forests or taiga

These occupy a total area approximately equal to that of the tropical rain forests, extending over more than 9×10^6 km² generally situated between the latitudes of 45° N and 57° N. They flourish on thin leached podzolic soils of glacial origin.

Their species diversity is always low. In Europe, the dominant species going from west to east are the Scots pine, the spruce (*Picea excelsa*) and in Siberia *Picea obovata*, *Abies sibirica*, and various larches. In North America, climax forests consist of various species of pine and spruce in Alaska and British Columbia; of spruce, fir and larch to the east of the Rocky Mountains; and finally of pine and hemlock spruce (*Tsuga canadensis*) as far as Labrador. It is important to realize that mountain forests in temperate or even in subtropical countries (in the Himalayas for instance) correspond to a high-altitude form of taiga consisting of coniferous forests at the subalpine stage whose dominant species are the Swiss mountain pine and the cembra pine, together with spruce and larch.

The primary productivity of taiga is low, amounting to only half that of temperate forests as far as timber is concerned. It often takes as long as a hundred years for the woodland to mature sufficiently for commercial felling.

Potential timber production in forests and woodland

The various types of forest and woodland have quite different capacities as far as the production of timber is concerned. The differences arise from the nature of the tree populations (broadleaved or coniferous) and from their productivity as regards usable wood: in particular, the volume of tree trunk available per hectare. In addition, it is obvious that economically, 1 m³ of rosewood is not exactly equivalent to 1 m³ of poplar or spruce. . . .

continent to the east of the 100° meridian up to the latitude of the St. Lawrence. In Europe, such forests consist mainly of beech or oak, depending on the amount of rainfall, together with less common species such as lime, maple and so on. In North America, the species richness of deciduous forests is distinctly greater than in Europe. Numerous types of woodland occur: mixed forests of beech and sugar maple at the northern limit; oak and chestnut to the east and south-east; oak and hickory (*Carya*) to the west and south-west of the Appalachians. Alongside these dominant species are others such as *Liquidambar* sp., tulip trees (*Liriodendron* sp.) and so on.

Table 6.3   Plant biomass of various types of woodland (from Ovington, 1965)

Dominant species or woodland type	*Pinus nigra*	*Pinus sylvestris*	*Betula verrucosa*	*Quercus borealis*	*Picea abies*	*Nothofagus truncata*	*Pseudotsuga taxifolia*	Evergreen gallery
Location	North-east Scotland	Eastern England	Moscow, USSR	Minnesota, USA	Sweden	New Zealand	Washington State, USA	Thailand
Status	Plantation	Plantation	Natural	Natural	Natural	Natural	Natural	Natural
Age of tree in years	48	55	67	57	58	110	52	—
Tree height in metres	14	16	26	17	17	21	17	29
Number of trees per hectare	1112	760	—	800	924	490	1157	16 200
Oven-dry weight in t ha^{-1}								
Tree leaves	5.6	7.2	2.8	3.5	9.1	2.7	12.0	19.0
Tree branches	11.2	12.3	11.3	49.5	14.3	42.0	17.9	50.0
Tree trunks	95.1	96.7	156.7	111.9	85.2	224.8	174.8	225.2
Roots	34.0	34.1	43.1	15.0	60.0	39.2	12.3	88.5
Dead branches on trees	10.0	10.0	2.0	21.9	2.6	1.1	11.2	—
Shrubs and herbs	7.0	2.6	2.0	0.6	1.0	0.0	0.1	0.2
Litter	22.0	45.0	3.0	36.7	78.0	16.7	117.3	3.0
Total biomass	184.9	207.9	220.9	239.1	250.2	326.5	345.6	385.9
Ratio of weight of trunk to total above-ground biomass	0.78	0.77	0.91	0.60	0.77	0.83	0.81	0.77

Table 6.3 gives figures for the biomass of the various parts of trees. Some of these trees occur in temperate forests and some in tropical forests. The most important figure from the point of view of timber production is that in the last row, the ratio of the biomass of the trunk to that of the whole tree, and it is noticeable that this shows considerable variations. In particular, it appears that monospecific plantations of conifers do not have any obvious advantage over natural woodlands in this respect.

A final point to notice is that tropical forests do not have a significantly greater productivity of usable wood than temperate forests, contrary to a commonly held opinion. Duvigneaud (1967) has already rightly pointed out that an exact comparison of timber production from temperate and tropical origins does not bear out the feeling that tropical rain forests are a veritable 'land of plenty' as regards wood supplies. The average values for productivity in tonnes of usable wood per hectare per year are as follows:

Temperate regions:	various broadleaved trees	5.1
	various conifers	5.3
Tropical regions:	various broadleaved trees	13.1
	various conifers	12.6

The net primary productivity in the tropics is reduced considerably by the higher temperatures that increase respiration and by the existence of a more or less prolonged dry season.

6.2   The Exploitation of Forests

The durability of natural forests and their resources depends on the manner and the intensity of their exploitation. Broadly speaking, there are two ways of exploiting such resources: conservatively or destructively. The conservative method observes certain safeguards in the use of forests and does not significantly change the structure or the productivity of the biocoenoses or biotopes concerned. Silvicultural management here is based on selective felling of trees, a moderate cropping of animal life by hunting, some recreative and tourist use, and a certain amount of grazing of the undergrowth. The creation of nature reserves, or the use of woodland to counteract soil erosion in areas of high relief or to provide water supplies for a drainage basin, all clearly represent conservative measures as well.

Destructive forms of exploitation, on the other hand, can often be observed. An example is the felling of trees for timber by the practice of clear-cutting (see below) in forests growing on land with high relief or fragile soil structure, as is often the case in tropical areas. Overgrazing also leads to an irreversible degradation through the removal of young plants in such quantities as to prevent any eventual regeneration. Other practices that lead to the destruction of forests are that of itinerant farming where the rotation is too rapid and, obviously, that of complete clearance of woodland for agricultural or other purposes.

Table 6.4   *Compatibility of the various ways of using forest and woodland resources*

Type of use	Nature reserve	Protection of soil against erosion	Regulating the water cycle	Production of timber	Hunting and food gathering	Tourism and leisure	Pasture	Agriculture
Nature reserve		+++	+++	—	—	+	—	—
Protection of soil against erosion	+++		+++	+	+++	++	+	—
Regulating the water cycle	+++	+++		++	+++	+++	+++	+
Production of timber	—	+	++		++	+++	++	—
Hunting and food gathering	—	+++	+++	++		++	+++	—
Tourism and leisure	+	++	+++	+++	++		+++	—
Pasture	—	+	+++	++	+++	+++		—
Agriculture	—	—	+	—	—	—	—	

+++ Completely compatible.
 ++ Compatible if some precautions taken.
  + Compatible, but only with great precautions.
 — Incompatible.

The various ways in which forests and woodlands can be used exhibit differing degrees of compatibility with each other, as is shown in Table 6.4.

6.2.1   The production of timber

Human exploitation of woodland occurs mainly through the production of timber, which is growing continuously and is the prime cause of deforestation. Forecasts made in 1952 of world demand for wood estimated that it would be 700 million m^3 by 1970 (USSR excluded). In fact, the volume had already reached 1900 million m^3 by 1959, and rose to 2185 million m^3 in 1969 and to 2497 million m^3 in 1973.

There was an estimated 85 per cent increase in the annual world production of wood between 1950 and 1979, but it was very unevenly distributed. In Europe and North America it amounted to a mere 20–25 per cent, but it was considerably greater in tropical areas: more than 200 per cent in Asia, for example. The deforestation 'record' is held by Malaysia where timber production has grown eight times in little over 10 years.

Table 6.5 and Figure 6.3 show how the world production of timber is distributed and how it has grown since 1950. They also show broadly the principal uses made of it, both for industrial purposes and even today as an important source of energy. The use of wood as a fuel has been steadily decreasing in developed countries, but in the Third World, where it often provides the only available source of energy, such use has been growing in line with the population growth. About half the total world production is used in this way, and in certain countries like Brazil such consumption amounts to over 1 tonne per person per year (Adams *et al.*, 1977). The resultant deforestation is having disastrous consequences in the majority of Third World countries, particularly in areas of marked relief or with a prolonged dry season.

In developed countries, wood finds many industrial uses—in furniture-making or as a construction material in buildings, for instance—but there is a particularly high consumption for paper-making, for cardboard containers and for the production of various wood fibre conglomerates. These types of use have more than doubled since 1955: whereas in 1950 only 18 per cent of the timber produced was converted into pulp, the proportion in European countries today is almost 60 per cent.

Although there has recently been a marked slowing down in the rate of increase of paper and cellulose-fibre production, the amounts consumed in developed countries are very high: about 250 kg person^{-1} year^{-1} in the USA and more than 100 kg person^{-1} year^{-1} in most European countries. Such a profligate use of paper and other wood products—the ridiculous

Table 6.5   *World production of timber in 1973 and its principal uses* (from Skogsstyrelsen, in Bramryd, 1979)

	Conifers			Broadleaved species			
	Industrial uses	Fuel	Total	Industrial uses	Fuel	Total	Grand total 1973
Volume without bark in 10^6 m^3	949.7	174.1	1123.8	401.5	972.2	1373.7	2497.5

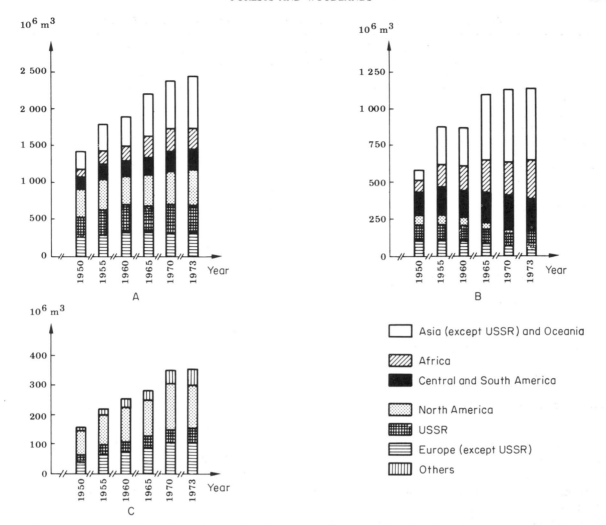

Figure 6.3   A, world production of timber by region; B, volumes of wood used as fuel; C, volumes of wood used by paper and cellulose industries. Other uses are represented by the difference between A and B + C. (From Bramryd, 1979; reproduced by permission of SCOPE (Scientific Committee on Problems of the Environment))

accumulation of advertising leaflets in letter boxes is only one aspect of this, although a striking one—is putting a disquieting pressure on world forest resources.

The effect of silviculture systems on the conservation of forest resources

Exploitation of forests for timber can be carried out using various techniques for cutting and harvesting the wood. There are three principal systems of felling which have quite different long-term effects: the clear-cutting, shelterwood and selection systems (Figure 6.4).

Clear-cutting is by far the most widespread system practised throughout the world today. Regrettably, it is even coming more and more into use among European foresters who justify it by shortages of manpower and by a decidedly technocratic concept of profitability. In this system of management, all trees are simultaneously felled over an area that is generally quite

sizable. The trunks alone are removed and the remainder of the trees, the branches, twigs and other debris, is left on the ground and often burnt, even though it could be used as fuel if removed. With some species, nearly 50 per cent of the biomass of the tree that is above ground is wasted in this way.

In areas of limited extent, regeneration of the vegetation can take place spontaneously from neighbouring land, but in general it must be brought about through the planting of young trees from a nursery.

Because the soil is left bare by this system of cropping, there is risk of erosion in areas with high relief and of soil changes such as laterization in tropical regions, which rapidly prejudices any chances of regeneration. This is precisely what has happened, and both in mountainous and Equatorial regions the practice of clear-cutting has devastated huge forest areas whose productivity could have been maintained quite easily if more rational methods of management had been adopted.

Figure 6.4   The main systems of silviculture: A, clear-cutting with natural regeneration; B, clear-cutting with planting; C, the shelterwood system; D, the selection system. (From Spurr, 1979, p. 74)

Even if clear-cutting does not induce erosion that is sufficiently catastrophic to prevent eventual regeneration, it does nevertheless profoundly disturb surface waterways. Streams and rivers through the forest are blocked with branches and become increasingly choked and silted up from the large amounts of fine soil carried into them by run-off at the time of heavy rain. Not only that, but the forest soils themselves lose large quantities of essential nutrients through increased lessivage and a higher rate of mineralization of organic matter (Figure 6.5). Thus, it is quite common to find after clear-cutting that surface waters in the area contain more than 10 p.p.m. of nitrates, a concentration that can lead to heavy eutrophication.

Selection systems of silviculture, on the other hand, leave the diversity and productivity of forests almost unchanged if they are carried out with the necessary safeguards. The methods are particularly suitable for natural or semi-natural woodlands where trees covering a range of species, ages and sizes all coexist. Selected mature trees are felled and particular care is taken during this operation not to damage the younger ones left standing. An excellent example of the way the selection system allows the resources and structure of a climax forest to be preserved is provided by mixed plantations of beech and fir.

The *shelterwood system* is practised in circumstances where there is a need to reduce competition

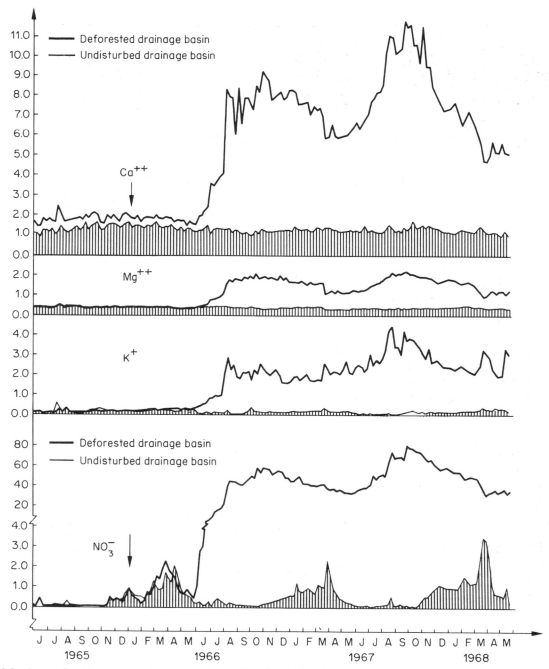

Figure 6.5   Loss of nutrients through increased run-off and leaching following a clear-cutting operation completed on the date indicated by the arrows, compared with that from undisturbed forest. The vertical axes give the measured concentrations in streamwater, with a change of scale in the case of nitrates. Note the size of the losses in nitrates, calcium and potassium. (From Likens *et al.*, 1970, p. 34)

between individual trees belonging to the same or to different dominant species. The idea is to encourage the growth of the best specimens and for this purpose harvesting is carried out on two or more occasions spaced out over several years. One cutting removes most of the young trees but leaves the more mature specimens to provide seeds for further seedlings and to provide shelter to enable the remaining young trees to become established. A second cutting then removes the mature trees at an appropriate time. This type of

silviculture has also been used to regenerate forests where excessive exploitation to provide firewood has caused them to regress to mere coppices with low productivity.

Timber production and the maintenance of productivity

The exploitation of forests for timber causes a continuous loss of nutrients contained in the biomass of the trees and their removal to the outside world. The

Table 6.6   Average amounts of various nutrients lost due to removal of plant biomass (from Rennie, in Duvigneaud, 1974)

Ecosystem	Amounts lost in kg ha^{-1}		
	Cal-cium	Potas-sium	Phos-phorus
Pine forest*	424	168	38
Mixed coniferous forest*	890	466	74
Broadleaved forest*	1930	483	106
Agricultural production (with 4-year rotation)†	2420	7400	1060

* For the case of a 100-year cycle.
† Oats, hay, potatoes, turnips.

amounts removed depend on the kind of woodland and its rate of exploitation, as well as on the character of the soil. Table 6.6 shows the average amounts of calcium, potassium and phosphorus lost per hectare of land due to the removal of plant biomass in various types of forest. Comparison with losses of the same elements through agricultural production in the last row shows that considerably less is removed by the clearing of forests. However, the quantities are by no means negligible and it is thus not impossible that timber production could cause a significant long-term reduction in the primary productivity of forests through exhaustion of the soil. Such a risk becomes greater when the wood itself contains a higher proportion of nutrient elements and when the recycling time of the nutrients is shorter.

Table 6.7 shows that the time taken for a given element (in this case calcium) to complete its cycle varies with the type of forest, and that the more rapid the turnover the greater the proportion immobilized in the biomass. The extreme case is that of tropical rain forests, where the greater part of the total nutrient supply is to be found in organic matter. These forests have developed on some of the poorest soils in the world, and their high productivities are the result of the large take-up and rapid turnover of nutrients. Thus, the calcium cycle in such forests takes only about 10 years to complete, whereas it is over 50 years in a temperate forest.

The proportion of the total nutrient content that is located in the biomass of a tree also varies with the species considered. Conifers, particularly pines, hold the smallest proportions and this means that there is less loss during harvesting. For that reason, these are often the best species with which to restock degraded or impoverished soils.

It is possible to consider compensating for the loss of nutrients through tree-felling by using fertilizers, a practice that is currently being developed in Scandinavia, although only on a limited scale. However, not only is the cost of the process relatively high in relation to the small returns from forestry, but the possible microbiological and other effects of adding fertilizers to forest soils is still largely unknown: more research is needed to assess its long-term validity.

Monoculture plantations and their consequences

The pursuit of short-term profitability at any price, justified by the technocratic concept of 'consumer needs', is leading to the replacement of natural woodlands and climax forests by plantations consisting of row upon row of trees belonging to a very small number of different species. These species are chosen for their speed of growth and consist mainly of conifers, poplar or eucalyptus. The situation was summed up by J. Dorst (1965): 'The tendency in modern forestry is to regard woodlands as tree farms for the satisfaction of human needs, with "fields of trees" just as there are fields of wheat: both of these are artificial in the eyes of the biologist watching the disappearance of natural communities, about which statistics can give a false impression.'

Tree populations containing a single species all of the same age make a mockery of ecological principles. The complete lack of diversity in these so-called 'forests' makes them particularly vulnerable to attack by pests: that is, by pathogenic insects and fungi. Proliferation of the bark-beetle Scolytidae, for example, generally haphazard and localized among natural conifers, assumes disastrous proportions in spruce plantations that have spread so extensively over France during the last few decades (Chararas, 1980).

Table 6.7   The calcium cycle in various types of forest (from Jordan and Kline, 1972)

Location	Type of forest	Time taken for cycle (years)				
		Soil	Wood	Litter	Foliage	Total
Puerto Rico	Mountain rain forest	3.0	6.4	0.9	0.2	10.5
Ghana	Equatorial rain forest	8.2	6.8	1.5	0.2	16.7
Weve, Belgium	Oak forest	184.9	21.5	0.4	0.6	207.4
Washington, USA	Douglas fir forest	57.4	20.2	5.5	10.2	93.3
New Hampshire, USA	Broadleaved forest	14.0	10.8	0.8	34.8	60.4
Central USSR	Spruce taiga	7.6	18.3	3.2	13.6	42.7

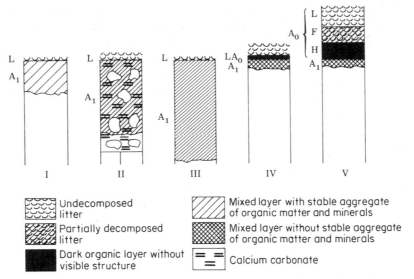

Undecomposed litter

Partially decomposed litter

Dark organic layer without visible structure

Mixed layer with stable aggregate of organic matter and minerals

Mixed layer without stable aggregate of organic matter and minerals

Calcium carbonate

Figure 6.6   Various types of forest humus: I, acid mull; II, calcareous mull; III, andosolic mull; IV, moder; V, mor. (From Duchaufour, 1980, p. 179)

The frantic planting of conifers that has been taking place for some years provides a perfect illustration of the harmful consequences of such an anti-ecological practice, yet these consequences were underlined in an EEC report published nearly 20 years ago (Noirfalise, 1964).

One of the results of the systematic but thoughtless introduction of coniferous plantations is a significant alteration in soil structure, with the production of a mor humus (Figure 6.6) and fulvic acids that tend to encourage podzolization. Other degrading effects connected with the introduction of conifers include a reduction in porosity of upper soil horizons, making them impermeable (Duchaufour, 1979). The water cycle is also modified: there is a two-fold or three-fold increase in run-off, and loss by evaporation increases because pine needles retain a large amount of rainwater. Coniferous forests are therefore drier than broadleaved woodlands, other conditions being equal. Where there is a pronounced dry season, the situation is obviously much worse.

Soil fauna are also affected. The number of earthworms is smaller by a factor of between 100 and 500 than under broadleaved trees, an impoverishment that

partly explains the transformation of humus to mor. The diversity of arthropod populations in general is reduced, but there is a corresponding increase in some groups (Acaridae and Collembola). Decomposers are also greatly affected and a marked reduction in bacteria occurs, particularly in the micro-organisms involved in the nitrogen cycle.

The simplified plant community formed by a monospecific and uniformly aged tree population provides a greatly impoverished habitat for bird life, whose density and diversity is considerably reduced. Blondel (1976 and later) has shown that the species diversity H_α of bird life is lower in the reafforested Swiss mountain pine area of Mont-Ventoux than in natural pinewoods of the same species, and even smaller in relation to that of mixed populations of conifers and broadleaved trees or that of Mediterranean oaks (Table 6.8).

The final disadvantage of uniform conifer plantations, and by no means an insignificant one, is their high susceptibility to fire. The replacement of the original sclerophyllous oak forests in Mediterranean countries by a disclimax of pines has rendered the remaining woodlands extremely vulnerable. Even in

Table 6.8   Features of bird populations in various types of forest (from Blondel, 1976)

Type of woodland	Total number of species	Average species richness	Species diversity H_α
Beech and fir	28	10.6	4.29
Mixed oak (holm-oak, pubescent oak)	28	9.9	4.38
Pubescent oak, Scots pine	22	9.3	4.05
Swiss mountain pine (natural)	21	10.2	4.03
Swiss mountain pine (reafforestation)	15	6.9	3.58

regions of taiga the fire risk is by no means negligible and considerable areas are destroyed each year in the high latitudes of Lapland, Siberia and Alaska. During the 1960s, for example, more than a million hectares of taiga were burnt down in the USSR each year. Although this is an insignificant area in relation to that of all the Siberian forests, it indicates the size of the risk associated with inflammable coniferous woodlands and underlines a major disadvantage of planting them without proper consideration.

Taking all these factors into account, the current practice of reafforestation with conifers in temperate, and even in tropical, regions would seem to constitute an astonishing perseverance with mistaken methods on the part of the authorities responsible for forests in the various countries involved. All the research and experimentation that has been carried out up till now has shown that such practices lead to harmful results even if priority is given merely to increasing productivity.

The German experience in forestry has long been a prime example of the absurd consequences of introducing monospecific coniferous plantations. Around 1840, the Germans began to practise a very artificial type of silviculture, involving the deliberate destruction of natural forests by clear-cutting and their replacement by pine or spruce plantations of a single species and of uniform age. The idea behind this was to achieve the greatest output of wood per hectare through the elimination of competition from the shrub and herbaceous strata. In fact, the effects produced were a change in the soil structure and a disturbance of the cycles of the various nutrient minerals. Losses increased through attacks by pests and trees brought down by storms. After the second or third generation of the pine or spruce, the output declined because the rate of growth of individual trees slowed down and because the quality of the wood deteriorated. As a result, the German forestry authorities decided to return to a more natural type of silviculture after 1918. Replanting was carried out with mixed forests of appropriate species, clear-cutting was replaced by selective felling, and methods were adopted for the removal of logs that involved the least possible damage to the remaining trees. All this has resulted in an increased productivity in timber and an improvement in forest soils.

It is ironic to have been forced to watch the same errors being committed for several years now in France as were perpetrated by their neighbours a century earlier.

6.2.2   Other uses of forests and woodland

One way in which forests influence human activities indirectly but strongly is through the part they play in *regulating the water cycle*. The maintenance of dense and uniform woodland, particularly over hilly areas, is a major factor in guaranteeing water supplies during the dry season and the best way of preventing floods in drainage basins at other times. Countries with Mediterranean climates and tropical countries, especially India and the Far East, have long experienced disastrous floods produced by unwise deforestation of the slopes above river valleys. Such flooding becomes gradually more frequent as afforested areas covering the hills are cleared for cultivation. The destruction of tropical rain forests, on the other hand, can cause great aridity. In the Congo basin, for example, it has been shown that 90 per cent of the rainfall is due to evapotranspiration from plant life and that only 10 per cent is brought by oceanic depressions. It follows that destroying vast areas of the forest would transform the land into a new Sahel, as has already happened in north-eastern Brazil over the last century. . . .

Animals grazing in forests generally jeopardizes their durability, partly by the removal of nutrients through secondary productivity which impoverishes the soil, but mostly by the damage done to young plants by herds of excessive size, so that regeneration of the tree population is prevented. In most regions where forest grazing is practised, the animals are brought to feed on the undergrowth during the summer, and this dries out the land which is trampled on. Cattle, sheep and goats each exploit their own particular layer of vegetation, and by nibbling off young shoots they shorten the life of the forest. Goats are the bane of Mediterranean countries in their readiness to climb trees to get at the foliage, and the state to which the forests in Crete have been reduced during the last few centuries by overabundant flocks of sheep and herds of cattle is well known. Another damaging feature linked to animal grazing is the widespread practice of firing the undergrowth during any pronounced dry season so as to encourage the regeneration of the herbaceous layer when the rains come. Even if the fires are handled with great care, young trees are destroyed and any eventual regeneration of the tree population is compromised.

Large animals that have the forest as their natural habitat, particularly deer, are often looked upon as a nuisance from the point of view of timber production. However, while it may be true that the absence of natural predators can allow excessive populations to grow, the level is kept low enough by hunting, certainly in France, so that the tree population can support it without much damage.

A significant role played by forests in developed countries is that of providing *areas for recreation* and leisure activities, particularly near cities. However, this type of use is only compatible with other functions if its capacity for coping with visitors is not exceeded.[1]

---

1. The forest of Rambouillet near Paris had more than 5 million visitors in 1979!

Beyond a certain limit, even walking can prevent any regeneration of vegetation (see Plate XVII, Figure 2). If that is the case, what is to be said about the opening up of national parks and forests to the so-called 'trail-bikes'?

6.3   The Destruction of Forests

World forests have been steadily receding ever since the earliest Palaeolithic times, but the pace of retreat has accelerated since the Neolithic age and has become particularly rapid over the last 100 years as populations have exploded. Deforestation has arisen from four principal causes, often in combination with each other: excessive felling of trees for timber, overgrazing, fire, and clearance of land for cultivation and pasture.

A major factor in the disappearance of forests since the beginning of the Neolithic age has been the search for new agricultural land, which in historical times has mostly been conducted by clearance of woodland. In general, the most fertile soils—at least in temperate regions—were originally covered by forests, so that farming from its very start was a major cause of their disappearance. It is no accident that the most disastrous levels of deforestation have occurred in areas that witnessed the birth of the earliest civilizations: the Mediterranean basin, the Indus valley, China, Mexico. . . .

Nine-tenths of Europe in Roman times was covered by the vast Hercynian forest, and clearance in central and northern regions only became significant towards the end of the first millennium AD. Communities of monks began increasingly to colonize areas they had cleared during the early Middle Ages, soon followed by the rest of the population, so that in the end only the most inaccessible belts of forest remained reasonably intact. However, even at the beginning of the nineteenth century the armies of Napoleon could march from Paris to Madrid almost entirely in the shade of trees, something that would certainly seem improbable to the present-day traveller journeying across the denuded plateaux of Aragon and Castile. Today, no more than 30 per cent of the land surface of Western Europe is afforested, and even that is very unevenly distributed, with the largest continuous areas in Scandinavia.

Of all the great forest ecosystems in the world, those of the Mediterranean region and of China have been among the most devastated by human activity. It is not only clearance that has dealt fatal blows to these woodlands, but overgrazing and fire as well. In Mediterranean Europe and Turkey, today's forests occupy only 5 per cent of their original area. The situation is even worse in North Africa and the Middle East, where the desert is encroaching on land formerly covered by immense forests. As an indication of the changes that have occurred, it is noteworthy that Hannibal's elephants, which belonged to the subspecies *Loxodonta africana cyclotis* and which lived in the African forest, were captured in southern Tunisia! Again, it is possible to find in Israel boundary stones which mark the limits of forests once situated on hilltops, which today are completely eroded and where soil scientists have estimated that at least a metre of soil has been carried away to leave bare chalk and clay.

In the Far East and in the Indian sub-continent, both

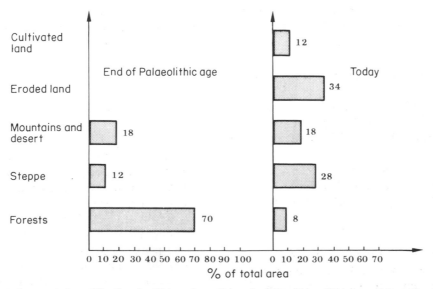

Figure 6.7   Deforestation and desertification in China since the end of the Palaeolithic age. More than 70 per cent of its territory once covered in forests has been reduced to a mere 8 per cent today, while 30 per cent has been destroyed by erosion and/or deserts because of deforestation

areas where ancient civilizations flourished, woodlands
have similarly suffered incalculable damage. The
Chinese have been among the worst enemies of trees
for millennia, and current estimates that barely 8 per
cent of their territory is afforested should be compared
with the 70 per cent at the end of the Palaeolithic age
(Figure 6.7).

Temperate North America has suffered an even
more brutal deforestation, the USA having lost more
woodland in 200 years than Europe did in 2000
(Figure 6.8). The first colonists arriving in the
Mayflower found a continent completely wooded from
the Atlantic coast to the Mississippi by an eastern
forest that originally occupied some 170 million
hectares. Today, only about 10 million hectares remain
and, moreover, climax forests constitute barely 5 per
cent of US woodland areas.

There has also been immense destruction through
forest clearance in tropical and subtropical regions.
The Thar desert of north-west India, for instance, now
covers more than 150 000 km² of land where 2000
years ago there was an impenetrable jungle. The
tropical forests of Africa and Latin America have
similarly suffered severely through cultivation of the
land, and the disastrous influence of European settlement there can never be overemphasized. Industrial
crops, introduced indiscriminately and sometimes
with a cynical contempt for the natural resources of the
countries in which they were developed, have
devastated millions of km² of both tropical rain forest
and deciduous forest. In Nigeria, 250 000 hectares of

forest were turned over to cacao plantations each year
over the period 1945 to 1965, and were quickly
ravaged by erosion. In Ghana, the growing of cacao
trees and other industrial crops has reduced the
afforested area of the country by 85 per cent.

Although they still cover 20 per cent of the continental surface, the South American forests have also
suffered irreparable damage through human activity.
The largest country, Brazil, has lost 45 per cent of its
forests since European colonization began. The last
great virgin forest in the world, that of the Amazon
basin, is today threatened with annihilation: the construction of several trans-Amazonian roads which has
been going on for some 20 years is a prelude to
agricultural exploitation. In just a few years, some
areas alongside the developing road network have
already become deserts even though they are near
waterways.

The irresponsible attitude of the Brazilian authorities
is even worse in that the policy has been adopted in the
face of clear warnings by scientists of the inherent
ecological risks in destroying so much of the
Amazonian forest (for example, Goodland and Irwin,
1975). Moreover, recent studies combining soil
research *in situ* and remote sensing methods
(Hammond, 1977) have shown that only 3 per cent of
the land in the Amazon basin is suitable for cultivation!
Yet the Brazilian Government has recently decided to
turn over more than 1 million km² of forest to growing
crops that will produce 'gasohol', a substitute for
petrol.[1]

The same country has also suffered much deforestation through the development of coffee plantations and
those of other industrial crops. This has been the
principal cause, for example, of the reduction in
woodland in the state of São Paulo from 250 000 km²
in 1910 (60 per cent of its total area) to less than
25 000 km² today. Again, the monoculture of coffee in
the state of Parana has been increasing at the rate of 3
per cent per year since 1953 and has reduced the forest
area from 65 000 km² to 29 000 km² (Adams *et al.*,
1977).

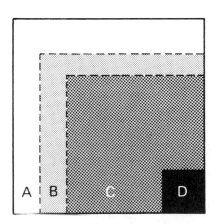

A Original forest area 370 million hectares
B Remaining forest area 288 million hectares
C Forest area still exploitable 184 million hectares
D Unmodified or climax forest areas 17.6 million hectares

Figure 6.8 Extent of deforestation in the USA since
European settlement. It should be pointed out that the
remaining 184 million hectares of forest still exploitable
consist mainly of artificial woodland and secondary forests,
heavily modified by excessive felling and bearing no relation
to the original climax ecosystems. (From US Forest Service)

Destruction by pasturage

The use of forests for pasturage has been one of the
major causes of their recession. In Latin America and
other tropical regions, for example, significant
afforested areas are transformed annually into
pastureland after clear-cutting. In dry tropical forests
and in the Mediterranean area, forest grazing has been
practised from the earliest times and has had similar

---

1. It seems outrageous to me that cultivable land should be used
to provide motor fuel at a time when malnutrition in Third World
countries is so widespread.

disastrous consequences. In the Sudan, in northern Nigeria, in Mali and Senegal 'the uncontrolled wanderings of tribesmen and their herds have had effects on the vegetation as serious as those caused by the advent of a much drier climate . . . yet there is no evidence of any such fundamental climatic change in this region for 4000 years. There is, on the other hand, plenty of evidence of the destruction of plant life by mankind, particularly by the herds of nomadic tribes' (Fishwick, 1970).

It is the same in Somalia, where a region extending to the Gulf of Aden at Hargeisa was described by some nineteenth-century naturalists who travelled through it as a paradise of open forests populated by a rich fauna of large ungulates. Today, following the destruction of woodland by overgrazing, the forests and the wild life have disappeared: in less than a century, new areas of desert have advanced over hundreds of kilometres.

Temperate countries have also lost significant areas of woodland from the same cause. The whole of the Atlantic coastal regions of Europe, from Galicia in north-western Spain to Pomerania in northern Poland, where deciduous forests once flourished, are now covered by vast areas of heath land degraded by sheep.

The disastrous consequences for forest ecosystems of uncontrolled grazing is particularly evident in the semi-arid zones of the world. The plant formations which flourish there, the dry forests or coniferous woodlands, are more fragile than those growing in areas where the rainfall is more evenly distributed throughout the year. Overgrazing has been fatal to Mediterranean forests, which now cover no more than 5 per cent of their original area. Spain, peninsular Italy, the Balkans, Greece, Turkey, North Africa and the Middle East have all lost the major part of their original legacy of immense sclerophyllous forests to what is virtually desert, and what remains has been profoundly altered by sheep grazing.

Excessive feeding on undergrowth by sheep or cattle inhibits any regeneration since the young plants and shoots are eaten during the first year of growth. In order to allow saplings to reach a height that would put their foliage out of the reach of animals, it would be necessary to fence off the forest for anything between 4 and 12 years, depending on the severity of the summer drought.

Domesticated livestock do not have any adverse effects on the fully grown trees in woodland, except for goats which tend to climb them to reach the foliage. However, an unfortunate practice of local populations in North Africa, Turkey and the Middle East is that of topping conifers to feed their animals during periods of exceptional drought. This gradually causes a thinning out of the forest and allows desert to spread, as it has done in the African Sahel, where the practice is widespread.

The paradox in all this is that, when grazing is carefully controlled, not only does it have no adverse effect on forests but it can actually increase the food value of their herbaceous strata. Work carried out in Tunisia has shown that in an open wooded area of Aleppo pines, where grazing is continuous and permanent, the production of fodder is 46 standard units per hectare per year. When a similar area is fenced off, the productivity under otherwise similar conditions amounts to 380 units, but if carefully controlled grazing is allowed it reaches 568 units (Quezel, 1980).

Furthermore, the extension of coniferous woodlands at the expense of deciduous forests should generally be seen in relation to possible overgrazing, since conifers are able to withstand the depredations of domesticated livestock better than sclerophyllous or broadleaved species whose seedlings and saplings are more vulnerable. In addition, the herbaceous stratum is more productive under conifers than under deciduous trees, so that there is a greater degree of self-regulation encouraging a higher production of fodder.

Destruction through felling for timber and fuel

Another major cause of the destruction of forests is their excessive exploitation for the production of wood. Over-frequent felling, clear-cutting in fragile woodlands, and in general the practice of removing more of the biomass than is replaced by growth, are leading to the annihilation of large afforested areas each year.

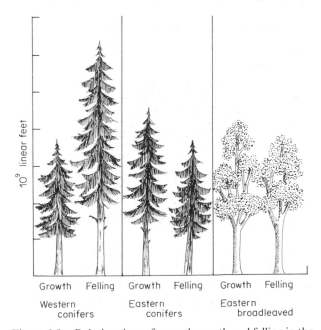

Figure 6.9   Relative sizes of annual growth and felling in the two great forest areas of the USA, the western and eastern. Notice how much western forests are being overexploited with a rate of cutting greatly exceeding the net annual productivity. (From the 7th report of the Council for the Quality of the Environment, Washington, 1976)

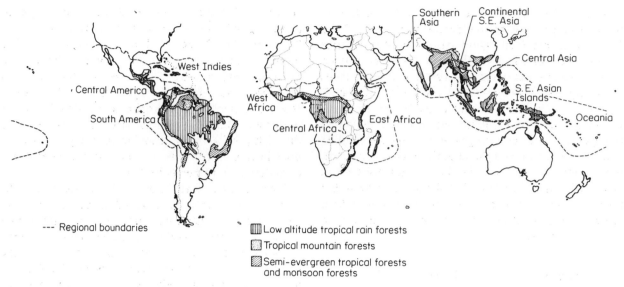

Figure 6.10   World distribution of tropical forests. It is estimated that they now occupy two-thirds of their original area at most, and that they will have totally disappeared by the middle of the next century if no measures are taken to protect them. (From IUCN, 1980b)

In the North American continent, the huge northern coniferous forest of Canada now has a standing biomass barely equal to a third of its original value, while in the USA felling in the western forests has consistently exceeded the annual growth for 20 years or more (Figure 6.9). In Siberia, the so-called production forests which constitute 87.4 per cent of the total woodland areas of the USSR are being exploited at a rate that exceeds the average annual growth by 150 per cent. Over 10 million hectares cleared at the beginning of the 1950s were still showing no signs of regeneration 15 years later (Pryde, 1972).

However, the types of woodland suffering most overexploitation are the various categories of tropical forest (Figure 6.10), which are being cut back at an extraordinary rate for the production of timber for export or of wood for fuel. The reduction in area of

tropical rain forests throughout the world during the 1970s is estimated to have amounted to 110 000 km² per year: at this rate, they will be completely destroyed by the middle of the next century (Table 6.9).

In Africa, much of the overexploitation has been due to the activities of foreign companies. Whereas 60 per cent of its total land area was afforested 2000 years ago, the figure today is barely 25 per cent and more than half of that is severely degraded woodland. The great rain forest areas of West Africa and the Congo basin have suffered particularly badly since the Second World War. Thus in the Ivory Coast, some 400 000 hectares of virgin forest have been cut down since 1955 and only a few thousand hectares have been replanted.

The use of wood as a fuel is one of the major causes of overexploitation in tropical forests: numerous wood-fired industrial installations have been set up in Brazil,

Table 6.9   The destruction of tropical forests by the year 2000 (from IUCN, 1980b)

| | Predicted losses in area between 1975 and 2000 | | | |
| | Dense forests | | 'Exploitable' deciduous forests | |
Region	In 10³ ha	As % of 1975 area	In 10³ ha	As % of 1975 area
West Africa	6 600	47.1	6 600	54.7
Tropical Asia under centralized development	6 300	29.1	6 000	35.3
Southern Asia	16 400	23.0	13 600	27.9
East Africa	3 300	17.8	3 200	50.4
South-East Asian Islands	21 600	16.5	20 000	26.3
Tropical South America	64 200	12.0	7 300	13.3
Continental South-East Asia	4 100	10.6	4 000	13.3

in Zaire and in other African countries. The extension of tobacco-growing by multinational firms in Third World areas provides a good illustration of this source of much deforestation. Whereas the drying of tobacco leaves is carried out in industrialized countries in ovens heated by gas-fired or oil-fired burners, the corresponding plant in the Third World is fuelled by wood. A United Nations study pointed out some years ago that this use of wood, encouraged as it is by the tobacco industry, posed a major threat to the environment. At a time when 55 per cent of the world's tobacco is produced in the Third World, it should be realized that for each hectare of Virginia tobacco that is oven-dried in developing countries a hectare of forest goes up in smoke (Land, 1980).

Destruction by fire

Fire is another significant cause of deforestation throughout the world. It poses a permanent threat to dry forests, particularly those consisting of conifers, which are vastly more inflammable than other species because of the resins and various terpenes they secrete. Even northern forests of this type are vulnerable to fire during the summer: in Alaska, for instance, some 1.7 million hectares of coniferous forest were burnt down in 1968 alone.

However, the greatest inroads have occurred in Mediterranean and tropical forest areas having a prolonged dry season. Mediterranean people have long used fire as a means of transforming woodland into pasturage or preparing land for cultivation. Autumnal firing of forests, designed to encourage the growth of pasture, is still widespread in the Mediterranean basin (in Corsica, for instance), even though it is forbidden in most of the coastal countries since much woodland can be destroyed if herdsmen start the fires too early in the year.

The vast increase in tourism is also responsible for many of the forest fires today, particularly through cigarette ends thrown from vehicles.[1] This is an aspect of smoking that is rarely recognized, but which is a significant cause of forest fires in European countries of the Western Mediterranean.

Whatever the reason, the original southern European forest has suffered great depredation from fire in recent years. In Provence and Languedoc alone the area destroyed since 1960 is estimated at 350 000 hectares, and in the single year of 1979 more than 50 000 hectares of forest and garrigue were burnt down (Figure 6.11). On the average, 4 per cent of the forest area of the French Mediterranean coast is

destroyed by fire each year, so that the mean expectation of life of woodland in this region is no more than 25 years. As Quezel (1980) has pointed out: 'The tourist explosion has been fatal for the French and Spanish Mediterranean forests, and is in the process of becoming so for Turkey.' Over the whole Mediterranean basin, approximately 200 000 hectares of forest are burnt down each year, and it is obviously the highly inflammable conifers that pay the heavier price in comparison with sclerophyllous and deciduous trees that are much less affected.

Mediterranean forests, under the combined attacks of fire and other destructive influences, are now so reduced and transformed that they are only a pale reflection of their condition at the beginning of human history. Hardly any remnants now remain of the original forest which consisted largely of pubescent oaks and/or various other species of sclerophyllous oak: of *Quercus ilex* (holm-oak) in the western Mediterranean areas of Europe, of *Quercus mirbecki* (Canary oak) in Morocco, and of *Quercus calliprinos* (Palestine oak) in the Middle East (Lemée, 1967).

The original coniferous areas of the Mediterranean mountains, although less degraded by human activity than the broadleaved forests at low altitudes, have nevertheless suffered considerably over the last few hundred years. What is left today, for example, of the *Abies pinsapo* forests of the Andalusian sierra; of the cedar forests stretching through the Lebanon and other Middle Eastern areas; of the forests of *Pinus brutia* in Crete (Figure 6.12); or, finally, of the *Abies nebrodensis* in Sicily?

The enormous areas of garrigue and maquis in southern Europe, amounting to tens of millions of hectares, exist because of fire. The original forests have been gradually wiped out and replaced by degraded ecosystems with smaller and smaller biomass: *garrigues* in limestone areas, *maquis* in siliceous soils. These two terms designate formations of evergreen shrub species that enjoy arid conditions but do not cover the ground completely: on the contrary, large patches of land are left bare of vegetation, allowing the soil to be eroded and the underlying rock to show through (Plate XIX).

Fire has also been a controlling factor in the replacement of the climax woodlands of Mediterranean oaks by secondary coniferous forests, and in the elimination of many species of herbs and shrubs in favour of woody formations more resistant to repeated fires (Figure 6.13). Among trees, only the cork oak (*Quercus suber*) and some species of *Pinus* are able to resist fire reasonably well because of the thickness of their bark: the Aleppo pine (*Pinus halepensis*) is an example of a species that is quite resistant in spite of the high degree of inflammability associated with all conifers. A fire that spreads rapidly through

---

1. In California, this practice is discouraged by the imposition of fines and prison sentences, a long way from the laxity of French law in this respect.

Figure 6.11   Charred forest and garrigue in the south of France. (Photograph F. Ramade)

Figure 6.12   Crete is one of the Mediterranean countries that has suffered the greatest loss of forest cover due to human activity dating from the earliest antiquity. This photograph shows a view of Mount Levka on the west of the island. Although formerly covered in dense coniferous forests, Cretan mountains are today largely bare. (Photograph F. Ramade)

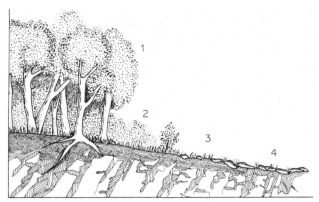

Figure 6.13 Successive stages in the regression of a climax forest of holm-oak due to overexploitation and repeated fires in a limestone area of Mediterranean France. 1: the original forest of holm-oak (*Quercus ilex*). 2: garrigue with kermes oak (*Quercus coccifera*), the dominant fire-resistant species in these degraded woody formations. 3: land with covering of false brome grass (*Brachypodietosum ramosi*). 4: overgrazed pasture containing *Euphorbia characias*. (From Braun-Blanquet, in Lossaint and Rapp, 1978)

Forests in tropical regions where there is a pronounced dry season have also been greatly reduced by fire: in many areas of East and West Africa, for instance. The practice of burning brushwood has led to a continual transformation of forest to savanna, and the encroachment of this type of grassland to the detriment of the forests is almost universal in the tropics. This is why only a few isolated patches of the original sclerophyllous and deciduous forests of Kenya and Tanzania remain: in their place, vast areas of the Sudanese savanna have encroached and are traversed each year by brushwood fires.

In Malagasy, fire has been responsible for the destruction of much of the dry deciduous forest that once covered more than a third of the western half of the island. For some decades now, this forest has been continuously reduced in size by fires lit by herdsmen: several thousand hectares are destroyed every year, leaving the land denuded and vulnerable to erosion or alteration of the soil. Eastern Malagasy is covered by a tropical rain forest in which fire is used by the inhabitants as a preliminary to the clearance of land for farming. During the 1960s, some 150 000 hectares per year were destroyed in this way, in a forest that covered a total of 6 million hectares at the beginning of that period (Chauvet, 1972).

undergrowth may leave some large trees standing to become efficient disseminators of seed. The cones of some species are indehiscent (do not burst open), thus protecting the seeds during the passage of the fire and allowing them to be released later. All these factors, together with their greater ability to tolerate arid conditions, have favoured various species of Mediterranean pine (Figure 6.14) at the expense of climax populations of pubescent and other oaks. Forests of these latter species in southern Europe, for instance, have been burnt down and replaced by conifers (Aleppo pines, cluster pines or Corsican pines), whose rapid growth and production of acid humus retaining less moisture has prevented almost any regeneration of the original oak populations.

Forest fires are a real ecological disaster in regions with an intense summer drought because the soils become impoverished and their structure degraded. The temperatures reached in the surface layers of the ground are very high and this completely burns up not only the litter but the humus as well, and to a considerable depth. In addition, the decomposition of organic and even inorganic compounds of nitrogen by the heat causes an extra loss of nutrients. Enrichment of the soil by burning organic matter is only apparent, since the ashes are often either blown and dispersed by the wind

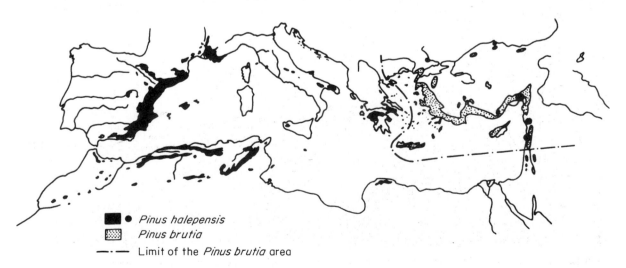

Figure 6.14   Geographical distribution of *Pinus halepensis* and *Pinus brutia* in the Mediterranean basin. (From Quezel, 1980)

or are severely leached by the first violent rainstorms after the fire such as occur in Mediterranean and tropical climates.

6.4   The Protection of Forests

Forests and woodlands are renewable resources important to the human race in many different ways. If they are managed rationally, taking due account of all the basic ecological factors involved, there is no reason why they should not provide permanently all the wood needed for purposes where it is indispensable. In addition, forests play a major part in the conservation and

regeneration of soils and, by protecting areas of high relief against erosion, they save drainage basins both from flooding and from becoming badly silted up. They also smooth out climatic fluctuations and increase rainfall. Finally, they act as genetic reservoirs not only for all existing tree species but for the whole community of living organisms that inhabit them. The destruction of the coniferous forests in tropical America, for instance, would have serious consequences for the future of world forestry since the various species of pine that grow in them constitute the genetic raw material for reafforestation programmes in several other tropical countries.

In many regions of the world where forests have

Figure 6.15   A stand of giant sequoia (*Sequoiadendron giganteum*) in the Sequoia National Park of Central California. The complete protection of these woodlands from fire seems to hinder the regeneration of this species, whose young stock are fire-resistant. (Photograph F. Ramade)

Plate XIX   Mediterranean forests

1   Garrigues and the remnants of Aleppo pine forests in the eastern section of La Sainte-Baume. This region, which has suffered three large fires during the last 25 years, was formerly covered with dense woodlands of pine and holm-oak. Now it has degenerated to garrigue, a degraded formation characteristic of limestone areas and consisting of plants that can survive repeated fires. In the distance, the massif of La Sainte-Victoire (Bouches-du-Rhône) can be seen.

2   Pubescent oaks (*Quercus pubescens*) in the national forest of La Sainte-Baume. This species, together with other sclerophyllous oaks, formed the principal tree populations of the Mediterranean basin at the beginning of the Christian era.

3   Aleppo pines (*Pinus halepensis*) in the Chaîne de l'Etoile near Marseille. This species has replaced the original forests of Mediterranean oaks because of its more rapid growth and to a certain extent because of its ability to survive fire.
(Photographs F. Ramade)

become badly degraded, 'protection' implies regeneration and enrichment. Immense efforts need to be made, for example, in the rehabilitation of woodlands throughout all countries bordering the Mediterranean. Such an operation involves not only a quantitative aspect because of the very size of the areas concerned, but a qualitative one as well: that is, one promoting the reconstitution of broadleaved forests whose trees would have a much greater fire resistance than the disclimax species of pine which have replaced the original oak.

In a similar way, but more generally, protection of forests implies that priority is given to the growth of climax species better adapted to local ecological conditions and thus more resistant to attacks by insects and disease. As Chararas pointed out in 1964, the degeneration of the cluster pine in Provence undoubtedly stems from its artificial extension into a low-altitude Mediterranean environment to which it is ill-adapted.

In most Third World countries, the protection of forests would necessitate the abandonment of wood as fuel and the development of new sources of energy to replace it. A short-term measure to alleviate the wastage of forest resources in this way would be to establish plantations specifically for firewood in order to preserve natural woodlands. However, the high rate of population growth in developing countries makes this an inadequate response to their energy needs and it is in fact only technological and thus economic development that will make forest protection possible. An illustration of this is the introduction of compressed gas in the rural areas of Spain and Algeria which has led to the abandonment of wood as fuel there.

Mountain forests, more than any others, are worthy of special protection because of the essential part they play in the fight against erosion and in the regulation of the water cycle. It is imperative that they should all be declared as nature reserves and that any modifications making them more vulnerable should be prohibited: I am thinking here particularly of road construction that cuts savage gashes through them, generates erosion and eventually degrades the whole woodland area.

Protection against fire is a permanent concern in forests growing in climates with a pronounced dry season. Developments like the creation of fire-breaks and the installation of water points and so on are certainly desirable measures, but the best method of inhibiting fires is the reestablishment of broadleaved plantations as climax ecosystems. As far as conifers and other highly inflammable species are concerned, it has been suggested in the USA that controlled fires should be used outside the dry summer period to eliminate the undergrowth from which disastrous uncontrolled fires generally start.

However, it must not be thought that fire in forests is always undesirable. There is a term 'pyroclimax' that describes precisely those formations in which the dominant species of tree are favoured by the passage of fire, so long as it does not occur at too frequent intervals of time. Thus the Californian forests of giant sequoia (Figure 6.15) cannot be regenerated unless they are occasionally traversed by an undergrowth fire. If this does not occur, the young sequoia stock is challenged by other less fire-resistant species with more rapid rates of growth which would end up by smothering them (Dasmann, 1976).

Similarly, fires which are spontaneous (and not lit accidentally or deliberately by humans) and which break out every 200 or 300 years, are a natural phenomenon in northern coniferous forests. Indeed, such fires accelerate the mineralization of woody matter and other plant detritus, which is a very slow process under sub-arctic conditions, and thus form a natural feature of the biogeochemical cycles in taiga. Recognition of this effect has led to a decision in the USA that spontaneous fires in the Yellowstone National Park should not be fought. In Mediterranean areas, too, fire appears to benefit the regeneration not only of Aleppo pine, whose current extent is largely the result of human interference, but of species like *Pinus brutia* and the cedars.

In certain aspects, therefore, fire appears to be a natural feature in the dynamic development of some types of forest. Both its total elimination and its excessive frequency due to accidental or deliberate human action would seem capable of producing undesirable ecological disturbance in the ecosystems involved.

Chapter 7

Natural Rangelands: Savanna, Steppe and Other Grasslands

Wherever the growth of trees is hindered by insufficient rainfall or other climatic factors, forests give way to different types of biome in which the dominant vegetation is herbaceous, with a ground cover principally of grasses. This is the case at all latitudes: in tropical regions the biome occurs as savanna, in semi-arid areas as steppe, while in cool temperate countries there are alpine meadows in mountainous territory beyond the tree limit. This last type of biome is a high-altitude variant of tundra, the typical ecosystem of sub-arctic regions, characterized by a vegetation that is predominantly of grasses and Cyperaceae with various associated lichens.

Hardly any of these areas are capable of being transformed into arable land because of the shortage of water, the poor soil structure and/or the low temperature. For that reason, they are mainly used for extensive rearing of livestock, a practice that has been in existence in the Old World from the beginning of the Neolithic age.

Ecosystems of this type occupy more than 30 million km^2 of the world's land mass (Table 7.1), although the distribution is somewhat uneven, being concentrated rather more in South America, Africa and Asia than in other continents. However, the total areas now covered by the various grassland formations are by no means all natural ecosystems—far from it. In Europe, India and Africa, and in tropical and temperate America, vast expanses of supposedly 'natural' grassland are in fact human formations. As soon as Palaeolithic hunters discovered fire, they disrupted plant communities in many regions of the globe. The first to be affected was Africa: the original plant cover of dry deciduous forest, so easily destroyed by fire, has probably been giving way to savanna from the Sudan and Guinea for hundreds of thousands of years. There is certainly archaeological evidence in Central Europe of the same sort of destruction of woodlands by fire in the Palaeolithic age, and during the same period tropical regions of the Old World undoubtedly witnessed similar destruction of vast areas of primitive forest which were thus prevented from being ultimately regenerated.

Since Neolithic times, natural vegetation has been deliberately degraded in various parts of Africa, Asia and South America with the idea of creating large areas of grassland capable of supporting a much greater population density of ungulates than the original forests were able to. The extension of savanna in both Africa and Asia beyond their original limits has been due to the deliberate firing of climax woodlands. The North American Indians extended the prairie in a

Table 7.1   *Areas of permanent pasturage in various regions of the world* (from FAO, 1971)

Region	Areas in 10³ km²		Pasturage as % of total
	Total land surface	Pasturage	
Europe	4 930	930	18.8
USSR	22 402	3 740	16.7
Asia	27 531	7 350	26.7
(of which, China	9 561	1 770	18.5)
North and Central America	22 410	3 720	16.6
South America	17 840	4 130	23.2
Africa	30 300	8 440	28.1
Oceania	8 510	4 640	54.5
World total	133 923	30 950	22.8

similar way some 10 000 years ago so as to provide ranges for their buffalo, with which they lived in a virtual symbiosis.

Be that as it may, permanent pasturage in one form or another covers 23 per cent of the total land surface of the world at present, and 47 per cent is potentially of use only as grazing land for farm or wild animals. According to the FAO, 75 per cent of the world's domestic livestock population of more than 3.8×10^9 head (1.1×10^9 cattle, a similar number of sheep, 6.7×10^8 pigs and 3.8×10^8 goats) are range-fed on these open grasslands.

7.1 Grassland Ecology

The ecology of grasslands is complex, yet in spite of the large variety of biomes used for range-feeding they have a number of characteristics in common. For instance, although the vegetation can include some trees and shrubs, it is grasses that overwhelmingly pre-

dominate both in total standing biomass and in numbers of individual plants.

In the Guinean savanna of Lamto, grasses (mainly from the genera *Loudetia*, *Hyparrhenia*, *Andropogon* and *Imperata*) form between 75 and 99 per cent of the biomass of the herbaceous plant cover (Lamotte, 1978). In North America, 90 per cent of the prairie herbage consists of grasses, predominantly from species belonging to the genera *Buchloe*, *Bouteloua*, *Stipa* or *Andropogon*. A natural meadow of *Holcus lanatus* and *Poa* sp. in Normandy contains 65 per cent grass in its total plant cover (Ricou *et al.*, 1978).

If there is no human interference, grasses form a tight continuous network of vegetation over the soil of almost all the grassland formations. The plants fall into one of two distinct groups: annual (therophytes) and perennial, mainly tufted grasses and geophytes. Perennial grasses belong to two morphological types (Figure 7.1), some, as in Figure 7.1B, branching out over the soil to produce suckers that form new root systems, thus considerably increasing the area covered by each

A   B   C

Figure 7.1   Growth forms of typical range grasses: A, annual (therophyte); B and C, perennials. In B is shown a species that spreads by putting out suckers, in C a species that forms tufts. (Reproduced with permission from Dasmann, *Environmental Conservation*, Wiley, 1976)

Table 7.2　Primary productivity of the principal types of grassland (from Lieth, 1975)

Formation	Total area (10^6 km^2)	Net primary productivity in g of dry matter $m^{-2} \text{ yr}^{-1}$		Net production (10^9 t yr^{-1})	Energy fixation $(10^3 \text{ kcal } m^{-2} \text{ yr}^{-1})$
		Normal range	Mean		
Tropical savanna	15.0	200–2000	700	10.5	2.8
Semi-desert scrub	18.0	10–250	70	1.3	0.3
Temperate grassland	9.0	100–1500	500	4.5	2.0
Tundra	8.0	100–400	140	1.1	0.6

plant. Other species, as in Figure 7.1C, form tufts. Most of the grass crop on ranges consists of perennial varieties that can withstand grazing well and thus protect the soil quite efficiently against erosion.

Another characteristic of grasslands is the considerable development of root systems that often penetrate to great depths. Because of this, the above-ground biomass B_e of the plants is always much less than that below ground (B_h), even in the absence of grazing. In the Lamto savanna, most of the crop is perennial grass with a B_e/B_h ratio generally less than 0.4, and it is even as low as 0.2 for many geophytes (Lamotte, 1978). Ricou *et al.* found a B_e/B_h ratio of 0.35 for natural meadowland in Normandy, while a Siberian shrub tundra yielded a value of 0.32 (Rodin and Basilevitch, 1964). In all cases, most of the biomass is concentrated in the top 15 centimetres of the soil.

The soils on which natural grasslands are established show considerable variations, with structures that depend both on the type of vegetation and on the climate, just as with other ecosystems. In cold wet conditions, plant cover is not enough to prevent podzolization. If, on the other hand, the rainfall is relatively light, and if there is also considerable evapotranspiration, mineral salts migrate to the upper soil horizons where the combination of calcium carbonate and humus produces chernozem (black earth), which is characteristic of temperate grassland with a slightly alkaline pH.

The primary productivity of grasslands varies with latitude and amount of rainfall, although it is always lower than that of forest biomes (Table 7.2). Its value is greatest for tropical savanna and falls off with increasing distance from the Equator as the rainfall decreases: then, to the north of the subtropical desert belts in the Northern Hemisphere, the productivity begins to increase once more, rises to another peak in the temperate grasslands and finally declines again as tundra is reached.

In general, the primary production of the epigeal plant layers (those above ground) is greater than or equal to that of the hypogeal layers, although there are exceptions. One example is that of a prairie of *Andropogon* in Missouri whose annual epigeal production has been estimated at 1962 kcal m^{-2} while the corresponding hypogeal production is 2380 kcal m^{-2} (Kucera *et al.*, 1967). More in accordance with the general pattern are the figures from a permanent meadow in Normandy produced by Ricou *et al.* (1978): here the primary productivities were 3910 kcal m^{-2} for the aerial parts of the plants, 1173 kcal m^{-2} for the litter and 2200 kcal m^{-2} for the root systems.

That part of the secondary productivity that is useful to humans is, unlike the total primary productivity, greater in grasslands than in forests. The reason for this is that most of the primary production in forests is taken up by consumers that are of no interest to

Table 7.3　Secondary productivity of the principal land-based biomes (from Whittaker and Likens in Lieth and Whittaker, 1975)

Ecosystem type	Area (10^6 km^2)	Animal consumption (10^6 t yr^{-1})	Secondary production (10^6 t yr^{-1})	Secondary productivity $(\text{kg ha}^{-1} \text{ yr}^{-1})$	Animal biomass (10^6 t)
Tropical rain forest	17.0	2600	260	152.9	330
Tropical seasonal forest	7.5	720	72	96.0	90
Temperate deciduous forest	7.0	420	42	60.0	110
Taiga	12.0	380	38	31.7	57
Tropical savanna	15.0	2000	300	200.0	220
Semi-desert scrub	18.0	48	7	3.9	8
Temperate grassland	9.0	540	80	88.9	60
Tundra	8.0	33	3	3.8	3.5

humans as food. In temperate deciduous forests, the mammalian secondary productivity amounts to only a few kilograms per hectare per year, and although the total secondary productivity in tropical forests seems very high, it is mainly due to various inedible insects and other arthropods. In grasslands, on the other hand, the majority of the available primary production forms the basic foodstuff for most of the wild and domesticated ungulates. An additional factor in savanna is the presence of scattered trees which increase the primary productivity and the diversity, thus accounting for their exceptionally high secondary productivity (Table 7.3).

No overall study has been undertaken of the energy flow through all the populations of large herbivores in a grassland ecosystem. The best-known synecology as far as savanna is concerned is that of Lamto in the Ivory Coast, but this is not a very good subject for study since large mammals form only a small proportion of the total animal population: buffalo and antelope alone have escaped an almost complete massacre of large fauna in the country.

In Uganda, Buechner and Golley (1967) have studied a population of some 15 000 kobs (*Adenota kob thomasini*) in the Toro reserve to the south-west of Lake Albert, and estimate the animal biomass as being about 2.17 tonnes km^{-2}, quite a high value for wild hoofed mammals. These authors have also calculated the energy transfer through the food chain of these animals and obtain the following results (expressed in kcal m^{-2} year^{-1}):

Primary production consumed	74.1
Energy flow	62.4
Metabolism	61.6
Secondary productivity (growth)	0.81

The ecological efficiency is thus of the order of 1 per cent and the secondary productivity is about 577 kg m^{-2} year^{-1}. This is quite a low value, although the energy flow is among the highest known for any wild mammals, the reason being that they alone consume 10 per cent of the primary production in their savanna homeland.

Energy flows and secondary productivities much greater than those for wild ungulates are found when domesticated livestock form the main consumers on a natural grassland, and they are greater still when the grassland has a high primary productivity through good agricultural management. Ricou *et al.* (1978), for instance, have determined the general pattern of energy flow and productivity for permanent meadowland in Normandy. This ecosystem, in spite of its great age, is not a primitive formation but originated from the clearance of deciduous forests that covered the area until the end of the first millennium AD. The primary productivity is particularly high because of the

excellent condition of the soil and the favourable climate. It amounts to 920 g of dry matter m^{-2} year^{-1} for the epigeal growth consumed by cattle: the equivalent of 3910 kcal m^{-2} year^{-1}. The density of the herds (2.2 milking cows per hectare) is quite high, since they consume 782 g of dry matter m^{-2} year^{-1}, the equivalent of 3323 kcal m^{-2} year^{-1} or 85 per cent of the epigeal primary production (Figure 7.2), of which 2420 kcal are assimilated. The secondary productivity amounts to 518.6 kcal m^{-2} year^{-1}, a much higher figure than that for wild herbivores, or even for domesticated livestock, living on natural grassland. Even if the *total* primary productivity of 7323 kcal m^{-2} year^{-1} is used, the efficiency of energy conversion is 7 per cent, a very high value for permanent pasturage. Thus, cattle raised in California on grassland with a net primary productivity of 1410 kcal m^{-2} year^{-1} give a net secondary productivity of 69 kcal m^{-2} year^{-1} or an ecological efficiency of 0.5 per cent.

The biogeochemical cycles of biogenic elements show a remarkable stability in grasslands. Rearing of livestock, unlike arable farming, plays only a minor part in the removal of nutrient minerals, at least where the rainfall is abundant. On a permanent pasture in the Pennines of northern England, for example, it has been shown that the loss of phosphorus from the grazing of sheep intended for the meat market is barely 1/200 of that from run-off due to rainfall in the area studied.

7.2 Savanna

Tropical savanna covers some 15 million km^2 of the total land surface of the globe, mainly in the Northern Hemisphere. Geographically, it succeeds tropical rain forests either after a transitional area of dry deciduous or sclerophyllous forest or else quite abruptly with no intervening stage. Sometimes fingers of gallery forest may prolong the rain forest into drier regions along the banks of waterways where the soil and humidity are suitable, while savanna occupies the plateaux beyond.

Various types of savanna occur, depending on the amount of rainfall in the region, and this produces a remarkably regular succession in Africa in going north from the Equator. Guinea savanna occupies areas where the annual rainfall exceeds 1200 mm and is sufficient to allow the growth of many species of tree. It is sometimes called preforest savanna since it occupies areas in contact with tropical forests, but is generally known as tree savanna because there is a significant tree population in spite of the fact that the predominant ground cover is grass (Figure 7.3).

Further to the north, the grassland becomes Sudan savanna, containing various species of trees and shrubs that are resistant to the bush fires that cross the region

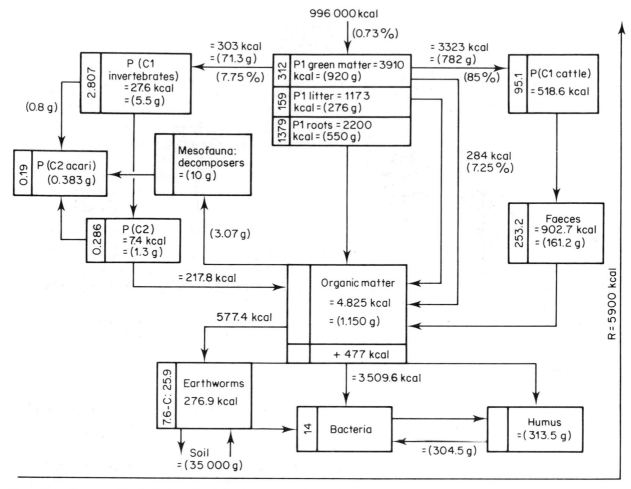

Figure 7.2   Pattern of biological productivity for permanent meadowland in Normandy. Transfer of matter is indicated by arrows with the biomass written vertically. The figures giving productivities are expressed either in g of dry matter m^{-2} year^{-1} or in kcal m^{-2} year^{-1}. P1 = producers; C1 = primary consumers (herbivores); C2 = secondary consumers (carnivores); R = respiration; P = productivity. (From Ricou *et al.*, 1978)

each year and destroy the grass cover. Woody plants belong mainly to the myrobalan family (Combretaceae), but whether the grassland is tree savanna or shrub savanna depends on the rainfall and the frequency and intensity of the fires. This type of savanna has fewer woody plants than the Guinea type, but even so they can reach a density of 1000 trees or shrubs per hectare.

Further north still is the Sahelian savanna, forming an intermediate stage between the above types and the semi-desert scrub that leads to desert biomes. It contains far fewer species of tree than either the Guinea or Sudan savanna and at a much lower density. For instance, in the Sahelian savanna of Fété Olé in northern Senegal, 98 per cent of the trees belong to only six species: *Guiera senegalensis, Balanites aegyptiaca, Grewia bicolor, Commiphora africana, Acacia senegal* and *Boscia senegalensis* (Bourlière, 1978).

Many of the woody species in the Sahelian savanna are thorny and possess other characteristics of plants well-adapted to arid conditions. The density of trees and shrubs is much less than in other types of savanna and rarely exceeds 250 individuals per hectare: at Fété Olé, for example, a count of woody plants taken before the great drought at the beginning of the 1970s showed a density of only 133 per hectare having a height of 1 metre or more, and after 1972 the species *Acacia senegal* and *Guiera senegalensis* were found to have suffered a mortality rate of 60 per cent (Poupon, 1979).

As a general rule, the abundance of trees and shrubs in the African savannas decreases with increasing distance from the Equator and, as has already been pointed out, the major part of the plant biomass throughout these regions is provided by grasses. These are predominantly perennial species, often exceeding 1 metre in height at maturity. The dominant genera in Africa are *Andropogon, Loudetia, Imperata, Hyparrhenia, Aristida* and *Pennisetum* (elephant grass).

Grasses in the South American savanna consist primarily of *Pennisetum, Trichloris* and *Setaria*, while

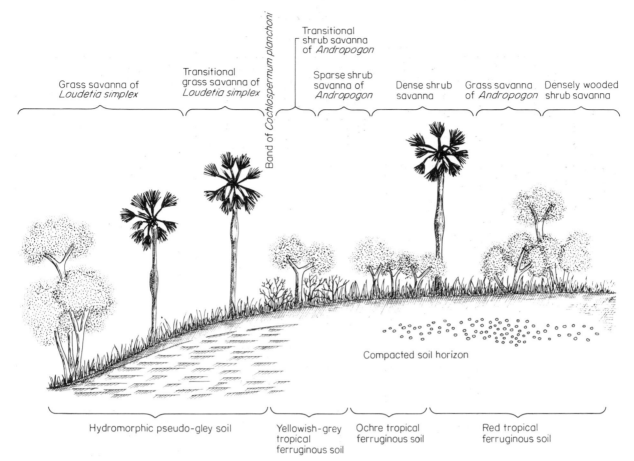

Figure 7.3   Influence of topography and soil structure on the character of a Guinea savanna in the Lamto area of the Ivory Coast. The term *shrub savanna* is used here in the same sense as *tree savanna* in our text. (From César and Menant, in Lamotte, 1978)

in the tropical savanna of northern Australia the main grass crop is of *Themeda*, *Astrebla* and *Trioda* (Lemée, 1967).

The high primary productivity of grasslands encourages proliferation of larger ungulates: various Bovidae, antelope and zebra in the African savanna and large kangaroos (*Macropus*) in Australia. In fact, the Sudan and Sahelian savanna support the largest mammalian biomass of any terrestrial ecosystem: in no other region of the world is there now such a great concentration of large mammals. An illustration of this is the migration that takes place each year of nearly 2 million wildebeests (*Connochaetes taurinus*) between the Serengeti National Park in Tanzania and the Masai Mara Reserve in south-west Kenya (Figure 7.4).

The character of the vegetation in tropical savanna has been greatly modified by the grazing of both wild and domesticated livestock and by the widespread practice, among the herdsmen and farmers of these regions, of seasonal firing. In Africa and tropical Asia, bush fires have led to a considerable extension of secondary savanna at the expense of forests. Evidence that savanna is a secondary formation due to human interference rather than a climax ecosystem is provided

by the work of Lamotte *et al.* (1967). By protecting an area of preforest savanna at Lamto in the Ivory Coast from fire, they found that a transition to dense secondary forest began to develop quite rapidly. In reality, the opposite transformation, from forest to savanna, is taking place quite generally throughout the tropical regions of the world because of the bush fires that constantly encroach on the afforested areas.

Just as in the Mediterranean garrigue, the climax vegetation of these tropical regions has been replaced by species that are much more fire-resistant. The grasses now consist either of species with rhizomes like *Imperata* or *Chrysopogon*, or of *Andropogon* whose young shoots are hidden in the centre of thick tufts and are thus protected when fire crosses the land. The shrubs that grow have a large underground structure of roots, whose biomass forms 30 per cent of the total plant. This is undoubtedly an adaptation linked to fire-resistance, since the new young growth develops very rapidly on the stumps of the burnt-out plants. As for the trees, the species that persist are mainly those whose trunks are covered by a thick fire-resistant bark. However, apart from a few large species like the doum palm of East Africa or the palmyra palm (*Borassus*

Figure 7.4   Migration of wildebeest (*Connochaetes taurinus*) between the Masai Mara reserve and the Serengeti national park in East Africa. (Photograph F. Ramade)

aethiopum) of the Ivory Coast, trees are in general sparsely scattered and in poor condition, with twisted trunks and stumpy lower branches as a result of repeated fires.

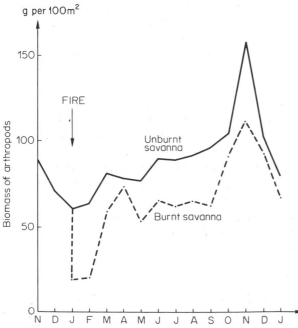

Figure 7.5   Effect of fire on the biomass of arthropods in the herbaceous layer of a Guinea savanna. (From Gillon and Gillon, 1967, 21–3)

The effect of fires on the structure and fertility of the soil in tropical savanna is still the subject of controversy. Research in the South African veldt suggests that it enriches the surface layers in nutrients and encourages the growth of *Themeda*, a grass much sought after by the wild and domesticated livestock that graze there. However, firing carried out at the height of the dry season produces such high temperatures at the surface that modifications of the soil occur similar to those observed after fire in the coniferous forests of the Mediterranean: loss of carbonaceous material, dissolving away of inorganic nutrients, an increase in the pH value due to the formation of basic compounds of the alkali and alkaline-earth metals, and, lastly, a reduction in the water-retaining capacity.

The consumption of litter due to bush fires also impoverishes the soil in humus: a shrub savanna in Kenya protected against fire for 10 consecutive years showed a very large increase in humus as compared with neighbouring unprotected terrain that was fired each year. Other studies in South Africa (Daubenmire, 1968) have demonstrated the great differences that can occur in the water content of soils during the dry season, where burnt areas showed a far greater degree of aridity.

A significant proportion of the net primary production is destroyed by bush fires. In the Lamto savanna, for instance, they consume 8.5 tonnes per hectare per year of the plant biomass: more than 30 per cent of the average total production estimated at 28 tonnes per hectare per year (Lamotte, 1978). However, such an adverse effect is only apparent, particularly for the case of herbivores like the large ungulates. The fire encourages the growth of species of grass with high nutritive value soon after it has passed. Certain antelopes like topi (*Damaliscus korrigum*) can even be observed feeding on very recently fired savanna, probably attracted by the mineral salts contained in the remnants of calcined plants scattered over the land surface.

Invertebrates in the savanna are affected to varying degrees by the passage of fire. Unlike mammals and birds, which can flee at its approach, or reptiles, many of which can bury themselves in the ground, most of the invertebrates perish, with the exception of soil-dwelling species[1] and those that can fly or jump well enough. However, the latter group are often eaten as they flee from the flames by the numerous insectivorous birds of prey attracted by the fire: Gillon (1971) has mentioned the presence of large numbers of kites capturing insects fleeing in advance of a fire, particularly those of the Acrididae family (crickets, locusts, etc.).

Arthropod populations are in general quite significantly reduced by fire. Gillon and Gillon (1967) found that their total biomass decreased by about 30 per cent, from an average of 9.2 kg ha^{-1} (330 000 individuals) in an unburnt savanna to 6.2 kg ha^{-1} (230 000) in one exposed to fire at corresponding times of the year (Figure 7.5). In fact, however, fire affects different taxonomic groupings in different ways: thus, Blattidae, which require moist conditions, are greatly reduced in numbers, their average density falling from 55 500 per hectare to 16 000 per hectare. On the other hand, Orthopterae are more numerous in a fired savanna than in an unburnt one.

Secondary productivity in savanna

The trees and shrubs scattered over the Sudan and Sahelian savanna contribute to a high secondary productivity both directly and indirectly. First of all, they enhance the primary productivity of the herbaceous layer by increasing soil fertility (Leguminosae like the acacias), by protecting the soil against erosion and by reducing evapotranspiration (Ovington *et al.*, 1963; Poupon, 1979). Also, their foliage provides an additional essential source of food for livestock during the dry season, for instance for the giant eland (*Taurotragus derbianus*). For some species of giraffe

and certain antelopes like the gerenuk (*Litocranius walleri*) this is practically the only source of food available.

Tropical savannas have an exceptionally high biological productivity considering the generally poor quality of their soils and the distinctly low rainfall in many of them. This apparent paradox comes about from two causes: first, from the particular structure of the vegetation, already discussed above, in which that part of the primary production consumable by large herbivores is especially favoured; secondly, from the ecological diversity of the ungulate populations, which enables optimum use to be made of the plant growth. It is therefore not surprising that, of all land-based ecosystems, tropical savannas (particularly those of Central and East Africa[1]) have the largest density of biomass and the highest secondary productivity as far as large mammals are concerned (Bourlière, 1963 and later; Table 7.4).

In the Albert National Park of Zaire, for example, a count taken several decades ago of antelopes alone yielded 24 kobs (*Adenota kob thomasi*), 12 topis (*Domaliscus korrigum*) and 2 reedbucks (*Redunca redunca*) per km^2: a biomass of more than 3.8 tonnes for every km^2 of savanna. The total mammalian biomass in the grasslands of this park amounted to over 200 kg ha^{-1}, whereas a rain forest in southern Ghana barely yielded 1 kg ha^{-1} even including very small mammals (Bourlière and Hadley, 1970). A similar count taken in the Serengeti National Park in Tanzania (1971) yielded some 42 000 buffaloes, 220 000 Grant's zebras (*Equus burchelli böhmi*), 330 000 white-bearded wildebeests (*Connochaetes taurinus*), 150 000 Thomson's gazelles and 20 000 topis (*Damaliscus korrigum*), to quote only the main species in an area of 23 000 km^2: in other words, 33 of these various animals per km^2. Large herbivores account for nearly 95 per cent of the mammalian biomass in such regions.

The large species diversity of ungulates in the African savanna arises from the multiplicity of ecological niches occupied by the various species. Each type of primary consumer has a particular diet throughout the year which is complementary to those of others occupying neighbouring niches. The most clear-cut example of this is provided by comparing the diets of giraffes (foliage and young shoots of trees), rhinoceroses (bushes) and gnus (almost exclusively grass).

---

1. Ants, for example, are affected little or not at all by the passage of fire (Lévieux, 1968).

1. In West Africa, and more especially in territories previously under French administration, the large fauna of the savanna have been virtually exterminated by hunters who are, as everybody knows, indispensable as regulators of animal populations and who are the best protectors of nature (at least according to statements made by certain official spokesmen)!

Table 7.4   Biomass of herbivores supported by various types of protected savanna in tropical Africa (from Bourlière and Hadley, 1970)

Location	Type of habitat	Number of herbivore species*	Biomass (in kg ha⁻¹)
Tarangire Game Reserve, Tanzania	Open savanna with *Acacia* sp.	14	11
Kafue National Park, Zambia	Tree savanna	19	13
Kagera National Park, Rwanda	Savanna with *Acacia* sp.	12	33
Nairobi National Park, Kenya	Open savanna	17	57
Serengeti National Park, Tanzania	Open shrub savanna	15	63
Queen Elizabeth National Park, Uganda	Open shrub savanna	11	120
	The same overgrazed	11	278–315
Albert National Park, Zaire	Open shrub savanna, overexploited	11	236–248

* Only herbivores with an appreciable biomass are included.

There are, however, more subtle dietary differences limiting the amount of interspecific competition that occurs (Talbot, 1966). Thus, a detailed study of the way certain of the ungulates feed on the grasses reveals an extremely high selectivity in their eating habits. For example, one particular species of grass, *Themeda triandra*, is shunned by most herbivores but not by gnus, topis and zebras: the gnus feed only on its young leaves less than about 10 cm in length, the zebras eat only young shoots and older leaves longer than 10 cm, and topis only touch the leaves of the plant when they are dry.

Talbot's work (1963) also shows that the secondary productivity of wild ungulates in the East African savanna is much higher than that of domesticated livestock of similar size raised by nomadic tribes in the area, such as the Masai. This can be seen from the comparisons between the biomasses and growth rates shown in Table 7.5. Not only that, however, but the quality of the meat from the wild animals is also higher than that of the domesticated livestock. The carcasses of cattle raised by the Masai contain on the average 60 per cent of lean meat and 20 per cent fat, while the figures for gnus are 76 per cent and 5.4 per cent, and for kobs are 82 per cent and 3 per cent (McCulloch and Talbot, 1965). A final factor to the advantage of the wild species is their natural resistance to several parasites, particularly to those causing trypano-

Table 7.5   Daily growth and individual biomass of wild and domesticated animals in the East African savanna (from Talbot, 1963)

Species	Daily weight increase (kg day⁻¹)	Duration of growth (months)	Mean weight of adults (kg)	Age of maturity (months)
Thomson's gazelle	0.06	10	18.6	18
(*Gazella thomsoni*)	0.04	15	24.0	
Impala	0.12	10	45.8	24
(*Aepyceros melampus*)	0.09	18	59.4	
Grant's gazelle	0.12	10	45.8	24
(*Gazella granti*)	0.10	18	56.2	
Topi	0.20	12	114.3	30
(*Damaliscus korrigum*)	0.15	24	132.4	
Red hartebeest	0.23	12	122.5	30
(*Alcephalus buselaphus*)	0.13	24	150.6	
Wildebeest	0.24	12	163.3	45
(*Connochaetes taurinus*)	0.20	24	208.7	
Giant eland	0.24	18	283.5	?
(*Taurotragus derbianus*)			376.5	
Domesticated sheep	0.05	18	20.0	18
			45.4	
Domesticated cattle	0.14	—	158.8	60
			453.6	

Plate XX   Fire in an Ivory Coast savanna

1   The start of a bush fire in the Lamto savanna. Notice the number of palmyra palms (*Borassus aethiopum*) and the grass
not yet burnt in the foreground.

2   The savanna as it appears when fire is passing through it. Bush fires have only a limited effect on the trees and shrubs and
on the soil structure if lit quite early in the year at the beginning of the dry season.
(Photographs Y. Gillon)

Plate XXI   Animal species of prairie and savanna

1   Pronghorn antelopes (*Antilocapra americana*) in the Montana Bison Range Reserve, USA. This species, along with bison, formed the main ungulate population of the North American prairies before the arrival of Europeans.

2   Impalas (*Aepyceros melampus*) and Grant's zebras (*Equus burchelli böhmi*) in the Masai Mara Reserve, Kenya. The secondary productivity and biomass per unit land area of the wild ungulates in the tropical savannas are unequalled in any other terrestrial ecosystem.

3   Reticulated giraffes (*Giraffa camelopardalis reticulata*) in a Sudan savanna (Samburu National Park, Kenya).
(Photographs F. Ramade)

NATURAL RANGELANDS: SAVANNA, STEPPE AND OTHER GRASSLANDS   191

somiasis, a disease that is responsible for terrible losses among the domestic animals and can even prevent any breeding at all in some areas.

7.3   Temperate Grasslands

Where the rainfall in temperate regions of the world is insufficient to sustain the growth of trees, forests give way to grasslands covering vast areas of the Northern Hemisphere: the North American prairie is an excellent example of this type of biome. The soils of these areas are very different from those of temperate forests at the same latitude and with similar geological sub-strata. In the first place, they are much richer in humus than the forest soils because the short life of herbaceous species produces a large accumulation of organic matter: humification thus takes precedence over mineralization. In addition, loss of water through evaporation is much more significant because of the low rainfall, so that nutrients are not leached out of the soil and some mineral salts, particularly those of calcium and potassium, become deposited in the upper soil horizons. Taking these factors into account, it is not surprising that some types of soil in temperate grasslands, chernozem for example, are among the most fertile on the planet.

The vegetation consists mainly of perennial grasses with deep and complex root systems, and above ground either growing in tufts (*Stipa*, for example) or extending over the surface by putting out suckers (*Agropyron*). In areas of lower rainfall, annual plants (therophytes) can become preponderant. Since shrubs and woody plants generally are quite scarce, there is virtually a single layer of vegetation. Nevertheless, the species diversity is fairly high, and particularly noticeable are the many varieties of Compositae and Leguminosae.

Temperate grasslands, although originally populated by large herbivores, have long been turned over to pasturage, or cereal-growing where rainfall is sufficient. Even in the Southern Hemisphere, profound changes have been brought about by human intervention over the last 200 years. The veldt on the high plateaux of Cape Province in South Africa and similar formations in Australia were long ago taken over by cattle and sheep, while the pampas of Argentina is more and more being devoted to intensive cereal-growing. In many cases, overexploitation of the land by the rearing of livestock or by cultivation is showing itself by an irreversible degradation into desert.

All these temperate grasslands once had a rich fauna of birds and mammals well adapted to the vegetation. Among birds, they were the habitat of families like that

Figure 7.6   Bison grazing in one of the few remaining regions of prairie not degraded by human action: the Waterton national park in Alberta, Canada. At the beginning of the last century, between 60 and 100 million bison inhabited the vast North American prairies. Uncontrolled hunting then almost exterminated them: no more than a few thousand were left in the whole of the USA and Canada when the first protective measures were taken in 1880, thus saving the species from imminent extinction. (Photograph F. Ramade)

of the Otidae. Among mammals, both rodents and ungulates were abundant. Of the many species of rodent that lived in these great areas before they were given over to intensive pasturage or cleared for cultivation, prairie dogs (genus *Cynomys*) in North America and sousliki (genus *Citellus*) in Eurasia were the dominant ones.

There were also vast herds of large herbivores: before the arrival of Europeans, the North American prairie was the home of between 60 and 100 million bisons and a larger number of pronghorn antelopes (*Antilocapra americana*) (Figure 7.6).

In Eurasia, hardly anything now remains of the legendary herds of horses, asses and saiga antelopes (*Saiga tatarica*) that roamed over the steppes at the end of the Palaeolithic age. The saiga antelope was nearly extinct at the beginning of the 1920s, but protection of the species by the USSR has allowed a sizeable herd to be reestablished in Central Asia, amounting to more than 2 million animals. Out of this population, some 300 000 can be slaughtered each year, ensuring the survival of the species but also providing annually 6000 tonnes of high-quality meat.

The large kangaroos of Australia are ecologically equivalent to the ungulates of the Northern Hemisphere and, although reduced greatly in numbers, they do persist. It is not at all the same, however, in the South African veldt, where the various native antelopes and zebras have mostly been exterminated: the quagga and the blue antelope (*Hippotragus leucophaeus*), for instance, both became extinct during the nineteenth century.

7.4 Tundra

Tundra is a sub-arctic biome that lies mainly beyond the polar circle and only extends below 60° N in Alaska and Labrador. Because there is so little land nearer the South Pole than 45° S, apart from Antarctica itself, tundra is almost entirely confined to the Northern Hemisphere.

It includes a variety of biocoenoses whose vegetation depends on the local climate and soil. The short growing season (about 60 days) and the low summer temperatures (always below 10 °C) are the principal limiting factors, while the correspondingly long and severe winters mean that the deeper soils are permanently frozen (permafrost). Only the upper few decimetres of soil thaw out during the summer, but the water produced by surface melting is prevented from draining away by the permafrost. This leads to the formation of raised polygonal areas, hummocks, pelsas and other peculiar structures characteristic of sub-arctic soils.

The vegetation exhibits only a low species diversity, but varies with latitude, rainfall, relief and so on. Very broadly, a distinction can be drawn between two main types: shrub tundra and grass tundra. The first is predominant in the southern belt adjacent to the northern limits of taiga and generally in the less cold areas. It is characterized by the presence of birch (*Betula tortuosa* in Lapland, for example), dwarf willows and heathers. Elsewhere, these species give way to non-woody plants in the grass tundra: that is, to herbaceous plants (mainly grasses and sedges) and to cryptogams (particularly lichens of the genus *Cladonia*).

The plant biomass of tundra varies according to the region from a few tonnes to about 30 tonnes per hectare, but most of this comes from the hypogeal (underground) part of the vegetation and is thus not available to herbivores. The primary productivity is very low, at best a few tonnes per hectare per year, and again is mainly hypogeal. Conditions like these can only support a small mammalian biomass, consisting principally of one of the Cervidae: the reindeer (*Rangifer tarandus*) or its ecological equivalent in Canada, the caribou.

The reindeer has long been domesticated by Lapps and other Finno-Ugrian groups living in sub-arctic Eurasia, and extensive rearing of this animal is the major method by which humans exploit the tundra. The species lives on lichens, grasses and the foliage of birches and other woody plants. Experiments in the protection of tundra have been carried out under the MAB programme at Kevo in Finnish Lapland and have shown that grazing by reindeer has a significant effect on the grasses and lichens (Ramade, 1975). The low primary productivity means that each reindeer needs a large area to feed itself, amounting in Lapland to an average of 8 km² per animal.

In the North American tundra, the musk ox (*Ovibos moschatus*) is another large herbivore that became adapted to the extreme conditions and, considering the low capacity of the environment to support a large population, was once relatively abundant. It has, however, now been drastically cut down by excessive hunting and no longer plays an important part in sub-arctic regions.

7.5 Overgrazing and Desertification

Overgrazing can be caused by an excessive density of domesticated livestock and sometimes of wild ungulates as well. In both cases, the overpopulation arises, either deliberately or inadvertently, from human activity.

In the national parks of East Africa, for instance, excessive concentrations of elephants and of many species of antelopes and other herbivores have been observed for some years. This has arisen in large measure from the pressure of hunting and from the reduction in areas of savanna outside the parks which the animals once used for seasonal migration. Increases in the human populations of Kenya, Uganda and Tanzania have led to the clearance and pasturization of vast areas that were formerly inhabited only by wild fauna. As a result, the national parks suffer severe overpopulation at the height of the dry season.

Research carried out in several of the parks, particularly that at Murchison Falls in Uganda (Spence and Angus, 1970) has shown that such excessive densities extend the herbaceous layer of the savanna at the expense of the tree layer. At Murchison Falls itself, for instance, large elephant herds, which are capable of felling trees several metres in height and of killing giant baobabs by eating their bark, have destroyed a tree savanna populated with *Combretum* sp. and *Terminalia* (Figure 7.7). When these same savannas are protected, the trees rapidly reestablish themselves. In a similar way, baobabs and mopanes have been largely destroyed in the Tsavo National Park in Kenya.

It therefore appears that in savannas—and in temperate grasslands as well, according to other studies—overpopulation by wild herbivores produces effects comparable to those of fire in favouring grasses and other herbaceous formations at the expense of woody plants.

Most grasslands that are used for extensive rearing of livestock now show many signs of overgrazing, no matter what region of the world is considered. Excessive populations of domesticated animals, because of their different behaviour, produce effects that are still more harmful than those caused by wild animals. Instead of spreading out over the land to feed, they remain in groups and so increase the amount of vegetation removed from the area they are occupying. The pasture is thus easily overgrazed. In addition, the trampling of such closely packed herds wreaks havoc among the vegetation and denudes the soil along the tracks they make around water holes and in the corrals, as well as along the paths made by their movements to and from them. Gullies are formed and gradually encroach on neighbouring land: in some

Figure 7.7   Trees destroyed by elephants in the Amboseli national park, Kenya. The considerable growth in human population has put an end to elephant migration because so much of the area they once travelled over has been cultivated. As a result, their density in some national parks becomes excessive during the dry season. They eat the tree bark to obtain certain trace elements they need at this time and so kill the trees. (Photograph F. Ramade)

grasslands with fragile soils, veritable ravines several metres deep can now be seen where cattle tracks once existed. Finally, sheep and goats are particularly harmful when too numerous because they tear away the tufts of grass as they graze and thus increase soil denudation.

In fact, a good proportion of the ecological changes that occur in grassland where there is extensive stock-rearing stems from the low diversity of the herds. A fauna consisting of a number of wild ungulate species is replaced by cattle and sheep with a much more limited dietary range. As a result, only the preferred species of edible plants are consumed, leaving the way open for the proliferation of varieties with little or no food value: thorny or toxic species, for instance, or those with limited size or a very short growing season. These would only be poorly represented in a normal undisturbed plant community because of interspecific competition.

This situation is quite different from that in the African savannas, even where the density of wild herds is high. Here, the distribution of plant species remains unchanged in the absence of human intervention because each type of herbivore has a diet complementary to that of others. The primary production of the various plant species is thus affected uniformly by grazing.

In grasslands subjected to grazing by herds of domesticated herbivores, the species composition of plants is profoundly modified under the combined effects of an excessive population and preferential grazing of certain plants. The ground becomes gradually invaded by vegetation rejected by the animals: examples are the proliferation of *Rumex* easily observed in high-altitude pastures (Figure 7.8) and of *Eryngium* and matgrass (*Nardus stricta*) in sheep pastures. Mediterranean grasslands, too, have been overrun by thorny legumes (*Ononis*) and by asphodels that are avoided by livestock.

In the Middle East, overgrazing has caused the disappearance of most of the species of tall grasses that formed the original vegetation (*Stipa, Aristida, Bromus, Agropyron, Andropogon*, etc.). Instead, there has been a proliferation of many geophytes (various Liliaceae, which are inedible) and of many sub-species of *Poa bulbosa*, which have become the principal fodder for the herds. Unfortunately, because of ridiculously intense grazing unrelated to the carrying capacity of the land, the *Poa* are too often eliminated by competition from sedges. The worst of these is *Carex stenophylla*, whose tight network of rhizomes and dense mat of ground cover make it a veritable couch-grass of the desert and prevent recolonization by forage plants (Pabot, 1980).

Figure 7.8   High mountain pasture overrun by *Rumex* near Vallouise, Hautes-Alpes, France. (Photograph F. Ramade)

Table 7.6   *Successional changes related to grazing in the North American prairie* (from Weaver, in Dasmann, 1976)

Succession	Grazing or protection		Dominant species
Climax	Grazing	No grazing	*Decreasers* Big bluestem (*Andropogon gerardi*) Little bluestem (*A. scoparius*) Lead plant (*Amapla canescens*)
Middle succession	Grazing	No grazing	*Increasers* Kentucky blue grass (*Poa pratensis*) Blue grama (*Bouteloua gracilis*) Yarrow (*Achillea millefolium*)
Early succession (pioneer species)			*Invaders* Western wheatgrass (*Agropyron smithii*) Plantain (*Plantago* sp.) Russian thistle (*Salsola kali*)

In the Western USA, dry grasslands overgrazed by cattle and sheep have gradually been overrun by various inedible shrubs. Sage-brush (*Artemisia tridentata*), for example, is rejected as fodder by livestock and covers vast areas in the cooler sonoran rangelands of the western states. For the same reason, mesquite (*Prosopis* sp.) is a leguminous shrub that flourishes extensively over Texas and the south-western states.

In some cases, the introduction of domesticated herds can completely destroy the original vegetation in grasslands formerly grazed by quite different species of large herbivore or possibly without any such species at all. In New Zealand, for instance, the plains of South Island were once covered by herbaceous plants, mainly grasses belonging to the genera *Poa* and *Festuca*. This ecosystem has been totally transformed with the introduction of sheep, deer, goats and rabbits by Europeans and with the use of fire as a method of managing the pasturage. The original vegetation has given way to species like *Holcus lanatus* and to fescues that provide less ground cover.

The same phenomenon can be observed in the heathlands and moorlands of Western Europe where selective grazing by sheep has encouraged invasion by shrubs and other woody plants like heathers. In addition, highly nutritious species that flourish in chalk or limestone soils have disappeared and been replaced by others like *Nardus stricta* and *Molinia coerulea*, which have become dominant but are hardly eaten at all by

Table 7.7   *Effect of overgrazing on the biomass and primary productivity of certain North African grasslands* (from Le Houérou, 1979)

A: In the Hodna area, Algeria (*Salsola* and *Anabasis* dominant)

Treatment of pasture	% of ground covered by vegetation	Above-ground biomass (kg dry matter ha^{-1})
Protected	25	2088
Moderate grazing (1 sheep per 8 hectares)	5	850
Overgrazing (1 sheep per 3 hectares)	3	515

B: In southern Tunisia (*Stipa lagascae* and *Rhanterium suevolens* dominant)

Treatment of pasture	% of plant cover	Primary productivity	Edible productivity	Edible productivity per mm of rainfall
		(in kg dry matter ha^{-1} yr^{-1})		
Light grazing, good condition	25	1069	820	2.61
Average grazing, mediocre condition	8	614	493	1.57
Heavy grazing, already overgrazed	4	415	293	0.78

sheep. Fire is often used to get rid of the woody plants but this encourages the spread of these same inedible species and at the same time impoverishes the soil. Finally, because the secondary animal production is taken out of the ecosystem, overgrazing contributes to a reduction in the concentration of soil nutrients.

Eventually, overgrazing profoundly modifies temperate grasslands by causing climax species to be replaced by plants low on the successional scale: on the other hand, reduction of grazing pressure often allows the reverse succession to take place (Table 7.6).

The primary productivity of grassland and thus its meat-producing potential is drastically reduced by overgrazing (Table 7.7). Each type of pasture has a limited carrying capacity defined as the maximum density of domesticated livestock—or, more generally, of herbivores—that it can support without risk of degradation (Figure 7.9). It is expressed either in sheep-units (1 head of cattle = 1 horse = 5 sheep or goats) or as herbivore biomass (kg ha^{-1} or t km^{-2}). Unfortunately, it is a parameter rarely taken into account in the exploitation of grasslands with the result that overgrazing is now endemic in most parts of the world.

In the USA, the territory to the west of the Mississippi is estimated to have had a carrying capacity of 112.5 million sheep-units at the time when European settlement began. By the 1930s, a study of the same area carried out by the US Forest Service showed that several decades of overexploitation had reduced the capacity by more than half to 54 million sheep-units. In spite of that, the land was still in fact supporting more than 86 million units and the damage was thus still continuing (Dasmann, 1976). Experts estimate that it would take over 100 years for the degradation caused by 60 years of overexploitation to be reversed and the grassland reestablished—and only then on the assumption of an immediate reduction of the animal population to an acceptable level.

Grasslands of the Middle East have also suffered great and possibly irreversible damage from centuries,

if not millennia, of unrestricted exploitation as pastureland. Herds of goats and sheep with densities far exceeding the carrying capacity of the land still occupy the high plateaux of Turkey and the semi-arid grasslands stretching from Syria to Western Turkestan and Afghanistan. In Iran, for example, where the total carrying capacity is at most 24 million sheep-units, the land is currently supporting more than 70 million units.

Tropical Africa, South America, and to an even greater extent India, are all suffering from the same damage to their rangelands from excessive densities of their herds of livestock. Mediterranean Europe as well: it is surely somewhat surprising to find clear signs of soil erosion in the high mountain pastures of Haute-Provence due to overgrazing by sheep? In view of that, however, what is one to think of the damage being done to the alpine tundra of the Parc National des Ecrins by an excessive number of flocks of sheep, tolerated because of individual attitudes over customary grazing rights?

The carrying capacity of grasslands, in fact, shows great variations between one ecosystem and another (Table 7.8). For instance, while several tens of hectares per head of cattle are needed to provide adequate pasture in certain semi-arid ranges in Oregon, less than one hectare is sufficient on a permanent meadow in Normandy. As a general rule, the carrying capacity of land is considered to have been reached when 40 per cent of the primary productivity is consumed by herbivores (Bourlière and Hadley, 1970).

The detrimental effects of overgrazing on the primary productivity of grasslands may be direct, but they can also be indirect and delayed in time. The effects of overexploitation on the make-up of the plant communities have already been cited, but the changes produced may not only be qualitative but quantitative as well, through the interspecific competition introduced. Perennial plants consumed by animals are gradually replaced by annuals and bulbed geophytes, which have shallower root systems and provide less ground cover, thus reducing soil stability. The regrowth of the vegetation preferred by herbivores is also hampered by overgrazing, which eliminates seed-carriers and prevents dissemination. Inedible woody or herbaceous plants tend to proliferate and not only reduce the primary productivity usable by livestock but tend as well to reduce the area available to edible plants by covering the ground with a layer of unsuitable litter.

By these processes, overgrazing reduces the density of the herbaceous ground cover: it becomes patchy and this in its turn creates new factors that further reduce the primary productivity. Denudation of the soil and increasing adaptation to dry conditions, for instance, encourage infestation by grasshoppers and crickets and by various species of rodent. All these pests contribute to a reduction in that part of the vegetation

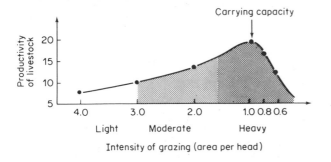

Figure 7.9 Relation between the intensity of grazing and the secondary productivity of domestic livestock. The productivity falls off for densities of animals above a certain value known as the carrying capacity of the land. (Adapted from Costin, 1964)

Table 7.8   Carrying capacity of various temperate and tropical grasslands (adapted from several authors in Dorst, 1965)

Type of pasturage	Grassland and location	Carrying capacity (herbivore biomass in kg ha^{-1})
Natural	Pampas, Argentina	140
	Prairie, Texas	110
	Veldt, South Africa	85
	Themeda savanna, Kenya	40
	Rangeland, Oregon	15
	Sub-desert scrub, Saudi Arabia	8
Artificial	Permanent meadow, Normandy	945
	Kivu, Zaire	650
	Meadow, Belgium	480
	Ituri, Zaire	340
	Prairie, Oklahoma	36

available to domesticated livestock. Thus, overpopulation by sheep and cattle in western USA has led to an increase in field mice, kangaroo-rats (*Diplodomys* sp.), hares, ground squirrels (*Citellus* sp.), gophers (*Geomys* sp.) and prairie dogs (*Cynomys ludovicianus*). Proliferation of all these rodents has been further encouraged by the elimination of their natural predators, wolves and coyotes. Tens of thousands of hectares of prairie have been laid waste by the increase in burrows and the earth thrown up around their openings, while the rodents themselves also compete with the domesticated livestock for the available food supplies. The black-tailed hare (*Lepus californicus*), for example, consumes considerable quantities: 12 of them eat as much as one sheep, 59 as much as a head of cattle. An allied species, the antelope hare (*Lepus alleni*), widespread in the vast semi-desert pasture of the south-west of the USA and north-west Mexico, is still more voracious: it only takes 8 of them to eat as much as a sheep and 41 as much as a head of cattle (Burt and Grossenheider, 1964).

The chronically overgrazed Sahelian savanna has experienced similar increases in rodent populations. In the Fété Olé region, for example, gerbils (*Taterillus pygargus*) and Nile rats (*Arvicanthus niloticus*) were observed in much greater numbers after the rains of 1974 that followed the great drought of 1972–1973 (Poulet and Poupon, 1978; Bourlière, 1978). The rats in particular, normally confined to the wetter areas of the Senegal river valley, increased to a density of more than 100 per hectare by the beginning of 1976, attacking the woody vegetation and making considerable inroads into it: during the dry season of 1975–1976, more than 97 per cent of the young growth of *Commiphora africana* was destroyed (see also Figure 7.10 for the effects of the 1972–1973 drought on the trees in this area).

Desertification has now become a permanent threat to all grasslands in semi-arid climates exposed to over-

grazing. In varying degrees, it currently affects some 60 countries with around 700 million inhabitants. Over 60 million of these live in semi-arid or even arid regions and rear livestock in grasslands so degraded that the smallest deficiency in rainfall immediately threatens them with severe famine.

The spread of deserts created by human activity is caused mainly by the overgrazing. Countries bordering the Mediterranean are among the oldest sufferers: the Iberian peninsula, Languedoc and Provence in France, the Italian peninsula, the Balkans, North Africa and the whole of the Middle East today bear the signs of the overexploitation of natural grasslands as well as the extension of pasture to areas where the original forest

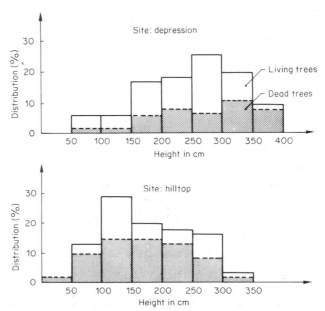

Figure 7.10   Mortality rates of a dominant tree species (*Guiera senegalensis*) in the Sahelian savanna at Fété Olé during the great drought of 1972–1973. (From Poupon and Bille, in Bourlière, 1978)

cover should have been uncompromisingly preserved. The regions most affected at the moment are North Africa (at the northern and southern boundaries of the Sahara), the Horn of Africa, the Middle East and the western part of the Indian sub-continent. Taking these as a whole, the area of desert quite recently created amounts to more than 5 million km², or one third of their total arid zones. A study of aerial records made in West Africa during the last few decades shows that the Sahara has advanced by an average of 2 km per year in the Sahel region to the south, while in the Sudan the desert has advanced 90 km in 20 years, affecting an area equal to 63 per cent of that of France.

The prime cause of such desertification is the destruction of the vegetation in these overgrazed ecosystems and the consequent increase in aridity. Sometimes the soil becomes covered with a 'sedge mat' consisting mainly of Cyperaceae such as the *Carex stenophylla* of subtropical and temperate grasslands, and this considerably reduces infiltration into the soil when the rainfall is low (Pabot, 1980). If the vegetation is completely destroyed, as it often is in the Sahel, any rain that falls on the bare soil does so at sufficient speed to disperse the finer fractions and clog the pores to form a superficial glaze. In other cases, some Cyanophyceae of the genus *Scytonema* have been observed to form colonies encrusting the surface of the bare earth with a continuous film that makes the soil almost completely impermeable (Le Houérou, 1979).

A final effect of the loss of plant cover is a greater fragility of the soil so that it quickly becomes eroded by water and/or wind. This causes a further increase in aridity through greater evaporation and run-off at the expense of infiltration.

In Africa, overgrazing has reduced the plant cover considerably during the last few decades. In Chad, the reduction amounted to 32 per cent between 1954 and 1974, at the same time as the area of eroded land increased by 28 per cent. In central Mali, Le Houérou *et al.* have shown that, out of a semi-arid area covering 60 000 km², the proportion of denuded soil grew from 2 per cent in 1952 to 26 per cent in 1975.

In the final analysis, the spread of desert in any region of the world is the result of excessive growth in the local human populations. During the last 20 years, for instance, the populations of the Sahelian areas, and of the pasturelands of the Middle East and India, have grown at a rate varying from 2 to 3.5 per cent per year depending on the region. As a response to the increased food requirements of the countries concerned, the number of livestock has grown at a comparable rate, but because of the low secondary productivity of the pastures some 30–40 sheep or goats, or 3–4 head of cattle, are needed to feed one person. In Africa today, some 55 per cent of the total domesticated herds are located in arid or semi-arid regions. The number of cattle in Nigeria rose from 4 million in 1925 to 10 million in 1970 (El Kassas, 1977), and for the whole continent the herds increased by 71 per cent between 1950 and 1973. The record is held by the Sudan with a growth of 189 per cent over that period!

Drilling bore holes for water, far from helping to raise large herds of livestock, is a major long-term cause of desertification and of a reduction in the numbers of wild ungulates. Whereas the traditional nomadic custom was to drive the herds away from the more arid regions during the dry season in search of better water supplies, the animals now congregate around natural or artificial water holes. The wild herbivores that used to occupy the abandoned areas and find sufficient food in them are now denied their use (Newby, 1980).

As a result, therefore, of a deliberate policy of constructing artificial water holes, extensive grazing now takes place at the height of the dry season in areas where the vegetation is already poor. The effect is made worse by a growth in human population that has reduced the extent of tribal lands. In Niger, Chad and the Sudan, a single water hole is used continuously day and night by tens of thousands of animals. Lack of water elsewhere means that cattle cannot stray further than 20 km from the water hole, so that the surrounding land becomes badly trampled and overgrazed. Vegetation around the hole is rapidly destroyed and a desert gradually spreads out from it. Moreover, as the pasturage becomes more and more degraded, cattle are replaced first by sheep and then by goats, which finally destroy the shrubs and grasses neglected by the other animals.

Desert can also be spread through the savanna in semi-arid regions as a result of bad farming methods. In areas of the African Sahel where rainfall is between 250 and 450 mm per year, particularly in the Sudan, a traditional form of agriculture has developed which alternates with the growing of acacias (mainly *Acacia senegal*) for their gum arabic. After the savanna has been cleared, the first crop to be grown is either barley, if the rains occur in the winter, or sorghum and millet, if they occur in the summer. This quasi-monoculture is continued for 5 or 6 years until the soil is exhausted by the repetition. The land is then allowed to lie fallow and is overrun by various shrubs and trees, especially by *Acacia senegal*. This semi-spontaneous plantation, which takes about 10 years to develop, is exploited in its turn by the extraction of gum arabic. After 10 years, however, this extraction exhausts the trees and they begin to die, but because the acacias are very thorny the dead trees lying around prevent grazing by domesticated livestock. A grassy layer thus starts growing among the fallen debris and becomes quite high and dense. After a few years lying fallow, the vegetation is burnt away and there begins another cycle of alternating cropping and gum harvesting (El Kassas, 1977).

During recent decades, the period over which the alternation takes place has shortened considerably under the pressure of population growth, and the traditional cycle has been abandoned. Because of this, the *Acacia senegal* does not regenerate itself so well and the increasing frequency of the fires at the end of the fallow periods degrades the soil and causes the plant cover to become patchy. The farmers have then moved south and developed the same practice in areas where the rainfall is rather more abundant and where tree savanna has become established. This type of ecosystem is then destroyed as it is replaced by the new method of cropping, the plant cover again begins to decrease and thus affords less protection to the soil. Desertification once more sets in as erosion by wind and water increases. . . .

Finally, the spread of desert into grasslands with inadequate rainfall is not the only threat posed by attempts to cultivate them. There is also the consequent displacement of the herds that would otherwise use them as pasturage towards land that is much more marginal, land that is then in its turn condemned to desertification before very long.

7.6 The Rehabilitation of Degraded Natural Pastureland

The rehabilitation of the vast stretches of natural grassland degraded by overgrazing is now becoming an ever more urgent problem. The FAO has predicted that the worldwide spread of desert and increased soil erosion will have caused some 25 million hectares of natural pasturage to disappear between 1972 and 1985, and similar estimates give a figure of at least 35 million hectares during the present decade.

In practice, the highest priority should be given to a more rational exploitation of pasturage that still exists and that is not yet irreversibly degraded. This would involve a reduction in the density of herds to a level below the carrying capacity of the land, and in some cases its partial protection to allow the regeneration of grasses and other species suitable for fodder: overgrazed herbage is prevented from producing and disseminating its seeds. To achieve this, the administrative authorities responsible for rural development should create protected zones of limited extent to allow the vegetation to reseed and also to demonstrate to local inhabitants the restorative effects of such a practice on a degraded pasture. Proof that such a policy has a beneficial effect on the land would certainly encourage herdsmen to reduce the density of

livestock and thus alleviate the pressure on the vegetation.

However, the implementation of policies involving the creation of protected zones and a reduction in animal population would cause problems that make the active participation of the governments concerned absolutely essential. Among other things, support would be needed for meat prices so as to encourage the sale of the extra livestock, and there would be a need for increases in agricultural production through improved methods of cultivation.

In the more arid regions, the spread of the desert can only be halted by the closure of artificial water holes coupled temporarily with help in food supplies and other forms of compensation. Finally, grazing at the head of drainage basins should be strictly controlled, and if necessary prohibited, because of the risks of flooding and erosion that follow the destruction of plant cover.

In some cases, however, the degradation of natural grassland is so great that grasses and other fodder plants can no longer recolonize it without positive human intervention rather than mere protection. It is then very often necessary to turn over the upper layers of the soil by disc-harrow, or even to plough them up, so as to destroy the thick mat of inedible plants that has formed over the surface. In Iran, for example, attempts to rehabilitate a very degraded pasture overrun by a sedge mat of *Carex stenophylla* with its tight network of rhizomes and roots have had encouraging results. At the end of the 1960s, after mechanical working of the soil, several tens of hectares were sown with a local grass (*Secale montum*) or with a species of astralagus (*Astralagus siliquosis*) which have provided fodder on the previously degraded land without irrigation (Pabot, 1980). It is even possible to envisage the reconstitution of the original vegetation by reseeding with a suitable mixture of native species.

However, another method of rehabilitating degraded pasture invaded by inedible woody plants is by the use of defoliants specific to dicotyledons, and about this there are strong reservations. In the USA, vast stretches of dry prairie rangelands overrun by sagebrush (*Artemisia tridentata*) have been treated with 2,4,5-T[1] in order to destroy it and encourage the restoration of grassy vegetation. The toxic risks to the environment associated with this method are insufficiently known and potentially too great for its safe use. Even in the USA, where it has been popular until recently, it now seems in the process of being

---

1. 2,4,5-T is a herbicide derived from phenooxyacetic acid much used as a defoliant for clearing land in silviculture and agriculture. It includes an extremely toxic trace impurity, dioxin, which was the cause of the disaster at Seveso in Italy in 1976.

Figure 7.11   A herd of *Oryx beisa* in an *Acacia tortilis* tree savanna, Samburu national park, Kenya. The oryx was originally tamed by the ancient Egyptians and attempts at its domestication are now being made in Kenya. It is a species much better adapted to semi-arid pasturage than traditional livestock. (Photograph F. Ramade)

abandoned under the pressure, among others, of the authorities responsible for the protection of the environment.

In some circumstances, the best way of exploiting natural grassland would lie in a greater use of the secondary production from wild herbivores: that is, by their domestication. In many cases this is the only possible alternative to severe degradation through excessive rearing of cattle, sheep and goats. In fact, it is really rather surprising to see the determined attempts made to raise cattle in the African savanna, when wild antelopes, with their diversified diets, use the primary production so much more efficiently and, unlike cattle, are resistant to most animal diseases, particularly trypanosomiasis. In addition, as already indicated, antelopes have a greater secondary productivity than cattle and their flesh contains a higher proportion of lean meat. Finally, the 92 or so species of wild ungulate in the African savanna are capable of living comfortably on arid pasture that is at best marginal for cattle, and they need much less water.

At the moment, it is possible to envisage the domestication of giant eland (*Taurotragus derbianus*) because of successful experiments carried out in Zimbabwe and in the south of—the USSR! The female eland has been shown to provide as much or more milk than African cattle of comparable weight, while the species is content with fodder of lower quality and eats shrubs that normal domesticated livestock cannot feed on.

The African buffalo (*Syncerus caffer*) is also a potential candidate for domestication and is another species that can live on plants rejected by cattle. Finally, the oryx, another species of antelope adapted to semi-arid pastureland, appears to be next on the list: it has a high secondary productivity and recent attempts at domestication carried out in Kenya have proved highly successful (Figure 7.11).

As in many other areas, overgrazing of grasslands stems mainly from the pressue of the socio-economic structures of pastoral societies. Such societies, from the Masai of Equatorial Africa to the Lapps of the Scandinavian tundra, are associated with a nomadic way of life and they certainly permit the exploitation of ecosystems whose plant production could not otherwise be used. However, it is a way of life incompatible with excessive growth in population and incompatible, too, with the literacy needed to make human progress possible. In the end, the only way of ending the degradation of natural grasslands in many regions of the world today is through a new type of rural development based on conservation.

Chapter 8

The Protection of Threatened Ecosystems: National Parks and Other Nature Reserves

The need to preserve certain natural resources has been recognized by various human societies for much longer than is generally supposed. Classical writers, particularly Plato, were referring even in the fourth century BC to the preeminent role of forests in the regulation of the water cycle and the protection of soils against erosion. However, the great variety of motives which led past civilizations to take protective measures were often very different from those of the modern world, although that has not prevented some forests, lakes, and animal species from enjoying effective protection for centuries, even in some cases for more than a thousand years.

Religious taboos have often been at the root of such measures: in many primitive societies, trees and water are worshipped as symbols of natural forces among other things. It is not surprising, therefore, that sacred forests exist in almost every part of the world where trees have flourished. In Europe, for example, gifts of forest areas made to the Church by native populations or barbarian invaders at the time of their conversion to Christianity have often allowed fragments of the vast Wurmian forest to be preserved in their original state. In Africa and Asia, the association of trees with religious rites was accompanied by a marking off of 'sacred woods' which amounted in effect to converting them into nature reserves.

The increasing scarcity of certain animal species through excessive hunting has in the past also been responsible for the introduction of protective measures. Animal sanctuaries set up by local chiefs have long existed in Africa. Itshyanya in Rwanda, for instance, has for centuries been strictly forbidden territory closely guarded by pygmies, to which the country's ruler, the Mwami, retires once every year for the ritual slaughter of an elephant (Harroy, 1970).

The oldest animal reserve in the world is undoubtedly the forest of Bialowieze in Poland. This was made into a completely protected area during the fourteenth century by King Jagellon to save certain species which had been threatened with extinction from the beginning of the second millennium AD: these were the urus or aurochs, a wild ox sometimes known as the European bison, and the tarpan or wild horse.

However, no matter how important these early attempts at conservation may have been, it is essential to realize that their effectiveness was relatively limited and that they sprang from ideas very different from those which have recently given such a decided boost to the protection of natural resources. These recent moves have arisen from the fact that there has been an exponential growth in the degradation of biocoenoses and the devastation of entire ecosystems since the dawn of the modern era, in parallel with population increase and technological progress. Western thought, whatever the period or philosophy considered, has been dominated by the Judaeo-Christian archetype of the human being as master of creation, an attitude also taken up by Islam. It is a creed that sees progress as symbolized by an uncontrolled exploitation ('squandering' would be a better word) of natural resources with ever greater areas coming under human domination. Such an aggressive and outdated conception of the place of the human species in the biosphere is nevertheless a very long-standing one: it has lasted from the *Horror sylvanum* of the monks in the Middle Ages who cleared the forests, to the determined promotion by contemporary technocrats of such measures as the draining of marshland, the regrouping of *bocages*, the opening up of forests and alpine valleys, and even the 'control' of animal populations by hunting.

The population explosion of the twentieth century and the predictable exhaustion of some resources have led biologists and, more recently, certain economists (see Passet, 1979, for example) to question such a dogmatic and mistaken view of humanity as independent of ecological factors. However, there were philosophers and scientists in previous centuries who, out of aesthetic or purely scientific considerations, founded the modern conservation movement. Among these were the eighteenth-century writers who were adherents of Rousseauism, above all the famous German biologist and geographer Alexander von Humboldt (1769–1859), one of the pioneers of present-day ecology. His ideas were permeated with a concern to bring out the relationship between human beings and their environment and made him the first theorist of nature conservation. We owe to him the expression 'natural monument' for areas of land of exceptional ecological importance.

8.1   The First Modern Reserves and National Parks

The first nature reserve of modern times was established, paradoxically, not by biologists but by a group of French painters known as the Barbizon School, who in 1853 secured the protection of part of the forest of Fontainebleau to preserve its natural beauty. We owe to them the creation by law in 1861 of the 'séries artistiques' covering some 124 hectares of beech grove (Jouanin, 1970).

However, the foundation year as far as the national park proper is concerned is traditionally fixed as 1872. A federal law promulgated in the USA on 1 March of that year created the first national park in the world, that of Yellowstone in Wyoming. The law prohibited any development that might alter its character in order to preserve the region 'as a public park or pleasure ground for the benefit and enjoyment of the people'.

In fact, the first decision aimed at protecting the natural wealth of the USA was taken before that. Following the advice of the famous American naturalist John Muir, a pioneer in conservation, President Lincoln had had voted unanimously through Congress in 1864 a law making the Yosemite valley and the neighbouring sequoia forest of Mariposa a nature reserve. Under the law, the state of California was entrusted with the preservation of the two areas in order to safeguard the magnificent forests of *Sequoia gigantea* which until then had been badly over-exploited. Both areas were eventually incorporated into the Yosemite National Park.

During the later part of the nineteenth century, only the British followed the American example and set out to transform vast tracts of their large empire into national parks. The Glacier National Park was created in Canada in 1886 and the Royal National Park in Australia a little later. Lastly, the Sabie National Reserve was established in South Africa in 1898 and this became the famous Kruger National Park in 1926.

In Western Europe, the conservation movement was well behind that of North America as far as concrete results were concerned. Sweden, a model nation in matters of environmental protection, was the first to create national parks: six were set up in 1909, among them those of Sarek and of Stora Sjöfallet in Lapland. This was followed after a short time by Switzerland which established a national park in the Engadine in 1915. This was the first area in which protective measures were justified by its scientific importance.

Although the British Government passed a law protecting sea-birds in 1869 and extended this to all wild fauna in 1880, it was not until after 1949, with the formation of the Nature Conservancy Council, that Great Britain had any national nature reserves.[1]

Metropolitan France, the largest of the Western European countries, has particularly lagged behind its Germanic, Anglo-Saxon and Latin neighbours in the matter of national parks and other nature reserves. Whereas between the two world wars such areas were multiplying all over the world, and not just in Europe, the only significant French creation was that of the zoological and botanical nature reserve in the Camargue in 1928. Even then, this was not set up as a result of government action but through that of a private scientific and philanthropic association, the Société Nationale de Protection de la Nature, one of the oldest such organizations in the world: established in 1853 under the name Société d'Acclimation! It was not until 1960 that a law on national parks was passed and it was 1963 before the first of them was created: the Parc National de la Vanoise.

8.2   Recent Developments in Protective Measures

Whatever the scale and importance of the earlier achievements in the field of conservation already described, they were always of relatively limited scope. The great upsurge of interest in the protection of nature did not occur until after the Second World War and it is only in the last two decades that it has become the concern of governments in most countries.

It has begun to be realized at the end of the 1940s that an international organization for nature conservation would be of considerable help to governments, to the United Nations and its specialist institutions, and to other interested organizations. In 1948, various international authorities, both official and private, founded the International Union for the Conservation of Nature and Natural Resources (IUCN). This has played a major part in the promotion of national parks and in the conservation of natural resources through its role as adviser to the governments concerned and through the assistance it has provided on the ground towards the implementation of numerous projects.

It was not until after 1945 that the majority of national parks were set up in Africa and Asia, first by the European colonial administrations and later by those of the newly independent nations who took over the work started by their predecessors and, in most cases, strengthened it. Similarly, although the first wild-life sanctuaries—the *zapoviedniki*—were set up in the

---

1. However, private reserves like those of the Royal Society for the Protection of Birds existed well before this.

Plate XXII   The protection of nature in the USA

1   View of Mount Rainier from the Nisqually Trail, Washington State. This extinct volcano, 4382 metres high, is the main attraction of the national park named after it.

2   Natural pillars of sandstone in the Arches National Park, Utah, standing among weathered rocks reminiscent of ancient ruins. The park owes its name to the numerous natural arches cut through by wind erosion. It protects communities peculiar to cool sonoran deserts.

3   Moose (*Alces alces*) in the Hayden valley, Yellowstone National Park. This is the oldest national park of modern times and was established in 1872.

4   Lake Jackson and, on the right, Mount Moran in the Grand Teton National Park, Wyoming. The extension of this park in 1950 to include the Jackson basin made it possible to arrest a worrying decline in the largest population of wapiti (*Cervus canadensis*) in North America.

5   Trumpeter swans (*Olor columbianus*) on the Madison River, Wyoming. This species was in decline but has been able to reestablish itself during the last few decades as a result of the strict protection it receives in the USA.

(Photographs F. Ramade)

Plate XXIII   The Parc National des Ecrins, Hautes-Alpes, France

1   View of the Massif de la Meije. This park covers an area of 91 800 hectares in the Massif des Ecrins, mainly of alpine and snow-covered mountains. Protection of fauna in the park is, somewhat regrettably, compromised by pressure from hunting under the pretext of a supposedly excessive density of ungulates.

2   An entrance to the park above Vallouise on the route over the Col de l'Eychauda.
(Photographs F. Ramade)

USSR during the first years of the Revolution,[1] it was once again only after 1945 that the greatest efforts began to be made, and it was the same in other countries of Eastern Europe.

The rate at which national parks were being created showed a marked increase from the middle of the 1960s, stimulated by various international conferences, particularly the 1968 Unesco conference in Paris, 'The Effect of Man on the Biosphere', and that of 1970 in Stockholm on 'Man's Environment'.

Five national parks have been established in France since 1963, in addition to that of La Vanoise: Port-Cros, Cévennes, Pyrénées, Ecrins and, in 1979, Mercantour, although here unfortunately various 'pressures' considerably reduced the area of exceptional ecological importance included within the boundaries from that originally planned.

During the last 15 to 20 years, many countries with a solid network of national parks have strengthened those already existing by extending them and reinforcing the laws protecting them, and have also created new ones. In the USA, for example, two great parks were established in 1968 and 1970 respectively: first, that of the Redwoods in Northern California aimed at the protection of *Sequoia sempervirens*, a tree that grows on the Pacific coast to heights of 120 metres; secondly, that of the North Cascade Range straddling the Canadian border in Washington state, a park covering some 230 000 hectares. Kenya, too, has established two parks: that of East Turkana in 1973 and the Mount Kulal nature reserve in 1978, the latter with an area of 700 000 hectares.

Over the same period, other areas that are already protected have been considerably extended: in Finland, for example, the Lemmenjoki National Park, created in 1956 with an original area of 38 500 hectares, was increased to 175 000 hectares in 1972.

According to the IUCN, around 1205 national parks or similar reserves could be listed worldwide in 1967, and among these Cury-Lindahl and Harroy (1972) have picked out just over 200 as being of major importance. An updating of this list published by the IUCN in 1982 included 2250 national parks and other reserves subject to laws guaranteeing absolute or partial protection, and these together cover a total area of nearly 400 million hectares. Such an increase is a witness to the considerable growth in efforts at conservation on the part of most countries in the world.

8.3   Threatened Species and Ecosystems

Such a significant increase in protective measures has unfortunately not been sufficient to end the worrying depletion of many animal and plant species, nor even to put a stop to the degradation of entire ecosystems amounting sometimes to their total destruction.

The most threatened communities at the present time are those inhabiting the following environments: islands generally, but particularly tropical and sub-tropical islands; tropical rain forests; arid and semi-arid regions; Mediterranean ecosystems; the sub-tropical mountain valleys of southern Asia and South America; and finally, the wetter zones, particularly fresh-water areas, mostly in Europe and eastern North America.

The ecosystems most exposed to the risk of short-term destruction are often those with exceptionally high diversities. Many tropical rain forests are threatened with complete extinction by the end of this century because they have been wantonly over-exploited, yet at the same time they contain more than 70 per cent of the total plant species in the biosphere, many of them still undescribed. In fact, among the most diversified ecosystems that need urgent protective measures going beyond the scope of strictly limited nature reserves, the IUCN (1980b) has picked out the tropical dipterocarp forests of Malaya, Borneo, Sulawesi, Sumatra and New Guinea; the tropical forests of the Philippines and Malagasy; and the rain forests of Central and South America (Figure 8.1).

Other ecosystems in urgent need of strict conservation measures are those of the Mediterranean area, those in South Africa and southern Australia, as well as some in island groups like New Caledonia and Hawaii, where there is both chronic degradation and a high species richness.

Finally, the IUCN also stressed in its report on a world strategy for conservation the exceptional importance of some aquatic ecosystems: the coral reefs, for instance, of Indonesia and Malaya, of the eastern and central Pacific, and of the Caribbean and south-west Atlantic. Again, sea areas like the Gulf of California, the coastal waters of the Gulf of Mexico, the Red Sea and the seas of Japan and China are all among oceanic ecosystems that should receive high priority in the protection of their great variety of fish species. The exceptional diversity of some fresh-water ecosystems was also emphasized, and here the report particularly mentioned the rivers of the Amazon basin, of Central and West Africa, of the Mississippi basin and of India; the lakes of Equatorial and tropical Africa; the Caspian Sea and the Aral Sea; Lake Baykal; and the many lentic environments in Indonesia.

The degradation of all these ecosystems through overexploitation and through the removal from them of excessive amounts of biotic material is threatening an ever greater number of plant and animal species with extinction. During the last 10 years or so, around

---

1. That of the lower Volga was created in 1919 by a decision of Lenin himself.

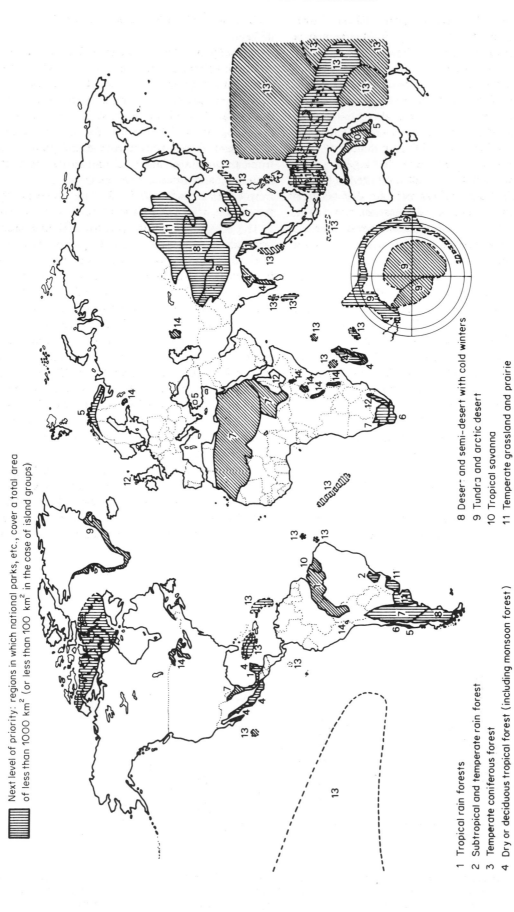

Figure 8.1   Regions of the world most urgently needing the creation of protected areas. (From IUCN, 1980b)

Highest degree of priority: regions without national parks or similar reserves

Next level of priority: regions in which national parks, etc., cover a total area of less than 1000 km² (or less than 100 km² in the case of island groups)

1 Tropical rain forests
2 Subtropical and temperate rain forest
3 Temperate coniferous forest
4 Dry or deciduous tropical forest (including monsoon forest)
5 Broadleaved temperate forest (and deciduous subpolar woodlands)
6 Mediterranean evergreen forest
7 Hot desert and semi-desert

8 Desert and semi-desert with cold winters
9 Tundra and arctic desert
10 Tropical savanna
11 Temperate grassland and prairie
12 Mountains and plateaux
13 Tropical and subtropical island groups
14 Lakelands

25 000 plant species (Lucas and Synge, 1978) and over 1000 species of vertebrate (IUCN, 1975) have shown signs of disappearing or at the very least have experienced a disquieting decline in numbers. In fact, these figures are only the visible tip of the iceberg, for the destruction of any habitat eradicates enormous numbers of invertebrates: arthropods on land and molluscs and coral in the seas. Cultivation and over-grazing of land, soil erosion, mining and quarrying, the construction of dams, ports and drainage systems, urban development, pollution: all these destroy much small animal life. Some writers estimate that between 500 000 and 1 million species of terrestrial, marine and fresh-water invertebrates will have disappeared by the end of this century through the destruction or altera-tion of their habitats.

The deliberate or accidental introduction of exotic plants or animals is often an important cause of the destruction or depletion of indigenous species, but one that is too seldom recognized. The disastrous effect of introducing rabbits on the marsupial population of Australia is well known (see Dorst, 1965, for example), but the goat and the rabbit have also destroyed the habitats of many plant, reptile and bird species in the islands of the Indian and Pacific Oceans.

A final major cause of the destruction of birds and mammals in every continent is that due to hunting and shooting by humans, whether authorized or not. A large number of species have been brought to the threshold of extinction through this activity: the black rhinoceros in Kenya, for example, fell in numbers from 11 000 to under 1400 during the 1970s, while Grevy's zebras in the same country decreased from

10 000 in 1971 to less than 1000 in 1978 (Ramade, 1979b).

In many developed countries, hunting far exceeds what the animal population can stand, and in addition it prevents the implementation of measures to protect threatened species. In France, for instance, there are more than 2.5 million huntsmen and, ever since the beginning of the Third Republic, electoral considera-tions have always prevented the politicians in power from passing legislation to reduce the number per-mitted to carry rifles and shotguns.

It is not only wild species that are in decline: there is also a disquieting decrease in the number of varieties among cultivated plants (Figure 8.2) and breeds of domesticated livestock. Among plants, a number of wild and cultivated varieties of food crops important to humans have already disappeared or are in the process of doing so: these include varieties of wheat, rice, potatoes, beans, bananas, tomatoes, yams, etc. It is the same with animals: out of 145 breeds of cattle in the European Mediterranean region, 115 are becoming extinct. In France, four breeds (the black and white French Friesian, the Normandy, the Charolais and the Montbéliarde) make up 80 per cent of the whole cattle population, while 20 of the 41 breeds of sheep form only 7 per cent of the total sheep population: for some breeds of sheep like the Roussillon Red there is virtually no hope.

The situation is even more alarming for Equidae. At the end of the last century, there were 380 000 asses in France, while today there are fewer than 30 000: the giant Poitou breed at present numbers no more than 44. In the same way, the population of draught horses

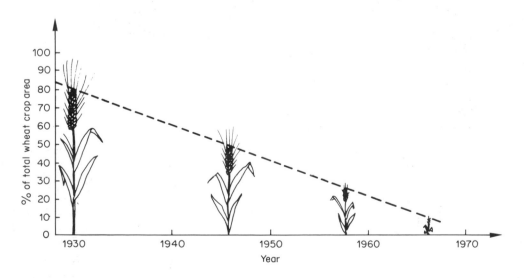

Figure 8.2   An example of the disastrous decline in the genetic diversity of crops: changes in the area covered by primitive varieties of wheat in Greece between 1930 and 1970 expressed as a percentage of the total area devoted to the wheat crop. Similar decreases have occurred in nearly every country over the same period. (From Fraenkel, 1973)

has fallen by 95 per cent since the end of the Second World War. The disappearance of breeds like the Percheron or the Boulonnais would be a real disaster if a new oil crisis forced farmers to turn once again to animal traction.

8.4   The Preservation of Genetic Diversity

The protection of plant and animal species and the establishment of nature reserves in ecosystems threatened by human profligacy are measures that have been undertaken with a variety of motives. As far as the general public is concerned, the basic criteria have until now have been largely aesthetic or even purely emotional: recent campaigns for the protection of 'baby seals' are perfect examples of this.

Governments in both developed and Third World countries, on the other hand, have principally been interested only in the importance of protected zones to the tourist industry. Many countries earn substantial amounts of foreign currency through the influx of tourists into their national parks: Kenya, for example, benefited from a capital inflow of more than 1 million dollars in 1978 from that source. This aspect of national parks has produced an ambivalent attitude towards them on the part of the administrative authorities running them. The success of some parks as a tourist attraction has been achieved with no regard for their carrying capacity and has undoubtedly led to erosion by visitors—and sometimes to their transformation into nothing more than a fun fair or amusement park.

However, the justification for setting up nature reserves ought not to lie in the aesthetic or picturesque character of an area nor in short-term economic values, but rather in scientific considerations of a genetic and agronomic nature: in other words, biological in the widest sense of the word. For, if world agriculture is to support a total human population that at best is going to stabilize at around 8000 million, preservation of the plant and animal genetic pool is essential. The imprudent spread of monoculture that is now taking place, coupled with the reliance of industrial stock-rearing on a very small number of domesticated breeds, could expose humanity to genetic disaster. It should not be forgotten that cultivated plants and domesticated livestock originate from wild stock that must be preserved for the improvement of the selected varieties and breeds.

As far as plants are concerned, strict preservation of the older wild varieties is the only insurance there is against the destruction of today's principal crops, either by a sudden attack from extremely voracious pests or by infection from a virulent mutant strain of pathogenic fungi. This is not mere speculation:

fered from a lowered resistance to devouring insects or to various cryptogamic and viral diseases, and this is now posing serious problems in agricultural production. Among many examples are an alarming spread of maize rust in the tropics and increasing attacks by pathogenic fungi on coffee and sugar-cane plantations, all producing a dramatic fall in yield.

The risk of a global agronomic disaster is increased by the excessive homozygosis in the varieties of crop on which food production is based. This is threatening developed nations as much as those of the Third World because of the almost exclusive monoculture of crops like Mexican wheat, IR8 rice and others, in what is alleged to be the 'green revolution'. Thus, the whole of the American soya industry relies on the progeny of six varieties coming from the same region of China. In France, a single potato variety, the Bintje, still made up nearly 90 per cent of the area devoted to this crop a few years ago, while in the USA only four varieties of potato account for 72 per cent of the total production. In Canada, a single cultivar accounts for more than half the national production of wheat.

Under these conditions, it is easy to see that such crops are extremely vulnerable to a sudden explosion in the population of a pest or parasite. If abnormally susceptible, the crop can be completely wiped out, a situation that is by no means fanciful. Many examples can be quoted: in the 1840s, for instance, the Irish potato crop was attacked by a pathogenic fungus, *Phytophtora infestans*, which caused a disastrous famine killing 2 million people and provoking a general emigration to the USA. Similarly, European vineyards were completely destroyed between 1860 and 1880 by an invasion of phylloxera: vine-growing was only reestablished by grafting European varieties on to rootstock introduced from the USA, using a species of American *Vitis* (*Vitis rupestris* or *riparia*, for example). If the habitats of these wild species had been destroyed, European vines might never have recovered.

Generally speaking, it is unselected wild stock that has the greatest genetic variation. This gives it the best chance of possessing some adaptive characteristic required by a selector. Plant stock of this sort cannot, however, be cultivated in fields or it rapidly loses its genetic variability. It is just the same with animals, where wild species closely allied to domestic breeds cannot be raised in enclosed land for similar reasons.

The maintenance of a worldwide genetic pool thus involves not just the preservation of individual specimens but also of whole populations of neighbouring species, the value of every gene then being established by the whole of the population within which it is exchanged. Such preservation is essential since any possibility of improvement in plant varieties or animal breeds depends on it. To take an example: merinos were a breed created by crossing domesticated sheep with mouflon (wild sheep), and this could not have been done if the latter had already been exterminated.

It has also often happened in the past that a species of plant or animal apparently of no importance, and sometimes on the verge of extinction, has suddenly found an extremely useful application. Several years ago, *Chirostoma estor* was a fish confined to a single Mexican lake and threatened with extinction through overfishing and deterioration of its habitat. It was then used successfully for the restocking of several large lake-reservoirs behind new dams and other stretches of calm water: it is now of major importance to the future of all the inland fisheries of Central America.

Turning to another area, more than 40 per cent of currently used drugs and medicines contain a natural substance as active ingredient, a substance which in two-thirds of cases has been extracted from a plant. If the ecosystems richest in plant species are not given adequate protection, there is nothing to prevent the disappearance of large numbers of phanerogams or cryptogams of great pharmaceutical and dietetic importance, some of which undoubtedly yield substances that are more effective than pyrethrum or penicillin, for example.

Another aspect is the surprisingly small number of plant and animal species (most of them domesticated since the beginning of Neolithic times) that currently provide all human food requirements. Only a few hundred plants and a few tens of animals are used, when there are more than 100 000 species of spermatophyte, some 3000 species of mammal and 10 000 species of bird known to exist. Most of the world's agricultural output is represented by some 20 or so cultivated plant species at most, and 8 of them alone (wheat, rice, maize, potatoes, barley, sweet potatoes, cassava and soya) make up more than 75 per cent of the total crop. Our descendants will surely learn how to make better use of the enormous potential residing in the vast number of unused wild species—as long as we have made sure that they survive.

The urgent need for protective measures is perfectly illustrated by the case of the large African ungulates. The local inhabitants and then the Europeans, persisted with the rearing of cattle poorly adapted to tropical conditions, as even indigenous breeds may be. Yet, as has already been pointed out in previous chapters, the various species of antelope are clearly superior to cattle in that they use the vegetation more efficiently and are more resistant to parasites. The gnu, kob and oryx, for instance, are much better meat producers than cattle and the giant eland is a better producer of milk as well. The protection of such species would seem vital for the future of animal production in Africa, if not in all tropical regions.

Many other arguments of an equally fundamental nature support the rigorous conservation of natural resources and of various types of ecosystem still relatively unaltered by human action. Every ecologist is well aware of the preeminent role of the diversity of an environment in guaranteeing its internal stability. The preservation intact of a large number of biocoenoses is essential as a counterbalance to the uniformity and levelling processes produced in living communities by contemporary civilization. If technological society does not change, it will continue to carry within itself the seeds of genocide by its reduction of everything to a common level and its brutal rejection or elimination of anything that is different.

8.5   The Cultural Significance of Nature Protection

However important the protection of wild life may be for its potential applications and long-term economic benefits, it is not to be justified on utilitarian grounds alone. In the first place, the preservation of plant and animal species and of unmodified ecosystems is also of prime importance to all the biological sciences. It is, moreover, essential to preserve for future generations some of the remaining communities only slightly affected by human interference or even some of those in their virgin state where they still exist: only then will it be possible to understand the differences which separate such ecosystems functionally from those that are exploited in one way or another by humanity. Without adequate protection, many living communities will have disappeared by the end of this century, before ecologists have even begun to study them.

Then there are aspects like the aesthetic character of many natural species and the exceptional splendour of certain landscapes, whether natural or man-made: such aspects are still as worthy of consideration as they were at the beginning of the modern conservation movements and justify protective measures just as much as others. The creation of national parks and nature reserves is thus more than ever necessary today, not only because of their great importance in ensuring a lasting development of the communities within them, but also because of their aesthetic, educational and cultural role. What teacher would question the value of enriching contact with an unspoiled natural environment to the psychosensory development of the child and the full flowering of its personality?

The weekly stampede by the urban populations of industrialized nations to the countryside, riverside or mountains shows how much this contact is also necessary for the mental equilibrium and well-being of adults. When one thinks of the pernicious psychological effect of 'mineral landscapes, divorced from life, infinitely sad and ugly' to which Grassé (1970) compares modern cities, the conservation of nature would appear to be essential to the physical health and morale of urban populations.

8.6   The Organization of Nature Protection

The preservation of disappearing plant and animal species and the more general protection of all threatened biocoenoses can only be secured by the creation of a great many nature reserves. At present, 35 of the biogeographical zones into which the biosphere can be divided have no national park or similar type of reserve; 38 more of them, although containing at least one protected area, are not adequately covered (Figure 8.1). The establishment of an international network of nature reserves in the context of the MAB programme of the United Nations is far from being completed, with the 162 reserves created up till now being located in only 76 of the 193 biogeographical zones. So far, only 12 reserves have been established in tropical rain forests and there are fewer than 6 in each of the following types of ecosystem: subtropical rain forests, temperate coniferous forests, cold semi-deserts, temperate grassland, tropical savanna, and fresh-water environments (IUCN, 1980a).

The success or failure of any attempts at preservation quite clearly depends on the methods of organization adopted in the various countries concerned, on the degree of rigour in the laws governing nature reserves, and finally on the way in which the laws are applied. In fact, the general term 'nature reserve' covers protected areas of widely varying status under the law and with a great range of aims and activities (Figure 8.4).

According to criteria laid down by the IUCN, the status of *national park* or analogous reserve should only be granted to areas that enjoy protection within a legal framework: protection of their natural resources from all human exploitation[1] and defence against any other attack on their territorial integrity resulting from human activity. In addition, a national park should be equipped with sufficient material and human resources (personnel, finance for its administration and upkeep) to ensure that the laws are adequately applied. The final criterion laid down by the IUCN is that of a minimum area. A sufficiently large area is in itself, of course, some guarantee of conservation within it, just as too small an area makes it vulnerable, but the dividing line between the two will depend on the geographical location and the population density in the country concerned. There is, in addition, a minimum area for every living species and every community

below which they cannot reproduce even if they enjoy absolute protection.

Alongside national parks are other types of nature reserve which confer absolute protection either on the complete ecosystem within it or at least on certain components of it. *Wild-life sanctuaries* represent the most rigorous form of protection. Unlike national parks, in which tourism is not only permitted but often encouraged (at least over part of their domain), sanctuaries are forbidden territory as far as human visitors are concerned: only the wardens, and scientists carrying out research on some special biological feature, have right of access and even then it is only on foot.

Sanctuaries can be situated outside national parks, as is the case, for instance, with certain stretches of land and inland waters in the Camargue nature reserve in the south of France. On the other hand, the IUCN thinks that a national park ought always to include a certain number of sanctuaries within its boundaries. Although French law has anticipated this possibility, and although some of their national parks have been established for more than 18 years, none of them includes any such sanctuaries.

There are in addition *biosphere reserves* created at the instigation of Unesco from 1974 onwards. These are designed to protect ecosystems, whether natural or modified by human activity, in order to preserve ecological 'evidence' for the purposes of scientific research. Reserves of this type should satisfy 7 essential criteria (Maldague, 1984):

1. They should provide a network of protected terrestrial and coastal environments which form a coherent system on a world scale.
2. They should occur in each of the 193 biogeographical provinces of the world distinguished in the classification of Udvardy (1975), so as to exhibit the maximum genetic diversity.
3. They should show a complete range of the different types of human interference, from ecosystems untouched by any anthropic action to those which have been degraded by humans very severely and/or for a very long time.
4. Their structure and size should ensure the efficient conservation of the ecological systems they are designed to protect.
5. They should have sufficient resources available for ecological research to be carried out on the spot together with education and training in matters concerning nature conservancy.
6. They should if possible have geographical continuity with other types of protected zone.
7. They should receive adequate protection under the law with long-term safeguards.

In the original concept, biosphere reserves consist of

---

1. In this connection, it is strange to see that in some so-called nature reserves recently created in France the regulations authorize hunting and even exploitation of forests.

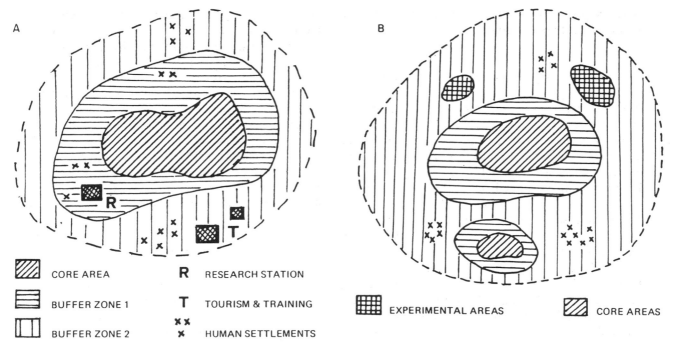

Figure 8.3 A, typical biosphere reserve; B, cluster type of biosphere reserve with two core areas and with integrated research and experimental areas. (From Batisse, 1982)

a central zone (the core area in Fig. 8.3A) with maximum protection and, in the case of virgin areas, complete exclusion of people other than those carrying out research. Outside the core are two concentric buffer zones in which controlled exploitation of natural resources is possible. This simple zoning system can be adapted to various geographical, ecological and cultural situations, including, for example, the case where animals migrate from one section to another or where two or more core areas must be protected, forming a cluster type of biosphere reserve (Fig. 8.3B).

Other areas have a status which generally only gives them partial protection, limited perhaps to some components only. Very small areas, sometimes known as national monuments, while subject to very restrictive regulations, are in fact too small to provide protection as adequately as national parks. Then there are *national forests* (or *forest parks*) having a very long regeneration cycle during which all exploitation is prohibited, and these thus effectively become botanic reserves. In a similar way, temporary or permanent *game reserves* are areas that give *de facto* protection to the fauna within them, at least to the extent that the authorities enforce restraint on activities there. In some cases, these more specialized reserves have achieved a highly satisfactory conservation of natural resources. The Masai Mara Game Reserve in Kenya, for instance, covers 181 000 hectares and forms the northern extension of the Serengeti National Park on the other side of the border with Tanzania (Figure 8.5). Within this area, the traditional grazing rights enjoyed by the nomadic herdsmen have in no way jeopardized

the continuing stability of the extraordinary wild ungulate population.

Nature reserves of all types play an economic role even outside the activities related to tourism. The preservation of plant cover in regions of high relief, whether trees or grasses, protects the soils of drainage basins against erosion and thus increases the importance of any reserves established in such areas. Again, certain game reserves in temperate countries where predators have often long been eradicated become overpopulated with ungulates some time after their creation when hunting is prohibited. As a result, excess animals can be removed without doing any harm to the remaining population and used, for instance, for restocking elsewhere. Generally speaking, the decline in carnivore populations is so great at present that a region put under protection experiences a growing proliferation of herbivores, reaching densities in excess of the carrying capacity of the environment.

Ecological problems connected with conservation

The ecological equilibrium in protected zones is sometimes disturbed, directly or indirectly, by the effect of human activity on the fauna. Damage to vegetation, for instance, has occurred in several such areas in Europe and North America because of excessive increases in the populations of Cervidae and other ungulates. In the Yellowstone National Park, proliferation of the mule-deer (*Odocoileus hemionus*) and the pronghorn antelope (*Antilocapra americana*) has led the US Fish

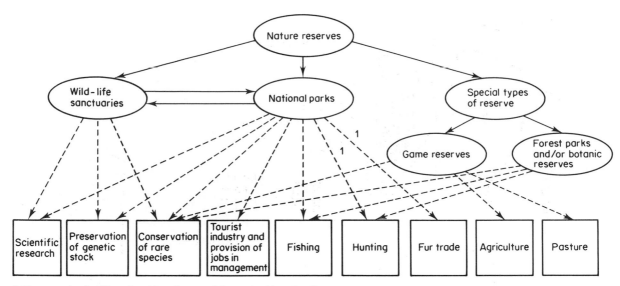

1 Excess animals either slaughtered or used for restocking elsewhere.

Figure 8.4   The various types of nature reserve affording protection to a whole ecosystem or to some components of it. The possible aims and activities shown at the bottom of the figure will vary according to the status of the reserve

and Wild-life Service to consider reintroducing wolves to limit their populations. In the same way, the wolves and cougars (*Felis concolor*) that once roamed the Kaibab plateau in Arizona were almost completely wiped out when the fauna reserve was established there. This generated such a proliferation of mule-deer in the 1920s that the woody and herbaceous vegetation was almost irreversibly damaged before starvation and disease caused a collapse in numbers in the mid-1920s (Figure 8.6).

A similar serious problem at present being encountered in the national parks of East Africa arises from the damage to trees caused by elephants. The former migration of these animals to areas outside the parks has been almost totally ended by the cultivation of the land or its use for other human activities. This has caused a much greater concentration of the elephants inside the protected areas, where they damage trees during the dry season by eating the bark and foliage (see Figure 7.7, p. 193). Overpopulation by baboons (*Papio olivaceus*) is another problem in some of the African national parks, this time caused by the almost complete extermination of their principal natural predator, the leopard, by poachers.

Certain types of human intervention can have a directly damaging effect in some nature reserves. Forest fires in the USA and Canada were once a case in point, but in coniferous formations these have no longer been fought systematically ever since it has been recognized that spontaneous fires play a positive role in the regeneration of woodland. The example of the development of young sequoias has already been mentioned. A similar effect occurs in national parks established in areas of savanna where it has been

observed that extinguishing fires and taking protective measures against them causes a development of woody plants at the expense of grasses. On the other hand, since too many fires cause catastrophic damage to trees and shrubs— even to pyrophytes—a programme of controlled firing is essential if the physiognomy characteristic of these areas before protection is to be maintained.

In countries where ancient civilizations have flourished, particularly in Europe, human interference with the environment has been on such a scale that very often it is not possible to eliminate the effects altogether after a decision to establish a protected zone has been taken. Thus, in the British and German parks that include heath and moorland, maintaining grazing at a reasonable level and instituting a programme of controlled firing seems in many cases to be indispensable: the vegetation is not in equilibrium and without such measures it would undergo a transition to shrubland and eventually to deciduous forest.

Even in the Camargue, an area seen by many Europeans as untouched wild land *par excellence*, the characteristic salt-adapted ecosystems are largely due to human intervention. In the middle of the last century, a dam was constructed so as to keep the sea separated from the complex system of lakes and lowlying land populated with halophilous flora and fauna. This ensures that the excessively salty conditions continue to exist as long as the dam itself is properly maintained. On the other hand, the discharge into this area of fresh water used for the irrigation of crops to the north of the delta can cause desalination and thus an alteration which ought to be quickly checked and reversed.

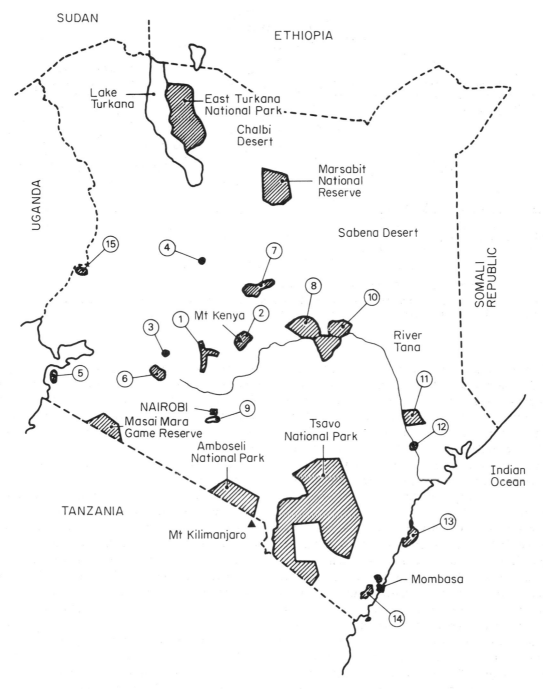

Figure 8.5 The network of national parks and nature reserves in Kenya is in many respects quite outstanding. In spite of difficulties in providing effective protection of the various areas, the measures taken by the authorities in this country set an example to the rest of the Third World and even to developed countries . . . like ours. (From Ramade, 1979b)

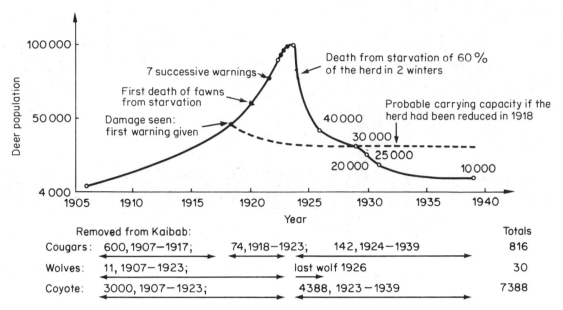

Figure 8.6   Effect of the deliberate extermination of predators (wolves, coyote, cougars) on the deer population of the Kaibab plateau, Arizona. After hunting had almost totally eliminated the carnivores, the mule-deer population reached such a density that it caused irreparable damage to the vegetation before collapsing to a stable level well below the carrying capacity of the environment. (From Leopold, in Kormondy, 1969)

It is thus apparent—and we should not lose sight of this—that the main problems in protecting some exceptionally important ecosystems arise from developments inside the reserved areas themselves or in their neighbourhood.

Protected fresh-water environments can suffer greatly from any changes in the normal rates of water flow: certainly from an excessive influx, as in the Camargue, but also from a diminution in outflow. The Everglades National Park in Florida, for example, which protects a complex area of low-lying ground and euryhaline marshland covering 540 000 hectares, is being threatened with quite a drastic modification of this type: fresh water from Lake Okeechobee, which used to feed the upstream part of the marshland in the park, is now being diverted to irrigate agricultural land. In a similar way, the drainage from crops that border the wetlands of a nature reserve can have disastrous effects on the protected areas whose water balance is upset.

Lastly, one of the greatest threats to national parks and other nature reserves arises from erosion due to an excessive influx of tourists. In many countries, and particularly in France, access to national parks is not regulated in any way (Figure 8.7). As a result, there is an invasion every year by visitors in numbers that are out of all proportion to any carrying capacity. Nearly 1 million tourists visit the Parc National de la Vanoise each year. In the Port-Cros national park over 300 pleasure boats are allowed to moor simultaneously in the sheltered creeks of the island in the high season, damaging the beds of *Posidonia* with their anchors and discharging rubbish that contaminates the sea-bed, an

environment that is theoretically protected. At the moment, this is the only French national park established in a sea area with the intention of protecting marine biocoenoses (see Plate XXV).

Turning to the USA, the total number of visitors to national parks and reserves was officially estimated as 95 million in 1960 and 239 million in 1975, and more than 300 million are expected in 1989 (from Environmental Protection Agency). Between 1960 and 1975, 102 new areas were classified as national parks or other types of nature reserve, and a system for regulating the number of visitors was instituted,[1] taking into account the carrying capacity of the various environments. In spite of all that, 'human erosion' in the form of damage to the grass cover has been increasing over this period, particularly around holiday residences and the most frequented sites. As a result, efforts have been made in recent years to develop a type of tourism with less harmful ecological consequences. First of all, plans to develop the hotel infrastructure have been abandoned (for instance, in the Yosemite National Park). Any extension of the sites where camping and caravanning is authorized has been banned. In addition, the US Secretary of the Interior, who is responsible for nature protection, has decided to close some of the roads to vehicular traffic,[2] although the

1. In the USA and Canada, nobody is allowed to stay for more than 10 days in one year in the same national park. In addition, the number of authorized daily visitors is not allowed to exceed the total number of places available in hotels and camp-sites.

2. Over the same period, new roads open to car traffic were being constructed in French national parks!

Plate XXIV   The protection of nature in Kenya

1   Buffaloes (*Syncerus caffer*) around a water hole in the Tsavo National Park. This park, created in 1948, covers nearly 21 000 square kilometres and is one of the largest in the world.

2   White rhinoceros (*Ceratotherium simum*) in the Meru National Park. This species was decimated by hunting and finally disappeared from Kenya for some decades. It was reintroduced during the 1970s but is unfortunately subject to massive poaching even in the national parks, and this is jeopardizing the chances of a recovery in its numbers.

3   Cheetahs (*Acinonyx jubatus*) in the Amboseli National Park.

4   Grevy's zebras in the Samburu National Park. This rare species of zebra was once also an inhabitant of southern Ethiopia and the Somali Republic, but has recently been finally exterminated there. At present, hardly more than 800 of them are surviving in the Samburu and Tsavo National Parks.

5   Grant's gazelles in the Samburu National Park.
(Photographs F. Ramade)

Plate XXV   The Port-Cros national park, Var, France

1   View of the north coast of Port-Cros. The national park embraces the whole 694 hectares of this small Mediterranean island, as well as part of the surrounding sea. It is supposed to protect the sea-bed whose greatest distinctive feature is one of the last barrier reefs of *Posidonia*, a large marine spermatophyte of the French Mediterranean coast.
(Photograph F. Ramade)

2   A bed of *Posidonia* in its natural state.
(Photograph J. C. Moreteau)

3   The pressure of tourism, mostly from pleasure boats which are permitted to moor in the waters of the park, is causing great damage to the sea-bed. This is not only from the rubbish thrown overboard but also from the anchors of the moored boats which destroy the plant life. This photograph is of part of the sea-bed where a dense growth of *Posidonia* was flourishing less than 15 years ago.
(Photograph J. C. Moreteau)

4   Gorgonia (*Paramuricia clavata* and *Eunicella verrucosa*) growing on a rocky substratum at a depth of 30 metres.
(Photograph J. C. Moreteau)

Figure 8.7   National parks in France are frequently subjected to excessive pressure from tourists, whose numbers are much above the carrying capacity of the environments the parks are supposed to be protecting. This photograph shows an over-crowded camp-site near Ailefroide in the Parc National des Ecrins. (Photograph F. Ramade)

area of American national parks accessible by car is relatively small: only 3 per cent of the Yellowstone National Park, for instance, can be visited in this way. In parallel with that development, there are moves to encourage tourists on foot to use the old tracks marked out by trappers.

Comparable tendencies in regulating access to national parks can be seen in other parts of the world such as Great Britain, the Scandinavian countries and Eastern Europe. Even in Kenya there is official concern over the disturbance of fauna by tourists in

cars (in 1978, I watched a family of cheetahs literally pursued for nearly an hour by half a dozen vehicles).

It is thus quite clear that excessive pressure from tourists can jeopardize the stability of the exceptional living communities which nature reserves are intended to protect. It is therefore essential that access to national parks and other reserves be regulated so that the number of visitors never exceeds the carrying capacity of the environments concerned and thus does not cause irreparable damage.

Conclusion

Two main factors are at the root of the current environmental crisis: one is the population explosion and the other is the wasteful consumption of energy and other natural resources by the developed countries of the world.

There can be no doubt at all that the continuing growth in human population presents a serious threat to the long-term stability of the biosphere. Yet, in spite of a worrying depletion of many mineral resources and irreparable damage to farmland from excessive population densities, governments and politicians in the vast majority of nations continue to deny the existence of a serious demographic problem and even persist in supporting a rising birth rate. This occurs as much in industrialized countries as in those discreetly described as 'developing'.

In the Third World, the most widespread attitude of the authorities is one of *laissez faire*. Under these conditions, human populations are going to exceed the carrying capacity of environments in accordance with ecological laws characteristic of r-type demographic strategies (see p. 16). In such countries, most of the economic development is not devoted to achieving a real improvement in the general standard of living, but is only just sufficient to maintain most of their inhabitants at a subsistence level.

The populations of industrialized countries, on the other hand, are now generally becoming stabilized through a more widespread use of contraception associated with a rise in average educational levels. Nevertheless, governments still promote policies that are decidedly favourable to higher birth rates, an approach that stems from a profoundly mistaken analysis of the interaction between fluctuations in population and the state of the economy. Population growth is still considered today to be an indispensable stimulus to economic activity by all technocrats and by all of the too-numerous specialists in the human sciences. In spite of the absurdity of such a proposition, demonstrated by many actual examples,[1]

authorities in developed countries persist in their attempts to stimulate higher birth rates through the offer of various material benefits.

In such a context, little consideration is given to the negative effects of the very same development: problems like those of overpopulation, urban disturbance, pollution, the losses of rural areas and natural environments generally. Worse still, when they are taken into account, it is only for their effect on the Gross National Product. According to this view, it is pointless to limit human populations as long as the relatively high standard of living assured by what is called the consumer society can be maintained— whatever the consequences for the human environment or even for the whole biosphere.

Until now, Western economic thought has been supremely indifferent to the scale of the human impact on natural resources and has even deliberately played it down. As Passet (1980) points out:

> So long as economic operations could not affect the regulating mechanisms of the biosphere, they could be considered as occupying a closed universe possessing its own logic and its own rules for optimization. On the one hand, there was humanity, producing and exchanging; on the other, there was nature, in principle unalterable and, to use Ricardian terms, spontaneously securing its own reproduction. It will, however, no longer be the same when larger populations, endowed with an effective technology, upset the environment supporting them and threaten to destroy it. Reproduction of the biosphere can then no longer be thought of as independent of the mechanisms governing the reproduction of natural and human resources.

In reality, the effect of humanity on the ecosphere since the dawn of the modern era has so increased that it has now become a dominant biogeochemical factor. One of the most tragic mistakes of Western economic thought lies precisely in the fact that it has not taken into consideration the true position of *Homo economicus* in the ecosphere: not outside it, as has long been believed, but an integral part of it and thus subject to all its biological uncertainties.

---

1. Since the beginning of the 1970s, the two Germanies and Sweden, whose economic performance is among the best in the world, have had a rate of population growth that has been nil and sometimes negative.

214

The extent of the present environmental crisis, and the disastrous consequences it leads us to expect, indicate the need for urgent implementation of a world strategy for conservation as the only guarantee of lasting development. The IUCN has recently evolved such a programme on behalf of the United Nations Organization, a programme that brings together the work of more than 700 specialists in general and applied ecology, in environmental planning and in related disciplines. Ecological models for the proper management of the Earth's surface and the rational utilization of the biosphere have been put forward by various ecologists for over a decade (see, for example, Labeyrie): only by applying these can a new, so-called post-industrial, type of civilization be conceived, free from the present difficulties.

It is with this aim in mind that the IUCN evolved its strategy, proposing that the main resources indispensable to humanity should be conserved through the application of the following principles:

(a) the maintenance of essential ecological processes and life-support systems (the regeneration and protection of soils, the recycling of inorganic nutrients, the natural purification of water);
(b) the preservation of the genetic diversity on which most of these processes and systems depend for their functioning: for instance, through programmes of selection to protect and improve cultivated plants, domesticated livestock, and micro-organisms; through scientific and technical progress; and through the future of many industries using living resources.
(c) the supervision and control of the long-term exploitation of ecosystems and species, particularly of the terrestrial and aquatic fauna upon which millions of rural communities and many industrial activities depend.

It is essential to realize that this programme should be implemented with the greatest urgency and its objectives achieved without delay. There should be no need, for example, to remind ourselves that several thousand million tonnes of soil are being destroyed each year by deforestation and poor agricultural practices, and that in the next two decades 30 per cent of the world's arable land will have disappeared through the spread of deserts alone.

Finally, it can never be repeated often enough that such a strategy can only succeed in the long term if associated with a rigorous policy of birth control on a world scale. A dramatic slowing down in the rate of population growth must be brought about by the end of the century. The leaders in some countries are aware of this, and in China, for instance, the authorities have recently launched a campaign encouraging couples to have only one child.

Many ecological data now available force us to the conclusion that humanity is already too numerous, in that the present size and density is incompatible with the maintenance of a satisfactory level of development, let alone its acquisition by all the Earth's inhabitants. The major problem that faces us is not securing the survival of the maximum number supportable by the ecosphere, but regulating the total population to an optimum value compatible with lasting development. To achieve such a solution, the world population, after reaching a maximum within a few decades, ought to decline and stabilize at a value distinctly lower than the present one. Many ecologists consider that no more than 2000 million people could be supported at a European standard of living without jeopardizing the future stability of the biosphere, and even this number presupposes an ending of all damage now being done to the environment.

Such ideas could appear utopian: in fact, they arise from a realistic evaluation of the constraints resulting from the way natural resources are being used in our technological society. Contrary to an opinion that is still widely held, the shocks that world economies have now been experiencing for a decade are by no means a passing phenomenon in an evolving industrial civilization. In reality, they are the forerunners of a much graver crisis that stems from the fact that our species is in the process of exceeding the carrying capacity of the biosphere.

The myth of indefinite quantitative growth based on a wastage of natural resources—which still remains the credo of governments, politicians and their technocrats—is completely defunct. If our civilization cannot comprehend and adopt a K-type strategy while there is still time, it will perish.

Bibliography and References

Because of the limited space available, it has not been possible to quote all the bibliographic references (nearly 2000) consulted during the preparation of this work. I ask for the indulgence of those authors and colleagues whom I have been forced to omit from this list.

Basic Texts on General and Applied Ecology

Blondel, J., 1979, *Biogéographie et écologie*, Masson, Paris, 173 pp.

Clarke, G. L., 1959, *Elements of Ecology*, 3rd edn, Wiley, 534 pp.

Dasmann, R. F., 1976, *Environmental Conservation*, 4th edn, Wiley, 436 pp.

Duvigneaud, P., 1974, *La Synthèse écologique*, Doin, 296 pp.

Ehrlich, P. R., Ehrlich, A. N., and Holdren, J. P., 1977, *Ecoscience: Population, Resources, Environment*, Freeman, San Francisco, 1051 pp.

Kormondy, E. J., 1969, *Concepts of Ecology*, Prentice-Hall, 208 pp.

Lamotte, M., and Bourlière, F., 1967, *Problèmes de productivité biologique*, PBI Publications, Masson, 246 pp.

Lamotte, M., and Bourlière, F., 1978, *Problèmes d'écologie: structure et fonctionnement des écosystèmes terrestres*, Masson, Paris, 345 pp.

Lemée, G., 1967, *Précis de biogéographie*, Masson, Paris, 958 pp.

Lemée, G., 1978, *Précis d'écologie végétale*, Masson, Paris, 285 pp.

Lieth, H., and Whittaker, R. H., 1975, *Primary Productivity of the Biosphere*, Springer-Verlag, 339 pp.

May, R. M., 1976, *Theoretical Ecology, Principles and Applications*, Blackwell Scientific Publications, 317 pp.

Moran, J. M., Morgan, M. D., and Wiersma, J. H., 1980, *Introduction to Environmental Science*, Freeman, 658 pp.

Odum, E. P., 1959, *Fundamentals of Ecology*, Saunders, 1st edn, 546 pp.; 1971, 3rd edn, 573 pp.

Odum, H. T., 1971, *Environment, Power and Society*, Wiley, 33 pp.

Pérès, J. M., 1976, *Précis d'océanographie biologique*, Presses Universitaires de France, 246 pp.

Ramade, F., 1978a, *Eléments d'écologie appliquée*, 2nd edn, McGraw-Hill, 576 pp.

Ramade, F., 1982, *Eléments d'Ecologie, Ecologie Appliquée*, 3rd edn, McGraw-Hill, Paris, 452 pp.

Ramade, F., 1984, *Eléments d'Ecologie, Ecologie Fondamentale*, McGraw-Hill, Paris, 408 pp.

Rickleffs, R., 1979, *Ecology*, Nelson, Sunbury, 966 pp.

Simmons, I. G., 1974, *The Ecology of Natural Resources*, Arnold, 424 pp.

Watt, K. F., 1973, *Principles of Environmental Science*, McGraw-Hill, 320 pp.

Whittaker, R. H., 1975, *Communities and Ecosystems*, Macmillan, New York and London, 385 pp.

Other Books, Review Articles and Original Articles Quoted in the Text

Acker, A., 1979, *Astronomie*, Masson.

Adams, J. A. S., Mantovani, M. S. M., and Lundell, L. L., 1977, Wood versus fossil fuels as a source of excess carbon dioxide in the atmosphere: a preliminary report, *Science*, **196**, no. 4285, pp. 54–56.

Aiken, S. R., and Moss, M. R., 1975, Man's impact on the tropical rainforest of peninsular Malaysia: a review, *Biol-Conserv.*, **8**, no. 3, pp. 213–229.

Ajtay, G. L., Ketner, P., and Duvigneaud, P., 1979, Terrestrial primary production and phytomass, in Bolin *et al.*, *The Global Carbon Cycle*, Wiley, pp. 129–181.

Allen, R. A., 1980, *How to Save the World. Strategy for World Conservation*, Kogan Page, London, 150 pp.

Amarger, N., Mariotti, A., and Mariotti, F., 1977, Microbiologie du sol: essai d'estimation du taux d'azote fixé symbiotiquement chez le lupin par le traçage isotopique naturel ^{15}N, *C. R. Ac. Sci.*, series D, **284**, p. 2179.

Amiard-Triquet, C., and Amiard, J. C., 1976, La Pollution radioactive du milieu aquatique et ses conséquences écologiques, *Bull. ecol.*, **7**, no. 1, pp. 3–32.

Amiard-Triquet, C., and Amiard, J. C., 1980, *Radioécologie des milieux aquatiques*, Masson, Paris, 191 pp.

Anonymous, 1977, *L'Energie solaire: de la recherche appliquée aux utilisations pratiques. Perspectives d'avenir*, Délégation aux Énergies Nouvelles, Ministre de l'Industrie, Actualités documents, October, 62 pp.

Batisse, M., 1982, *Environ. Conserv.*, 9, No. 2, p. 103.

Baumgartner, A., and Reichel, E., 1979, in S. Kempe, Carbon in the freshwater cycle, in Bolin *et al.*, *The Global Carbon Cycle*, Wiley, p. 319.

Bazilevitch, N. I., Rodin, L. Y., and Rozov, N. N., 1971, Geophysical aspects of biological productivity, *Soviet Geograph. Rev. Trans.*, **12**, pp. 293–317.

Bel, F., Le Pape, Y., and Mollard, A., 1978, *Analyse énergétique de la production agricole*, INRA-IREP, Grenoble, 163 pp.

Bell, H. V., 1971, A grazing ecosystem in the Serengeti, *Sci. Amer.*, **225**, no. 1, pp. 86–93.

Berg, R. R., Calhoun, J. C., and Whiting, R. L., 1974, Prognosis for expanded US production of crude oil, *Science*, **184**, no. 4134, pp. 331–339.

Bernhard-Reversat, F., 1975, Recherches sur l'écosystème de la forêt subéquatoriale de Basse Côte d'Ivoire, VI: les cycles des macroéléments, *La Terre et la Vie*, **2**, pp. 229–254.

Bernhard-Reversat, F., Huttel, C., and Lemée, G., 1978, *La Forêt sempervirente de Basse Côte d'Ivoire*, in

Lamotte and Bourlière, *Problèmes d'écologie*, Masson, pp. 313–345.

Berzin, A. A., and Yablokov, A. V., 1978, Number and structure of populations of the principal cetacean species exploited in world oceans (in Russian), *Zool. Jour.*, Moscow, **57**, no. 12, pp. 1771–1785.

Bethemont, J., 1977, *De l'eau et des hommes. Essai géographique sur l'utilisation des eaux continentales*, Bordas, 280 pp.

Beverton, R. J. H., and Holt, S. J., 1956, The theory of fishing, in Graham, M., *Sea Fisheries, their Investigation in the United Kingdom*, Arnold, London, pp. 372–441.

Blondel, J., 1976, L'Influence des reboisements sur les communautés d'oiseaux. L'exemple au Mont-Ventoux, *Ann. Sci. Forest*, **33**, no. 4, pp. 221–245.

Blondel, J., 1980, *Structure et dynamique des peuplements d'oiseaux*, in Pesson *et al.*, *Actualités d'écologie forestière*, Gauthiers-Villars, Paris, pp. 367–388.

Blondel, J., and Bourlière, F., 1979, La Niche écologique, mythe ou réalité? *La Terre et la Vie, Rev. Ecol.*, **33**, no. 3, pp. 345–374.

Bogorov, V., 1975, L'Homme, la société, et l'océan, in *Homme, société et environnement*, Acad. Sci. USSR, Institute of Geography, Moscow, pp. 284–312.

Bolin, B., Degens, T., Duvigneaud, P., and Kempe, D., 1979, The global biogeochemical carbon cycle, in Bolin *et al.*, *The Global Carbon Cycle*, Wiley, Chichester, pp. 1–56.

Borgstrom, G., 1969, *Too Many*, Collier–Macmillan.

Bourlière, F., 1962, Les Populations d'ongulés sauvages africains: caractéristiques écologiques et interprétations économiques, *La Terre et la Vie*, **109**, pp. 150–160.

Bourlière, F., 1963, Observations on the ecology of some large African mammals, in Howell, F. C., and Bourlière, F., 1963, *African Ecology and Human Evolution*, Aldine, Chicago, pp. 43–54.

Bourlière, F., 1973, The comparative ecology of main forest mammals in Africa and tropical America: some introductory remarks, in Meggers, B. J., Ayensu, E. S., and Duckworth, W. D., *Tropical Forest Ecosystems in Africa and South America*, Smithsonian Institution Press, Washington, pp. 279–292.

Bourlière, F., 1978, La savane sahélienne de Fété Olé, Sénégal, in Lamotte and Bourlière, *Problèmes d'écologie: structure et fonctionnement des écosystèmes terrestres*, Masson, pp. 187–229.

Bourlière, F., and Hadley, M., 1970, The ecology of tropical savannas, *Ann. Rev. Ecol. Systemat.*, **1**, pp. 125–152.

Bourlière, F., and Verschuren, J., 1960, *Introduction à l'écologie des ongulés du Parc National Albert*, Inst. Parc Nat. Congo Belge, Brussels.

Bramryd, T., 1979, The effects of man on the biogeochemical cycle of carbon in terrestrial ecosystems, in Bolin *et al.*, *The Global Carbon Cycle*, Wiley, pp. 183–218.

Brinkworth, B. J., 1975, Direct use of solar energy, in Lenihan, J., and Fletcher, W. W., *Environment and Man*, Vol. 1, *Energy Resources and the Environment*, Blackie, London, pp. 157–189.

Broecker, W. S., 1975, Climatic changes: are we on the brink of a pronounced global warming?, *Science*, **189**, no. 4021, pp. 460–463.

Brooks, D. B., and Andrews, P. W., 1974, Mineral resources, economic growth and world population, *Science*, **185**, no. 4145, pp. 13–16.

Brown, L. R., 1973, Population and affluence: growing pressures on world food resources, *Popul. Bull.*, **29**, no. 2.

Brown, L. R., 1978, La Terre mangée, *Le Forum du Développement*, United Nations, no. 49, November, pp. 1–2.

Brown, L. R., 1978, The world-wide loss of cropland, *Worldwatch Paper No. 24*, Worldwatch Institute Publications, Washington, October, 48 pp.

Bryson, R. A., and Wendland, W. M., 1970, Climatic effects of atmospheric pollution, in Singer, S., *Global Effects of Environmental Pollution*, Reidel, Dordrecht, pp. 130–138.

Buechner, H. K., and Golley, F. B., 1967, in Petrusewicz, K., *Secondary Productivity of Terrestrial Ecosystems*, Polish Acad. Sci., pp. 243–254.

Buechner, M. K., Morrison, J. A., and Leuthold, W., 1966, Reproduction in Uganda kob with special reference to behaviour, *Symp. Zool. Soc. London*, **15**, pp. 69–88.

Bunt, J. S., 1975, Primary productivity of marine ecosystems, in Lieth and Whittaker, *Primary Productivity of the Biosphere*, Springer-Verlag, pp. 109–183.

Burt, W. H., and Grossenheider, P. R., 1964, *A Field Guide to Mammals*, 2nd edn, Houghton Mifflin, Boston, p. 215.

California Department of Fish and Game, 1957, *44th Annual Report*, Sacramento.

Chapman, P. F., 1970, Energy production—a world limit?, *New Scientist*, **47**, pp. 634–636.

Chararas, C., 1964, *Le Pin maritime*, Lechevalier, 126 pp.

Chararas, C., 1980, *Ecophysiologie des insectes parasites des forêts*, published by the author, INA, 16, rue C. Bernard, Paris, 297 pp.

Chauvet, B., 1972, The forests of Madagascar, in *Biogeography and Ecology of Madagascar*, Richard-Vindard and Battestini, Elsevier, p. 191.

Chorley, R. J., 1969, *Water, Earth and Man*, Methuen.

Cieslik, S., 1976, L'Ozone stratosphérique, *La Recherche*, no. 68, June, pp. 510–519.

Cloud, P., 1971, Mineral resources in fact and fancy, in Murdoch, *Environment, Resources, Pollution and Society*, Sinauer, pp. 71–88.

Cloud, P., 1969, *Resources and Man, a Study and Recommendations by the Committee on Resources and Man*, Division of Earth Science, US National Academy of Sciences, Freeman, 259 pp.

Commoner, B., 1972, *The Closing Circle—confronting the environmental crisis*, Jonathan Cape, London, 336 pp.

Cook, E., 1975, Flow of energy through technological society, in Lenihan, J., and Fletcher, W. W., *Environment and Man*, Vol. 1, *Energy Resources and the Environment*, Blackie, London, pp. 30–62.

Costin, A. B., 1964, Grasses and grassland in relation to soil conservation, in Barnard, C., *Grasses and Grassland*, Macmillan, pp. 236–258.

Cravioto, J., Licardie, E. R. de, and Birch, H. B., 1966, Nutrition, growth and neurointegrative development, *Pediatrics*, **38**, no. 2, pt 2 supplement (August).

Curry-Lindahl, C., and Harroy, J. P., 1972, *National Parks of the World*, Vol. 1: *Europe, N. and S. America*, 217 pp.; Vol. 2: *Africa, Asia, Australia and Oceania*, 240 pp., Golden Press, New York.

Dajoz, R., 1971, *Précis d'écologie*, Dunod, 434 pp.

Damon, P. E., and Kunen, S. M., 1976, Global cooling? No, southern hemisphere warming trends may indicate the onset of the CO_2 greenhouse effect, *Science*, **193**, no. 4252, pp. 447–453.

Daubenmire, R., 1968, Ecology of fire in grasslands, *Adv. Ecological Res.*, **5**, pp. 209–266.

Detwyler, T. R., 1971, *Man's Impact on the Environment*, McGraw-Hill, 731 pp.

Dorst, J., 1965, *Avant que nature meurt*, Delachaux & Niestlé, 424 pp.

Dregne, H. E., 1977, Freiner l'avance des déserts, *Courrier de l'UNESCO*, July, pp. 14–17.

Duchaufour, P., 1965, *Précis de pédologie*, Masson.

Duchaufour, P., 1979, Bases écologiques de la conservation de la fertilité des sols, *C. R. Colloque CNRS Ecologie et développement*, Sept., CNRS no. 2-222-02778-0, pp. 101–105.

Duchaufour, P., 1980, Ecologie de l'humification et pedogenèse des sols forestiers, in Pesson *et al.*, *Actualités d'écologie forestière*, Gauthiers-Villars, pp. 175–201.

Duncan, D. C., and Swanson, V. E., 1965, *Organic-rich Shales of the United States and World Land Areas*, US Geological Survey, circular 523.

Dupas, A., 1975, Les Chlorofluorocarbones produits par l'homme sont-ils une menace pour l'ozone stratosphérique? *La Recherche*, **6**, no. 61, pp. 970–973.

Duvigneaud, P., 1967, *L'Écologie, science moderne de synthèse, vol. 1, Ecosystèmes et biosphère*, Ministère de l'Education Nationale et de la Culture, Brussels, 137 pp.

Duvigneaud, P., and Testement, P., 1977, *Productivité biologique en Belgique*, PBI-scope, Duculot, Gembloux, 617 pp.

Eddy, J. A., 1977, The case of the missing sunspots, *Sci. Amer.*, **236**, no. 5, May, pp. 80–92.

El Kassas, M., 1977, L'Avance des déserts et la complicité de l'homme, *Cour. Unesco*, **30**, July, pp. 4–6.

El Kassas, M., 1979, IUCN joins the desert war, *Bull. IUCN*, **10**, no. 9, p. 81.

Environmental Quality, Council of, 1976, *7th Annual Report*, US Government Printing Office, Washington, DC 20402, 378 pp.

FAO Production Yearbook, 1971, No. 25, table 1.

FAO, 1977, *The State of Food and Agriculture, World Review*, FAO.

FAO, 1979, *World Fisheries and the Law of the Sea*, summary of the EEZ programme, FAO, 30 pp.

Felden, M., 1976, *Energie: le défi nucléaire*, A. Leson, Paris, 379 pp.

Fishwick, R. W., 1970, The Sahel and Sudan zone of northern Nigeria, north Cameroon and the Sudan, in Karl, *Afforestation in Arid zones, Monographiae Biologiae*, Junk, La Haye, pp. 59–85.

Fraenkel, O. H., 1973, *Survey of Crop Genetic Resources in Their Centers of Diversity: First Report*, FAO/PBI. Additional reports in Plant Genetic Resources Newsletter.

Friedman, I., 1977, The Amazon Basin, another Sahel, *Science*, **197**, no. 4298, 1 July, p. 7.

Frontier, S., 1978, Interface entre deux écosystèmes: exemple dans le domaine pélagique, *Ann. Inst. Oceanog.*, **54**, no. 2, pp. 95–106.

Geistdorfer, P., 1975, Ecologie alimentaire des Macrouridae (Téléostéens gadiformes), thesis for the Doctorat-ès-Sciences, Paris, 315 pp.

George, J. C., 1972, The role of the Aswan Dam in changing the fisheries of the south-eastern Mediterranean, in Farvar, T. F., and Milton, J. P., *The Careless Technology*, Natural History Press, New York, pp. 159–178.

Gillon, Y., 1971, The effects of bush fires on the principal Acridid species of an Ivory Coast savanna, *Proc. Ann. Tall. Timber Fire Ecology Conference*, 22–23 April, pp. 419–471.

Gillon, Y., and Gillon, D., 1967, Recherches écologiques dans la savane de Lamto (Côte d'Ivoire): cycle annuel des effectifs et des biomasses d'arthropodes de la strate herbacée, *La Terre et la Vie*, **21**, no. 3, pp. 262–277.

Goethals, R., 1980, L'Energie éolienne, *La Recherche*, **11**, no. 109, March, pp. 263–271.

Goodland, R. J. A., and Irwin, S. A., 1975, *Amazon Jungle: Green Hell to Red Desert?*, Elsevier, New York.

Granat, L., Rodhe, H., and Hallberg, R. O., 1976, The global sulphur cycle, in Svensson and Söderlund, *N, P and S—Global Cycles*, Ecological Bulletin no. 22, pp. 89–122.

Gulland, J. A., 1968, The concept of marginal yield from exploited fish populations, *J. Cons., Cons. Int. Explor. Mer*, **29**(1), pp. 61–64.

Gulland, J. A., 1969, *Manual of Methods for Fish Assessment*, FAO, Rome.

Hammond, A. L., 1971, Solar energy: a feasible source of power?, *Science*, **172**, p. 660.

Hammond, A. L., 1977, Remote sensing II: Brazil explores its Amazon wilderness, *Science*, **196**, pp. 513–515.

Hammond, A. L., Metz, W. D., and Maugh, T. H., 1973, Energy and the future, *Amer. Ass. Adv. Sci. Miscell. public*, 1st edn, July 1973; 2nd edn December 1973, 184 pp.

Handler, P., 1979, *Energy in Transition 1985–2000*, final report of the Committee on Alternative Energy Systems, US Nat. Acad. Sci., Freeman, 677 pp.

Harroy, J. P., 1967, *United Nations List of National Parks and Analogous Reserves*, IUCN, Hayez, Brussels, 550 pp.

Harroy, J. P., 1970, Le rôle des parcs nationaux dans la conservation de la nature en Europe, in *Abst. Coll. Int. Parcs Nationaux européens*, FFSSN, Paris, pp. 9–27.

Hays, J. D., Imbrie, J., and Shackleton, N. J., 1976, Variations in the earth's orbit: pacemaker of ice ages, *Science*, **194**, no. 4270, pp. 1121–1132.

Hempel, M. G., 1977, 1978, Les Causes des fluctuations des ressources halieutiques d'interêt commercial, in *Abst. Coll. sur la conservation des ressources halieutiques dans la partie N.E. de l'Atlantique et en Méditerranée*, Malta, October 1977, Council of Europe, 1978, pp. 70–80.

Holm, L. G., Weldon, L. W., and Blackburn, R. D., 1971, Aquatic weeds, in Detwyler, *Man's Impact on the Environment*, McGraw-Hill, pp. 246–265.

Hopper, W. D., 1976, The development of agriculture in developing countries, *Scient. Amer.*, **235**, no. 3, pp. 196–205.

Hubbert, M. K., 1967, *Am. Ass. Petrol. Geol. Bull.*, **51**, 2207.

Hubbert, M. K., 1969, Energy resources, in *Resources and Man*, US Nat. Acad. Sci., Freeman, pp. 157–242.

Hubbert, M. K., 1971, Energy resources, in Murdoch, *Environment, Resources, Pollution and Society*, Sinauer, pp. 89–116.

Hudson, N., 1973, *Soil Conservation*, Batsford.

Huttel, C., 1975, Recherches sur l'écosystème de la forêt subéquatoriale de Basse Côte d'Ivoire, II: Inventaire et structure de la végétation ligneuse, *La Terre et la Vie*, **2**, pp. 178–191.

Huttel, C., and Bernhard-Reversat, F., 1975, Recherches sur l'écosystème de la forêt subéquatoriale de Basse Côte d'Ivoire, V: Biomasse végétale et productivité primaire. Cycle de la matière organique, *La Terre et la Vie*, **2**, pp. 203–227.

Hutter, W., 1976, Energie consommée pour la production de quelques cultures, *C. R. Ac. Agric.*, February, pp. 297–308.

IUCN, 1967, *Liste des Nations-Unies des parcs nationaux et réserves analogues*, Hayez, Brussels, 550 pp.

IUCN, 1975, *The Red Data Book*, IUCN, Gland, Switzerland.

IUCN, 1979 (in collaboration with MAB and Unesco), The Biosphere Reserve and its Relationship to Other Protected Areas, 19 pp.

IUCN, 1980a, *Liste des Nations-Unies des parcs nationaux et réserves analogues, 1980*, prepared by the IUCN Commission on National Parks and Protected Areas, Gland, Switzerland, 121 pp.

IUCN, 1980b, *World Conservation Strategy*, undertaken with the support of UNEP and WWF, Gland, Switzerland.

Jordan, C. F., and Kline, J. R., 1972, Mineral cycling: some basic concepts and their application in a tropical rain forest, *Ann. Rev. Ecol. Systems*, **3**, pp. 33–50.

Jouanin, C., 1970, Les Parcs nationaux dans le monde, in *Abst. Coll. Int. Parcs Nationaux Européens*, Paris, FFSSN, pp. 75–78.

Junge, C. E., 1963, *Air Chemistry and Radioactivity*, Academic Press, New York.

Kesler, S. E., 1976, *Our Finite Mineral Resources*, McGraw-Hill, 120 pp.

Klatzmann, J., 1975, *Nourir dix milliards d'hommes?*, Presses Universitaires de France, 267 pp.

Kovda, V., Smith, F., Eckardt, F. E., *et al.*, 1970, Conceptions scientifiques de la biosphère, in *Utilisation et conservation de la biosphère*, Unesco, pp. 13–21.

Kucera, C. L., Dahlmann, R. C., and Krelling, M. R., 1967, Total net productivity and turn-over on an energy basis for tall grass prairie, *Ecology*, **48**, pp. 536–541.

Labeyrie, J., 1976, La Datation par le carbone 14, *La Recherche*, **7**, no. 73, pp. 1036–1045.

Labeyrie, V., 1971, Communication to the 1st Salone International pour la Protection de la Nature (Protecna), Rouen, October 1971, *Actes du Colloque*.

Labeyrie, V., 1972, Modèles écologiques et aménagement de l'espace, *Experientia*, Basel, **28**, pp. 616–622.

Labeyrie, V. 1973, A propos de quelques conséquences écologiques de l'organisation des transports, *L'Espace géographique*, no. 1, pp. 5–20.

Labeyrie, V., 1974, Bases écologiques pour une prospective de l'environnement, in *Socio-économique de l'environnement*, Mouton, pp. 114–138.

Labeyrie, V., 1975, La Crise de l'environnement, l'économie de la nature et l'économie humaine, *Mondes en développement*, Editions Sociales, Paris, pp. 527–565.

Lacoste, A., and Salanon, R., 1969, *Eléments de biogéographie*, Nathan, Paris, 189 pp.

Lamotte, M., 1967, Recherches écologiques dans la savane de Lamto (Côte d'Ivoire): présentation du milieu et programme de travail, *La Terre et la Vie*, **21**, pp. 197–329.

Lamotte, M., 1970, La Participation au PBI de la station d'écologie tropicale de Lamto (Côte d'Ivoire), *Bull. Soc. d'Ecol.*, **1**, no. 2, pp. 58–65.

Lamotte, M., 1975, *The Structure and Function of a Tropical Savannah Ecosystem*, Springer-Verlag, pp. 179–222.

Lamotte, M., 1978, La Savane préforestière de Lamto, Côte d'Ivoire, in *Problèmes d'écologie: écosystèmes terrestres*, Masson, Paris, pp. 231–311.

Langlois, J. P., Ziegler, V., and Bertel, E., *et al.*, 1979, Réflexions sur l'approvisionnement mondial en uranium à l'horizon 2000 et perspectives ultérieures, *Abst. Colloque Int. sur l'Evaluation et les techniques d'extraction de l'uranium*, Buenos Aires, October, IAEA and OECD, pp. 3–11.

Lebreton, P., 1978, *Eco-logique*, Interédition, Paris, 239 pp.

Le Houérou, H. N., 1979, La Désertification des régions arides, *La Recherche*, **10**, no. 99, April, pp. 336–344.

Lenihan, J., and Fletcher, W. W., 1975–1977, *Environment and Man*, 6 volumes, Blackie, London.

Lévieux, J., 1968, Influence des feux de brousse sur le peuplement en fourmis terricoles d'une savane de Côte d'Ivoire, *Proc. 13th Int. Cong. Entomol., Moscow, 1968*, III, pp. 369–370.

Leyhausen, P., 1976, Erosion and the hippos, *Oryx*, GB, **13**, no. 3, pp. 303–304.

Lieth, H., 1975, Primary production of the major vegetation units of the world, in Lieth, H., and Whittaker, R. H., *Primary Productivity of the Biosphere*, Springer-Verlag, pp. 203–215.

Likens, G. E., and Bormann, F. H., 1974, *Science*, **184**, 1175–1179.

Likens, G. E., Bormann, F. H., Johnson, N. M., *et al.*, 1970, Effects of forest cutting and herbicide treatment on nutrient budgets in the Hubbard Brook watershed ecosystem, *Ecol. Monographs*, **40**, pp. 23–47.

Likens, G. E., Wright, R. F., Galloway, J. N., and Thomas, J., 1979, Acid rains, *Scient. Amer.*, **241**(4), 39–47.

Lorius, C., and Duplessy, J. C., 1977, Les Grands Changements climatiques, *La Recherche*, **8**, no. 83, November, p. 947.

Lossaint, P., and Rapp., M., 1978, La Forêt méditerranéenne de chênes verts (Quercus Ilex), in Lamotte and Bourlière, *Problèmes d'écologie*, Masson, Paris, pp. 130–185.

Lowe-McConnel, R. H., 1966, Man-made lakes, *Symp. Inst. Biol.*, Academic Press, no. 15, 218 pp.

Lucas, G., and Synge, H., 1978, *The IUCN Plant Red Data Book*, IUCN, Gland, Switzerland.

McCulloch, J. S., and Talbot, L. M., 1965, Comparison of weight estimation methods for wild animals and domestic livestock, *J. Appl. Ecol.*, **2**, pp. 59–69.

McNaughton, S. J., 1976, Serengeti migratory wildebeest: facilitation of energy flow by grazing, *Science*, **191**, no. 4222, pp. 92–94.

McNeil, M., 1964, Lateritic soils, *Scient. Amer.*, **211**, Nov., pp. 97–102.

Maldague, M., 1984, The biosphere reserve concept: its implementation and potential as a tool for integrated development, in Di Castri, F., Baker, F. W. G., and Hadley, M., *Ecology in Practice*, Part 1: Ecosystem Management, Tycooley, Dublin and Unesco, pp. 375–401.

Manabe, S., and Wetherald, R. T., 1967, Thermal equilibrium of the atmosphere with a convective adjustment, *J. Atmos. Sci.*, **24**, pp. 241–259.

Mariotti, A., 1977, Les Isotopes de l'azote: des indicateurs écologiques, *La Recherche*, **8**, no. 82, pp. 886–888.

Martin, R., 1978, *Les mammifères marins*, Elsevier, p. 75.

Maugh, T. H., 1979, The threat to ozone is real, increasing, *Science*, **206**, pp. 1167–1168.

Mayer, J., 1976, The dimensions of human hunger, *Scient. Amer.*, **235**, no. 3, pp. 40–49.

Menhinick, E. F., 1964, *Ecology*, **45**, 859–961.

Metz, W. D., 1974, Oil shales: a huge source of low-grade fuel, *Science*, **184**, 1271–1275.

MIT, 1970, *Man's Impact on the Global Environment*.

Mitchell, B., and Sandbrook, K. R., 1980, *The Management of the Southern Ocean*, Int. Inst. for Environ. and Development, London, 162 pp.

Mitchell, B., and Tinker, K. R., 1980, *Antarctica and its Resources*, Int. Inst. for Environ. and Development, Earthscan publication, London, 98 pp.

Mitchell, J. M., 1970, A preliminary evaluation of atmospheric pollution as a cause of the global temperature fluctuation in the past century, in Singer, S., *Global Effects of Environmental Pollution*, Reidel, Dordrecht, pp. 139–155.

Modha, K. L., and Eltringham, S. K., 1976, Population ecology of the Uganda kob (Adenota kob thomas) in relation to the territorial system in the Ruwenzori National Park, Uganda, *Jour. Appl. Ecol.*, **13**, no. 2 pp. 453–473.

Montbrial, T. de, Lattes, R., and Wilson, C., 1978, *L'Energie: le compte à rebours*, Report of the Club of Rome, J. C. Lattes, 317 pp.

Murdoch, W. W., 1971, *Environment, Resources, Pollution and Society*, Sinauer, 450 pp.

National Academy of Sciences, 1969, *Resources and Man*, Freeman.

Newby, J., 1980, Can addax and oryx be saved in the Sahel?, *Jour. of Fauna Preserv. Soc.*, London (Oryx), **15**, no. 3, April, pp. 262–266.

Noirfalise, A., 1964, *Conséquences écologiques de la monoculture de cônifères dans la zone des feuillus de l'Europe tempérée*, Council of Europe Report, 36 pp.

Odum, E. P., 1977, The emergence of ecology as a new integrative discipline, *Science*, **195**, no. 4284, March, pp. 1289–1293.

Odum, H. T., 1957, Trophic structure and productivity of Silver Springs, Florida, *Ecol. Monogr.*, **27**, pp. 55–112.

Olson, G. R., 1975, Preliminary studies of the colonization by benthic invertebrates of a new humped-storage hydroelectric reservoir, *Michigan Acad.*, **7**, no. 4, pp. 501–513.

Ovington, J. D., 1962, Quantitative ecology and the woodland ecosystem concept, *Adv. Ecol. Res.*, Academic Press, **1**, pp. 103–192.

Ovington, J. D., 1965, *Woodlands*, English Universities Press, London.

Ovington, J. D., Heitkamps, D., and Lawrence, D. B., 1963, Plant biomass and productivity of prairie grassland, savanna, oakwood and maize field ecosystem in central Minnesota, *Ecology*, **44**, pp. 52–63.

Pabot, H., 1980, Géophytes et hygrophytes dans les déserts d'Asie, *Bull. Soc. Lin. Lyon*, **49**, no. 3, March, pp. 148–152 and 201–216.

Passet, R., 1979, *L'Economique et le vivant*, Payot, Paris, 287 pp.

Passet, R., 1980, in Une Approache multidisciplinaire de l'environnement, *Cahiers du Centre de Recherche Panthéon-Sorbonne*, Série Science Economique, no. 2, pp. 5–6.

Pereira, H. C., 1973, *Land Use and Water Resources in Temperate and Tropical Climates*, Cambridge University Press.

Pérès, J.-M., and Devèze, L., 1963, *Océanographie biologique et biologie marine*, vol. II, Presses Universitaires de France, 514 pp.

Pesson, P. (ed.), 1980, *Actualités d'écologie forestière*, Gauthiers-Villars, Paris, 580 pp.

Pianka, E. R., 1970, On r and K selection, *Amer. Nat.*, **100**, pp. 592–597.

Pielou, E. C., 1966, Species diversity and pattern diversity in the study of ecological succession, *J. Theor. Biol.*, **10**, pp. 370–383.

Pierrou, U., 1976, The global phosphorus cycle, in Svensson and Söderlund, *N, P and S—Global Cycles*, Ecological Bulletin no. 22, pp. 75–88.

Pimentel, D., Hurd, L. E., Belloti, A. C., et al., 1973, Food production and the energy crisis, *Science*, **182**, no. 4111, pp. 443–449.

Pimentel, D., Terhune, E. C., Dysar-Hudson, R., et al., 1976, Land degradation: effects on food and energy resources, *Science*, **194**, 8 October, pp. 149–155.

Pinhorn, A. T., 1976, Living marine resources of New-

foundland and Labrador: status and potential, *Bull. Fish Res. Board*, Canada, no. 194, 64 pp.

Pomeroy, L. R., 1974, Cycles of essential elements, *Benchmark Papers in Ecology*, Dowden, Hutchinson & Ross, 369 pp.

Poulet, A., and Poupon, H., 1978, L'invasion d'*Arvicanthus niloticus* dans le sahel sénégalais entre 1975 et 1976 et ses conséquences pour la strate ligneuse, *La Terre et la Vie*, **31**, 161–193.

Poupon, H., 1979, *Structure et dynamique de la strate ligneuse d'une steppe sahélienne au nord du Sénégal*, thesis for the Doctorate-ès-Sciences, University of Paris-Sud, 20 September, 317 pp.

Pryde, R. R., 1972, *Conservation in the Soviet Union*, Cambridge University Press.

Pyefinch, K. A., 1966, Hydroelectric schemes in Scotland: biological problems and effects on salmonid fisheries, in Lowe McConnell, *Man-made Lakes*, Academic Press, London, pp. 139–147.

Quezel, P., 1980, Biogéographie et écologie des conifères du pourtour Méditerranéen, in Pesson, *Actualités d'écologie forestière*, Gauthiers-Villars, Paris, pp. 205–255.

Ramade, F., 1974, Crise de l'énergie et ressources, *Bull. Soc. d'Ecol.*, **5**, no. 3, pp. 185–206.

Ramade, F., 1975, Recherches écologiques en Laponie finnoise, *Bull. Soc. d'Ecol.*, **1**.

Ramade, F., 1978b, Consommation énergétique des agro-écosystèmes, in *Eléments d'écologie appliquée*, 2nd edn, McGraw-Hill, pp. 523–528.

Ramade, F., 1978c, Crise de l'énergie, ressources naturelles et production alimentaire, *Economie Rurale*, **124**, no. 2, pp. 30–38.

Ramade, F., 1979a, *Ecotoxicologie*, 2nd edn, Masson, Paris, 228 pp.

Ramade, F., 1979b, La Protection de la nature au Kenya—mythes et réalités, *Courrier Nature*, no. 64, pp. 1–12.

Ramade, F., and Garnier, C., 1974, Protection de la nature, crise de l'énergie et transports, *Courrier Nature*, no. 34, pp. 283–289.

Ramade, F., *Ecotoxicology*, Wiley, in press.

Rasool, S. L., and Schneider, S. H., 1971, Atmospheric CO_2 and aerosols: effects of large increases on global climate, *Science*, **173**, pp. 138–141.

Ricker, W., 1944, Further notes on fishing mortality and effort, *Copeia*, no. 1, April, pp. 23–44.

Ricker, W. E., 1945, A method of estimating minimum seize limits for obtaining maximum yield, *Copeia*, no. 2, pp. 84–94.

Ricou, G., et al., 1978, La Prairie permanente au Nord-Ouest français, in *Problèmes d'écologie, écosystèmes terrestres*, Masson, Paris, pp. 17–74.

Roberts, W. O., and Lansford, H., 1979, *The Climate Mandate*, Freeman, 197 pp.

Robinson, E., and Robbins, R. C., 1970, Gaseous atmospheric pollutants from urban and natural sources, in Singer, S., *Global Effects of Environmental Pollution*, Reidel, Dordrecht, pp. 50–64.

Rochlin, G. I., 1977, Nuclear waste disposal: two social criteria, *Science*, **195**, no. 4273, 7 January, pp. 23–31.

Rodin, L. Y., and Bazilevitch, N. I., 1964, *Doklady Akademii Nayk SSSR*, **157**, 215–218.

Rose, D. J., 1974a, Energy policy in the U.S., *Scient. Amer.*, **230**, no. 1, pp. 20–29.

Rose, D. J., 1974b, Nuclear electric power, *Science*, **184**, 15 April, pp. 351–359.

Ryther, J., 1969, Photosynthesis and fish production in the sea, *Science*, **166**, pp. 72–76.

Scrimshaw, N., and Young, V. R., 1976, The requirements of human nutrition, *Scient. Amer.*, **235**, no. 3, pp. 50–64.

Scudder, T., 1972, Ecological bottlenecks and the development of the Kariba lake basin, in Farvar, T. F., and

Schaeffer, M. B., 1954, Some aspects of the dynamics of populations important to the management of the commercial marine fisheries, *Bull. Inter. Amer. Trop. Tuna Commission*, **1**, no. 2, pp. 27–56.

Milton, J. P., *The Careless Technology*, Natural History Press, New York, pp. 206–235.

Sellers, W. D., 1965, *Physical Climatology*, University of Chicago Press.

Shapley, D., 1976, Crops and climatic changes: USDA's forecasts criticized, *Science*, **193**, 14 September, pp. 1222–1224.

Shell, 1978, Briefing service document, January.

Skaf, R., 1972, Le Criquet marocain au proche-orient et sa gréganisation, *Bull. Soc. d'Ecologie*, **3**, no. 3, pp. 1–325.

Slesser, M., 1975, Energy requirements of agriculture, in Lenihan and Fletcher, *Environment and Man*, vol. 2: *Food, Agriculture and the Environment*, Blackie, pp. 1–20.

Smith, F., Fairbanks, D., Atlas, R., Delwiche, C. C., *et al.*, 1972, Cycles of elements, in *Man in the Living Environment, a Report on Global Ecological Problems*, Wisconsin University Press, pp. 40–89.

Smith, G. H., 1971, *Conservation of Natural Resources*, Wiley, New York, 685 pp.

Smith, R. L., 1966, *Ecology and Field Biology*, Harper & Row, London, 686 pp.

Söderlund, R., and Svensson, B. M., 1976, The global nitrogen cycle, in Svensson, B. M., and Söderlund, R., *N, P and S—Global Cycles*, Ecological Bulletin no. 22, Stockholm, pp. 23–73.

Sofretes (Société française pour l'exploitation des techniques en énergie solaire), 1979, Actualités documents, *L'Energie Solaire*, June.

Sournia, A., 1977, Analyse et bilan de la production primaire dans les récifs coralliens, *Ann. Inst. Oceanog., Paris*, **53**, fasc 1, pp. 47–74.

Spence, D. H. N., and Angus, A., 1970, African grassland management—burning and grazing in Murchison Falls National Park, Uganda, in Jaffey, E., and Watt, A. S., *Scientific Management of Animal and Plant Communities for Conservation*, Blackwell, pp. 319–331.

Spurr, S. H., 1979, Silviculture, *Scient. Amer.*, **240**, no. 2, pp. 62–75.

Squires, A. M., 1972, Clean power from dirty fuels, *Scient. Amer.*, **227**(4), 26–35.

Steinhart, J. S., and Steinhart, C. E., 1974, Energy use in the U.S. food system, *Science*, **184**, April, pp. 307–316.

Stuiver, M., 1978, Atmospheric carbon dioxide and carbon reservoir change, *Science*, **199**, no. 4326, 20 January, pp. 253–258.

Stumm, W., 1973, *Water Research*, **7**, 131–144.

Suess, H. E., 1955, *Science*, **122**, 415–417.

Summers, C. M., 1971, The conversion of energy, *Scient. Amer.*, **224**, September, pp. 148–160.

Svensson, B. H., and Söderlund, R., 1976, *Nitrogen, Phosphorus and Sulphur—Global Cycles*, SCOPE (Scientific Committee on Problems of the Environment), no. 7,

Ecological Bulletin no. 22, Swedish Natural Science Research Council, Stockholm, 192 pp.

Talbot, L. M., 1963, Comparison of the efficiency of wild animals and domestic livestock in utilization of East African rangelands, IUCN, Arusha Conference, pp. 329–335.

Talbot, L. M., 1966, Wild animals as source of food, Bureau of Sport-Fish-Wildlife, Special Scientific Report no. 98, pp. 1–16.

Talbot, L. M., 1972, Ecological consequences of rangeland development in Masailand, East Africa, in Farvar, T. F., and Milton, J. P., *The Careless Technology*, Natural History Press, New York, pp. 694–711.

Teller, E., 1979, *Energy from Heaven and Earth*, Freeman, 322 pp.

Tett, P., 1977, Marine production, in Lenihan, J., and Fletcher, W. W., *Environment and Man*, Vol. 5, *The Marine Environment*, Blackie, London, pp. 1–45.

Thiriet, L., 1976, *L'Energie nucléaire*, Dunod, Paris, 253 pp.

Tivy, J., 1975, Environmental impact of cultivation, in Lenihan, J., and Fletcher, W. W., *Environment and Man*, Vol. 2, *Food, Agriculture and the Environment*, Blackie, pp. 21–47.

Udvardy, M. D. F., 1975, A classification of the biogeographical provinces of the world, IUCN occasional paper no. 28, Morges, Switzerland.

Verneaux, J., 1976, Fondements biologiques et écologiques de l'étude de la qualité des eaux continentales, in Pesson, P., *La Pollution des eaux continentales*, Bordas, Paris, pp. 229–285.

Vooys, C. G. N. de, 1979, Primary production in aquatic environments, in Bolin *et al.*, *The Global Carbon Cycle*, Wiley, pp. 259–292.

Weinberg, A. M., 1974, Global effects of man's production of energy, *Science*, **186**, no. 4160, October, p. 205.

Whittaker, R. H., 1972, Evolution and measurement of species diversity, *Taxon*, **21**, pp. 213–251.

Wilson, C. L., 1977, *Energy: Global Prospects 1985–2000*, Workshop on Alternative Energy Strategies, MIT, McGraw-Hill.

Wofsy, S. C., McElroy, M. B., and Dak Sze, N., 1975, Freon consumption: implication for atmospheric ozone, *Science*, **187**, pp. 535–536.

Wohletz, L., and Dolder, E., 1952, *Know California's Land*, California Department of Natural Resources, Sacramento.

Woodwell, G. M., 1970, The energy cycle of the biosphere, *Scient. Amer.*, September, pp. 26–36.

Woodwell, G. M., 1978, The carbon dioxide question, *Scient. Amer.*, **238**, January, pp. 34–43.

Woodwell, G. M., Whittaker, R. H., Reiners, W. A., *et al.*, 1978, The biota and the world carbon budget, *Science*, **199**, pp. 141–146.

Worthington, E. B., 1972, The Nile catchment—technological change and aquatic biology, in Farvar, T. F., and Milton, J. P., *The Careless Technology*, Natural History Press, New York, pp. 189–205.

Wortman, S., 1976, Food and agriculture, *Scient. Amer.*, **235**, no. 3, pp. 31–39.

Wyrtki, K., 1979, El Nino, *La Recherche*, **106**, December, pp. 1212–1220.

Glossary

Anthropogenic: arising from or generated by human activity.

Autotroph: an organism that can synthesize its organic requirements from inorganic materials, e.g. through photosynthesis. Adj: **autotrophic**.

Benthic: referring to the sea-bed. Organisms living on the sea-bed are known as **benthos**.

Biocoenosis: an interdependent community of plants and animals sharing a common environment.

Biomass: the total mass of living matter in a given population or ecosystem. Usually expressed either in living or dry weight, and sometimes per unit area.

Biome: a major ecological unit (a '**macro-ecosystem**') having characteristic climatic and other physical conditions, extending over a large area and supporting plant and animal communities adapted to these conditions. Examples are tropical savanna, desert, tundra and so on.

Biotope: the inert component of an ecosystem consisting of all its spatial, temporal and physico-chemical characteristics; or a habitat showing uniformity of such characteristics: a lake, a marsh, a mud-flat, for example.

BWR: boiling water reactor.

Climax: the ultimate stage in the evolution of plant communities in equilibrium with the climate and other environmental factors, which has developed in the absence of human interference. The plants themselves form a **climax community**.

Disclimax: short for disturbance climax. A stable community which is not a climax but is maintained by human interference.

Drainage basin: the catchment area from which water drains off into any given connected network of rivers, streams and lakes. (American usage calls this a watershed.)

Dystrophic: describing a lake with a high concentration of organic matter, particularly humic acid.

Ecosystem: a natural system consisting of living communities and their environment interacting to form a stable unit.

Edaphic factors: the physical and chemical properties of a soil that affect the growth of plants in it.

Epigeal (of plants): developing above the ground.

Eutrophic: describing a lake or river having a high concentration of nutrients and a high primary productivity.

Eutrophication of a water system is a process characterized by an increase in organic matter and a decrease in the concentration of dissolved oxygen. Eutrophic lakes are generally old and shallow.

Habitat: the locality having a particular type of environment in which a given organism normally lives.

Heterotroph: a living organism that needs an external supply of organic substances to synthesize its own organic requirements. All animals, fungi, and certain bacteria are **heterotrophic**.

Homeostasis: the capacity of a biological system or organism to maintain a constant internal environment as external conditions vary.

Humus: the organic constituent of soils, formed by the decomposition of dead plants and animals.

Hypogeal (of plants): developing below ground.

Lentic: associated with still water.

Lotic: associated with running water.

Oligotrophic: describing a lake poor in nutrients and organic matter, generally deep and relatively young.

Pelagic: describing organisms inhabiting a sea or lake which float or swim freely (as opposed to benthic organisms).

Phyllophagous: leaf-eating.

Productivity: the rate of production of organic matter by a living organism per unit area per unit time. **Primary productivity** refers to the production of organic matter by plants through photosynthesis. **Secondary productivity** refers to the production of organic matter by animals.

PWR: pressurized water reactor.

Saprophyte: an organism which can feed on dead or decomposing organic matter: such an animal is described as **saprophagous**.

Succession: the progressive or regressive evolution of a living community in a given biotope over a period of time.

t.c.e.: tonnes of coal equivalent.

Trophic level: one of the stages in a food chain to which a given organism belongs: occupied successively by **producers, primary consumers, secondary consumers** and so on.

Xerophyte: a plant adapted to arid conditions.

Xylophagous: wood-eating (of insects).

General Index

abiotic factors, 1
acidity of rainfall, 83
aerosol sprays, 80
Africa, afforested area, 162, 176
    aquatic weeds, 103
    bushfires, 187
    cultivated land area, 141
    deforestation, 174, 176
    desertification, 198
    food requirements, 140
    fossil fuel reserves, 30
    lithium reserves, 45
    malnutrition, 137
    oil shales, 32
    pastureland area, 182
    protected areas, need for, 204
    water power, capacity, 54
    water supplies, 99
agriculture, 138–142, 152–157
    energy consumption by, 154
    water consumption by, 99
agroecosystems, 134
aircraft, effect on atmosphere, 79
    efficiency as transport, 63, 64
Alaska, forest fires, 177
    metal deposits, 133
    oil reserves, 31
    tundra, 192
algae, 103
    coral reef, 115
algal beds, productivities, 12, 13
Allee's principle, 6
Amazon Basin, deforestation, 142, 174
    laterization of soil, 144
    origin of rainfall, 97
    tree density, 162
America, *see under* North America, South America, USA
amino acids, 135
anaemia, 137
animals, domesticated
    ecological efficiency, 138, 156, 190
animals, wild
    compared with domesticated, 190, 200
Antarctica, climatic changes, 86
    krill in seas, 127
aphotic zone, 113
aquatic weeds, 103
Argentina, cereal consumption, 139
Arizona, Kaibab Plateau, 209

Asia, afforested area, 162, 176
    bush fires, 187
    coal reserves, 35
    cultivated land, 141
    deforestation, 176
    fossil fuel reserves, 30
    malnutrition, 137
    oil shales, 34
    pastureland area, 182
    protected areas, need for, 204
    savanna, 185
    timber, 162, 166
    water supplies, 99
atmosphere, 69
    carbon cycle, 69
    nitrogen cycle, 104
    ozone cycle, 87
    sulphur cycle, 80
atmospheric concentration, CO, 69
    carbon dioxide, 24, 70
    ozone, 77
    sulphur dioxide, 79, 82, 83
atmospheric pollution, 57, 83
Australasia, cereal deficit, 138
    cultivated land, 141
    water power, capacity, 54
    water supplies, 99
Australia
    energy used in food production, 158
    irrigation needs, 101
    kangaroos, 187, 192
    rabbits, 205

bauxite, 68, 144, Plate VI
benthic zone, 112
beri-beri, 136
bilharziasis, 103
biomass, carbon content, 70
    distribution in trees, 165
    of forests, 71, 163
    of various ecosystems, 115
biomass/productivity ratio, 5, 15
biomes, land-based, productivities, 12, 184
biosphere, 1
    carbon content, 70
    productivity, 12, 13
biosphere reserves, 213
biotic factors, 1
biotic potential, 18
boiling water reactor, 38

boreal forest, *see* taiga
Brazil, deforestation, 142, 172, 174, 176
breeder reactor, 38, 42
broad-leaved forests, *see* deciduous forests

calcium cycle in forests, 170
California, coastal area, Plate V
    geothermal power, 47
    irrigation needs, 101
    Mojave Desert, Plate I
    oil pollution, 32, 58
    salinization of soil, 143
    sequoia forests, 181
Camargue, 202, 208, 211, Plate IX
Canada, acidity of lakes, 83
    Alberta tar sands, 33
    Candu reactor, 41
    cereal consumption, 139
    deforestation, 176
Candu nuclear reactors, 41
capacity of environment, 17
carbon cycle, 69
carbon dioxide in atmosphere, 24, 69, 75, 89, 93
carbon isotopes, 73
carbon monoxide, 69
carrying capacity, 17
    grasslands, 189, 196
    national parks, 211
cattle, ecological efficiency, 139
    and wild animals, compared, 190
    world population, 183
cereal deficits, 138
chaparral, 163
Chapman reactions, 77, 78
chemical elements, in ecosphere, 3
    in human beings, 3
    in sea-water, 133
chernozem, 145, 184
China, afforested area, 162
    cereal deficit, 138
    coal reserves, 35
    deforestation, 174
    erosion, 145
    malnutrition, 137
    timber, 162
clearance, forest, 173
    ecological effects, 167
clearance, land, 143, 173
    cost, 142

223

Taxonomic Index